Andreae
Farming, Development and Space

Bernd Andreae

Farming, Development and Space

A World Agricultural Geography

Translated from the German
by Howard F. Gregor

Walter de Gruyter · Berlin · New York 1981

Author
Bernd Andreae
Professor für Landwirtschaftliche Betriebslehre
Schweinfurthstraße 25
D-1000 Berlin 33

Translator
Howard F. Gregor, Ph.D.
Professor
Department of Geography
University of California, Davis
Davis, California 95616

CIP-Kurztitelaufnahme der Deutschen Bibliothek

> **Andreae, Bernd:**
> Farming, development and space: a world agricultural geography / Bernd Andreae.
> Transl. from the German by Howard F. Gregor. –
> Berlin; New York: de Gruyter, 1981.
>   Dt. Ausg. u.d.T.: Andreae, Bernd: Agrargeographie
>   ISBN 3-11-007632-2

Library of Congress Cataloging in Publication Data

> Andreae, Bernd.
>   Farming, development and space.
>   Revised translation of: Agrargeographie.
>   Bibliography: p.
>   Includes index.
>   1. Agricultural geography.
>   S495.A4713   338.1'09   81-4755
>   ISBN 3-11-117632-2   AACR2

© Copyright 1981 by Walter de Gruyter & Co., Berlin 30. All rights reserved, including those of translation into foreign languages. No part of this book may be reproduced in any form – by photoprint, microfilm, or any other means – nor transmitted nor translated into a machine language without written permission from the publisher. Typesetting: Verena Boldin, Aachen. Printing: Karl Gerike, Berlin. – Binding: Lüderitz & Bauer, Buchgewerbe GmbH, Berlin. – Printed in Germany.

Translator's Introduction

Professor Andreae introduces his volume as "a small textbook for students of geography and the agricultural sciences" and as the first German-language introduction to agricultural geography to be submitted by an agricultural economist. It is more. Andreae speaks from a land with a long and vigorous tradition in agricultural-geographic research, where agricultural economists antedate geographers in their particular concern for the spatial aspects of agricultural geography, and where the *lehrbuch* is usually more comprehensive and analytical than the kind that has given the textbook a dubious reputation in many circles elsewhere, particularly in the English-speaking world. But more specifically, Andreae draws on this inheritance and his own prolific work to offer us one of the few detailed pictures of world agricultural patterns as they are influenced by individual farm operations and agricultural development. Rather than proceeding from the region to the detail, the usual approach for geographers, Andreae is concerned with building the bridge (to use his metaphor) between agricultural economics and agricultural geography by proceeding from the farm to the comprehension of larger areas. He is also at pains to explain how farming practices and types change as population pressures increase and industrialization brings about increasing integration of agriculture into the national economy. His three-stage theory of farming development, monoculture→ diversification→ specialization, and its corresponding sequence of extensiveness→ labor-intensiveness→ capital-intensiveness, are themes that run though much of the book. These two distinctions, the emphases on farming operations and agricultural development insofar as they affect world agricultural spatial patterns, are expressed in the main title of this English-language edition of what originally appeared under the principal heading of *Agrargeographie**. Since that time, Professor Andreae has expanded his treatment of the ecological and economic boundaries of farming and the agricultural geography of the United States, and I have enlarged his already extensive bibliography to include especially more sources in English.

I am grateful to both author and publisher for their help and patience. No translator can do full justice to the original work, though I hope my role as both translator and colleague "at the other end of the bridge" will compensate in part.

Howard F. Gregor

Davis, California, March 1981

---

* *AGRARGEOGRAPHIE – Strukturzonen und Betriebsformen in der Weltlandwirtschaft.* Berlin 1977.

Foreword

Agricultural geography is a part of economic geography, and as such, is the science of the agriculturally transformed earth surface, with all its associated natural, economic, and social interrelations as reflected spatially. This surface is composed of agricultural zones and agricultural regions, which in turn are made up of farms, the building stones of agricultural geography.

Agricultural geography is distinctly a boundary science. Not only the economic geographer and the agricultural economist, but also the plant and animal geographers, the climatologist, the sociologist, the ethnographer, the cultural geographer, and others have contributed to the many causal relations in the world agricultural structure. Thus up to now no single and fairly exhaustive overall view has been presented by the various disciplines. That is the fate of any boundary science. Nor can the book offered here lay claim to a consideration of all aspects. Sociocultural and religious influences had to remain largely unconsidered.

Agricultural geography was for a long time accounted for scientifically almost exclusively by economic geographers, who for this purpose sought the help of agricultural economists and found it in rich amount, especially in the contributions of Engelbrecht, von Thünen, Aereboe, Busch, and Rolfes. In the last two decades agricultural economists have become notably more interested in the geography of agriculture as they have been called to all ends of the earth to advise, direct, and plan in areal development. They have thus been motivated to sharpen, through as geographically extensive a view as possible, their judgements on the agricultural problems of entire regions in Latin America, Africa, and Asia. Thus they have earnestly sought, from their side, associations with economic geography. The bridge was now being thrown across from both banks. The two disciplines were getting closer to each other.

As a result, I believed that the time had come to fulfill an oft-expressed wish by submitting from the side of agricultural economics the first German-language introduction to agricultural geography. Its aim is to be a small textbook for students of geography and the agricultural sciences, one that is embracing but still concise, motivating but still easily readable, and which seeks to provide understanding even more than information. The core of the book is based on my lectures on agricultural geography at the Berlin Technical University.

The economic geographer advances from the world agricultural economy or agricultural zones to the detail, whereas I, conversely, proceed from the farm to the comprehension of larger areas. It is obvious that the two different approaches must lead to different emphases, results, and conclusions. No value judgement, however, lies in this statement. I do not wish the reader to come away with an impression of "either-or" but rather one of "not only, but also," so that the bridgeheads thrown up from the two banks by economic geography and agricultural economics will finally coalesce into one unit.

I am grateful to the publishers for the encouraging trust with which they have received me into their author's circle, and for their appreciation of my wishes.

Respect for my most tried and true secretary and co-worker of more than sixteen years, Frau Elsbeth Greiser, née Goehle, requires that I cite her and all the other unnamed helpers whom she represents with warmest thanks.

I should like to dedicate this book to my wife, Gisela Andreae, née Freiin von Reibnitz. For more than three decades, she has supported me unfailingly in my work, sharing all sacrifices and privations. Without that support, the book would not have materialized. Now she can also share in the pleasure over its completion.

Berlin-Dahlem, January 1977 *Bernd Andreae*

Bernd Andreae
Farming, Development and Space

Errata

Listed here are corrections of errors, a number of which were made or whose corrections were overlooked in the final printing stages.

Page	Line from above	Line from below	For	Read
5		13	though	through
17		14	p. 16	p. 18
30		1	p. 15	p. 17
36	20		100 m	1,000 m
46	Table 2, D, Bahamas		Delete 0.1 under „% of".	
46	Table 2, E, col. 4		Shift 2.2, −, 0.4 to „Bush and Tree Crops".	
47	4		310,00	310,000
53	Table 3, col. 5		S. America S. L.	S. America 15° S. L.
			b) California	c) California
			b) Himalayas	c) Himalayas
			b) Tibet	c) Tibet
56	4		Sarfalvi 1970	Schweinfurth 1966
56	24		ley	grassland
58	Overview 6, col. 2		1,500	1,300
58		2	the	a
60		5	thus permitting	permitted by
67		8	Baltic	North
77		13	Central Schwerin Marsh	moderately heavy marsh
77		12	Baltic	North
77		2	land	period
78	3		patureland	pastureland
79	9		Hohnholz 1975	Raddatz 1954
84	11		central	southern
84	11		phenonemon	phenomenon
86	18		operating;	operating);
92	15		feed	humus
93	16		nautrally	naturally
96		9	mange	manage
101	Table 4, col. 5		Dairy-Feeder	Feeder-Dairy
119	Overview 9, title		German Farming	German Grain Farming
119	Overview 9, col. 3		wheat-catch	rye-catch
123	12		activities	activities
124	Table 6, cols. 3, 4, 8		−,−,−	·,·,·
124	Table 6, col. 2		1,449	1,440
125	17		digressions	degressions
139	Table 8, col. 2		100 kg	100 g
153	Table 9, col. 7		F, (P)	F
157	7		that	than
164	Table		Financial Input	Reimbursement share
178		8	46.5	46.8
178		6	cow	head of cattle
179	Table 12, col. 1		*Share of cropland*	*Share of CL*
179	Table 12, col. 2		$(DM)/AL^3$	$(DM)/ha\ AL^3$
186	20		constrasting	contrasting
202		13	Magedburg	Magdeburg

Page	Line from above	Line from below	For	Read
210				Delete last three lines.
212	1			Delete "ena. The".
217	15		cimpletely	completely
219	6		20	10
220		16	rainsing	raising
237		17	cultivated crops	cultivated fodder crops
238	Overview 19		conentrates	concentrates
239		11	coverted	converted
247		16	Of labor	Of course, by West German standards, a surplus of land and a scarcity of labor
247		13		Delete line.
249	Overview 21, col. 4		South	North
250	15		Turkish	Mongolian
253	Table 15, China PR, col. 6		...	—
255	Table 16, cols. 2, 3, 4		at least, ..., ...,	at least 60%, ... 70%, ... 80%
255	Table 16, cols. 2, 3, 4		40 to 60%, 30 to 70%, 20 to 80%	up to 40%, ... 30%, ... 20%
255		9	Plish	Polish
256		1	for fattening	and feeder
258	18		pure crop	pure leafy crop
258		19	are	is
261	9		must be	must not be
262	Overview 22, H, Central Russia		Flax (for fiber)	Hemp
262	Overview 22, J, Siberia 1.		Add "corn".	
264		14	back of	forward
264		6		Delete "much".
275		8	counries	countries
277		15	agricultura	agricultural
284		13	0.5	0.1
286		13	yield	supply
287	5		42%	40%
290		2	sishing	fishing
291		3		Delete period.
300		20	condideration	consideration
301		1	136	135
302		7	gegraophic	geographic
303		2–3	Vol. 2: Agriculture	Part XIII
311		3	Jentsch	Jentzsch
320		2	Woerman	Woermann
321	1		Woerman	Woermann
327	Baltic Sea coastal area		67, 77, 202	202

Walter de Gruyter · Berlin · New York 1981

Contents

Abbreviations		15
Glossary		17
Introduction:	Origins and Evolution of Agriculture – A Three-Stage Theory	27
I.	Agricultural Geography as a Science	30
	1. Definitions	30
	2. Objectives and Significance of Agricultural Geography	31
	3. Work Methods of Agricultural Geography	31
II.	The Climatic Zones of World Agricultural Space and their Significant Features for the Agricultural Economy	33
	1. Tropical Rainy Climates	33
	a) Rainforest Climate	33
	b) Humid Savanna Climate	34
	c) Tropical Highland Climates	36
	2. Dry Climates	37
	a) Dry Savanna Climate	37
	b) Shrub Savanna Climate	38
	c) Steppe Climate	39
	d) Semidesert Climate	39
	3. Humid Warm-Temperate Climates	40
	a) Subtropical Dry-Summer Climate	40
	b) Subtropical Warm-Summer Climate	41
	c) Marine Cool-Summer Climate	42
	4. Humid Cool-Temperate Climates	42
	a) Continental Warm-Summer Climate	43
	b) Continental Cool-Summer Climate	43
	c) Subarctic Climate	43
III.	The Delimitation of World Agricultural Space	45
	1. The Expansion of World Agricultural Space as a Current Problem	45
	2. Ecological Boundaries of Farming	48
	a) Polar Boundaries	48
	b) Altitudinal Boundaries	52
	c) Dry Boundaries	57
	d) Wet Boundaries	66

	e) Soil Boundaries	67
	f) Slope Boundaries	68
	g) Unfavorable Exposure	69
	3. Economic Boundaries of Farming	70
	a) Settlement and Industrial Boundaries	70
	b) Transport Boundaries	70
	c) Commercialization Boundaries	74
	4. Boundary Shifts with Economic Growth	76
	a) Mechanical Technological Advances as a Cause	76
	b) Biological Technological Advances as a Cause	78
	c) General Economic Growth as a Cause	79
	5. Contraction of World Agricultural Space as a Future Prospect	82
	6. Summary	85
IV.	Farms as Building Stones of the Agricultural Region	87
	1. Reasons for Diversified Farm Production	88
	a) Work Spacing	88
	b) Crop Rotations	90
	c) Fertilizer Balance	91
	d) Feed Balance	92
	e) Self-Sufficiency	93
	f) Spreading Risk	93
	2. Reasons for Spatial Differentiation of Farms	94
	a) The Physical Location of Production	94
	b) Population Density, Educational Level, and Personality of the Farm Operator	95
	c) Size of Farm, Ranch, and Plantation	96
	d) Location of the Farm for Transport	102
	e) The Diversified Farm in the Tension Field of Force Groups	105
	3. Reasons for Temporal Changes in Farms	106
	a) Price-Cost Development	106
	aa) Price Relations between Agricultural Products	107
	bb) Cost Relations between Agricultural Inputs	109
	cc) Price-Cost Relations between Agricultural Products and Inputs	110
	b) Technological Advances	111
	aa) Organic Technological Advances	111
	bb) Mechanical Technological Advances	111
V.	The Principal Farming Systems of World Agriculture	113
	1. Grassland (Grazing) Systems	114
	a) Nomadic Grazing	114

	b) Sedentary Extensive Grassland Farming (Ranching) .	115
	c) Sedentary Intensive Grassland Farming	115
	2. Annual-Cropping Systems	117
	a) Primitive Rotation Farming	117
	b) Ley Farming	118
	c) Grain Farming	119
	d) Hoe Crop Farming	120
	3. Perennial-Cropping Systems	122
	a) Gathering	122
	b) Bush and Tree Crop Farms	123
	c) Plantations	124
VI.	The Agricultural Geography of the Humid Tropics	127
	1. Regions of Rainfed Farming	132
	a) Rainfed Farming in the Tropical Rainforest Belt . . .	132
	b) Rainfed Farming in the Humid Savannas	137
	c) Rainfed Farming in the Tropical Highlands	140
	2. Regions of Irrigation Farming	141
	a) Farm Management Functions of Crop Irrigation . . .	143
	b) Irrigation Methods in Geographic Comparison	143
	c) Rice Growing as a Representative of Irrigated Cropping on Family Farms	145
	3. Regions with Predominantly Bush and Tree Crops . . .	149
	a) Farm Management Characteristics and Geographic Distribution	149
	b) The Principal Bush and Tree Crops and their Locations	152
	c) Peasant Farms or Plantations?	154
	aa) Typical Peasant Crops	155
	bb) Typical Plantation Crops	155
	cc) Crops Suitable for both Peasant Farms and Plantations	156
VII.	The Agricultural Geography of the Dry Areas	160
	1. Regions of Extensive Grassland Farming	165
	a) Zones of Nomadic Herding	166
	b) Zones of Sedentary Grassland Farming	167
	c) Seasonal Feed Balance as the Central Problem	169
	2. Regions of Dryland Crop Farming	174
	a) Wheat-Fallow Farming	176
	b) Millet-Sorghum-Peanut Farming	181
	c) Other Forms of Dryland Cropping	185
	3. Regions of Combined Extensive Grassland Farming and Dryland Crop Farming	187
	a) Risk as a Hindrance to Diversified Production	187

  b) The Example of Southwest Africa 187
  c) Modifications by Economic and Ecological Variants . 189
 4. The Macrospatial Structure of a Dry Area: The Example
  of Australia. 190
  a) Physicospatial Structure 191
  b) Agricultural Zones 191
  c) Water Reclamation Projects 195

VIII. The Agricultural Geography of the Middle Latitudes . . . 196
 1. The Agricultural Geography of Western Europe 197
  a) Place-Specific Characteristics 197
  b) Methodology of Delimitation and Statistical Derivation
   of Agricultural Regions 207
  c) Crop Rotation Regions 210
  d) Land Use Regions 214
  e) Livestock Regions 222
  f) Complex Agricultural Regions 227
  g) Summary 231
 2. The Agricultural Geography of North America 232
  a) Place-Specific Characteristics of the United States . . 232
  b) The Dairy Zone 237
  c) The Corn Zone 238
  d) Agricultural Regions with Mixed (General) Farming . 240
  e) The Cotton Zone 241
  f) The Wheat Regions 242
  g) Regions of Extensive Grassland Farming (Ranching) . 243
  h) Coastal Regions with Fruit and Vegetable Growing
   Dominant 244
  i) Industrialized Animal Production on Specialized Farms 246
  j) Regional Differences in the Factor Combination . . . 247
 3. The Agricultural Geography of the East Bloc Countries . . 250
  a) Place-Specific Characteristics 250
  b) Large Socialist Farms as a Regionalizing Feature . . . 252
  c) Problems of an Areally-Suitable Agricultural Production 252
  d) Stages of Socialization 253
  e) Agricultural Zones in the Baltic-Adriatic Area 256
  f) Agricultural Zones of the Soviet Union 260
  g) Agricultural Zones of the People's Republic of China . 266

IX. Structural Changes in World Agricultural Space with Economic Growth 271
 1. Forces for Development 271
 2. Changes in the Factor Combination 273

a) Factor Costs and Factor Combination in Sparsely Settled Countries	274
b) Factor Costs and Factor Combination in Densely Settled Countries	277
c) Differences in the Overall Trend of Change in the Factor Combination	279
d) Developmental Tendencies in Farm Size	280
3. Diversification and Specialization of the Production Program	284
a) One-sided Farms at the Beginning of Development	284
b) Diversification Tendencies in the Pre-industrial Era	286
c) Specialization Tendencies in the Industrial Era	287
d) Stages in Farming Diversification and Specialization in the March Toward Total Economic Integration	289
4. Changes in Farming Systems in Selected Climatic Zones	290
a) Tropical Rainforest Climate	291
b) Humid Savanna Climate	292
c) Dry Savanna and Steppe Climates	293
d) Marine Cool-Summer Climate	295
e) Continental Cool-Summer Climate	297
Outlook: The Agricultural Evolution Theory of Friedrich Aereboe in the Light of this Agricultural Geography	299
Bibliography	303
Figures	323
Index	327

Abbreviations

AC	Agricultural census (complete census)
AL	Agricultural land (FL + idle cropland and pasture + private parks, sod farms, ornamental gardens)
CC	Catch crop
CL	Cultivable (arable) land = land usually cultivated annually
CLLU	Large livestock unit (cattle)
°C	Degrees Celsius = 5(°F−32/9); °F = 9(°C/5) + 32
DM	Deutsche mark (West Germany) = $0.27 U.S., as of 1969
Dpf	Deutsche pfennig (West Germany) = $0.0027 U.S., as of 1969
EEC	European Economic Community (Common Market)
EC	European Community
FAO	Food and Agricultural Organization of the United Nations
FL	Farmed land = cropland, including kitchen gardens, permanent grassland, orchards, vineyards, hops, nurseries, and woodlots
FRG	Federal Republic of Germany (West Germany)
GDR	German Democratic Republic (East Germany)
GU	Grain unit, corresponding in nutritive value to 100 kg grain
ha	Hectare (2.47 acres) = 10,000 m^2; 100 ha = 1 km^2 (0.39 mi.^{2})
kcal	Kilocalorie (1,000 calories)
kg	Kilogram (2.2 lbs.) = 1,000 gm; 1,000 kg = 1 metric ton (2,204 lbs.)
K$_2$O	Potash fertilizers (pure nutrient)
kStU	Kilo starch unit (1,000 StU) = 1.408 Scandinavian feed units
LLU	Large livestock unit, corresponding to 500 kg liveweight
m	Meter (3.3 ft.) = 100 cm = 1,000 mm (40 in.)
MD	Man-day. In the tropics usually only 5.5 to 7 man-hours.
MH	Man-hour
MHP	Motor hp (horse power)
MHPh	Motor hp hour
MY	Man-year = equivalent of one full-time worker working at least 2,400 MH annually. In the tropics usually only 1,500 to 1,700 MH/year.
N	Nitrogen fertilizers (pure nutrient)
PFA	Primary forage area, used exclusively for feed production

Abbreviations

PLLU	Large livestock unit (productive livestock)
P_2O_5	Phosphoric acid fertilizers (pure nutrient)
% AL	% of agricultural land
% CL	% of cultivable (arable) land
% FL	% of farmed land
q	Quintal = 0.1 metric ton = 100 kg (220.4 lbs.)
RLLU	Roughage-consuming large livestock unit
S. beets	Sugar beets
S. gr.	Spring grain (spring plowing and planting)
TPh	Traction power hour
TPU	Traction power unit (equivalent of 1 horse; 5 tractor hp = 1 TPU)
W. gr.	Winter grain (fall plowing and planting)

Glossary

Basic agricultural-geographic concepts (after Manshard 1968, pp. 9 ff.)
- *agricultural geography* = Science of the agriculturally transformed earth surface.
- *social geography* = Science of the spatial distribution of geographically relevant social structures.
- *economic formation* = Common expression for economic form or life form.
- *agricultural space* = Portion of the earth surface used agriculturally in any way.
- *agricultural area* = Spatial unit with a clear farming dominance.
- *agricultural zone* = Agricultural area which can be associated globally with climatic and vegetation belts.
- *agricultural region* = Part of the earth surface with a clearly defined agricultural structure, thus with an assured unity.
- *farm* = Any type of operational unit in agriculture.
- *ranch* = Extensive livestock raising unit, usually of significant size.
- *plantation* = Large farms having their own processing facilities for their products (sugarcane mills, sisal factories, tea factories, etc.).

Rural land use concepts (alphabetical)
(see Fig. 1, p. 16)
- *Alpine farming* = Extensive livestock raising in high mountain areas, which often lie above timberline and can only be grazed about 90 days per year.
- *annual-crop farming* = Crop rotations with only annual crops.
- *cash catch crops* = Crops planted after the main crop in the same year, but still supplying a marketable product.
- *cereals* = Small grains (see leafy crops).
- *cover crop rotation* = Leafy crop – leafy crop – grain.
- *crop rotation* = Chronological succession of field crops.
- *cropland* = All land on which crops are regularly alternated (see Fig. 1).
- *cropland share* = Areal extent of crop types in % CL.

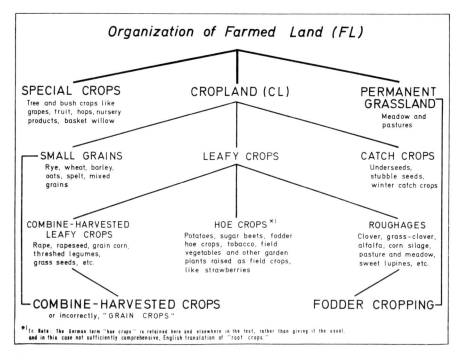

Figure 1

— *cropping system*	= Field- or crop-rotation system = form of cropping organization.
— *crop years*	= In ley farming (see *ley farming*) the cropping periods for annual crops; in contrast to *grass years*, cropping periods for perennial forage crops.
— *double-crop rotation*	= Leafy crop — leafy crop — grain — grain.
— *dry farming*	= Farming system in which fallow is inserted in the crop rotation to conserve water (e.g. 1. fallow — 2. wheat — 3. wheat).
— *extensive leafy crops*	= Threshed legumes, field forage crops, oil crops, fallow; in contrast to the *intensive leafy crops* (hoe crops).
— *five-course rotation*	= Leafy crop — grain — grain — grain — grain.
— *foreign compatibility*	= Compatibility of different crops; in contrast to *self-compatibility* = compatibility of similar crops.
— *four-course rotation*	= Leafy crop — grain — grain — grain.
— *grass-clover farming*	= Intensive form of ley farming, with only a two- to three-year period of fodder cropping.

Glossary

– *grassland cropping*	= Emphasizing rotation pasture, rotation meadow, grass-clover, alfalfa, grass-alfalfa.
– *grass years*	= Periods of perennial forage crops (see *crop years*).
– *"Hauberg" farming*	= Combined forestry and cropping on farms among lighter stands of oak forest in central Europe. Grain is planted between the trees, which are cut for tanbark.
– *intensive leafy crops*	= Hoe crops.
– *"Kunstegart" (Ger.)*	= Alpine ley farming (see *ley farming*) with seeded pasture.
– *leafy crops*	= All preparatory, support, and main crops other than the cereals in crop rotations: hoe crops, forage crops, legumes, oil and fiber crops, fallow.
– *ley farming*	= Based on crop rotations that include the harvesting and grazing of grass plots (leys) over a period of several years.
– *monoculture*	= Constant cultivation of the same crop, e.g. permanent grain cropping.
– *"Naturegart" (Ger.)*	= Alpine ley farming with volunteer pasture.
– *non-cash catch crops*	= Crops planted after the main crop in the same year, but which no longer yield a marketable product.
– *Norfolk rotation*	= Leafy crop – grain – leafy crop – grain.
– *primary crop rotation*	= Rotation nearer the farmyard or more important economically; in contrast to the *secondary crop rotation*, which is farther from the farmyard or less important economically.
– *production elasticity*	= Adaptability of farm organization.
– *production goal*	= Production program = range of farm enterprises.
– *production intensity*	= Amount of labor and materiel expenditures, plus tax payments, per hectare.
– *rotation component*	= Sub-rotation, consisting of a leafy crop and following grain crops.
– *rotation period*	= Length of the rotation in years.
– *rotation plan*	= Basic rotation type, in which only the two most important crop groups, i.e. leafy and grain crops, are distinguished.

- *secondary crop rotation* = See *primary crop rotation*.
- *self-compatibility* = Compatibility of a crop with itself (see *foreign compatibility*).
- *"Sennerei" (Ger.)* = Dairy farming on Alpine pastures.
- *three-field system* = Leafy crop – grain – grain.
- *useful economic life* = Span of years in which land can be cropped.

Intensity concepts
- *intensity* = Amount of labor and materiel expenditures, plus tax payments, per hectare. Thus one can speak of livestock-, labor-, or fertilizer-intensive farms. Of fundamental importance is the difference between
- *farming intensity* = *organizational intensity* = proportion of the farming structure in intensive enterprises like hoe cropping, raising of special crops, or milk production, and
- *specialization intensity* = *operational intensity* = expenditures of labor, fertilizers, etc., per hectare in one and the same farming enterprise.

Cost concepts (see Fig. 2.)
- Of fundamental importance are the concept-pairs:

    special costs — overhead costs
    variable costs — fixed costs
    average costs — marginal costs.

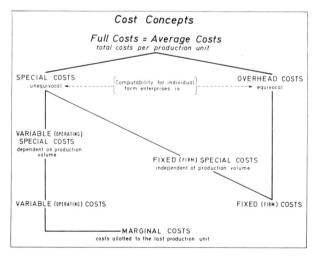

Figure 2

Glossary

Farm size classes

a) *Classification by market dependence*
   - subsistence farms: with operators selling less than 25% of production
   - weakly commercialized farms: with operators selling 25 to 50% of production
   - strongly commercialized farms: with operators selling 50 to 75% of production
   - fully commercialized farms: with operators selling more than 75% of production

b) *Classification by labor composition*
   - *large-scale wage-labor farms* are those whose operators do not work manually but can devote themselves exclusively to management problems.
   - *wage-labor family farms*, or *large family farms*, are those whose operators employ wage labor but participate both as manager and manual worker.
   - *land-rich family farms* are those with extensive farmland in relation to family labor potential.
   - *land-poor family farms* are those with insufficient farmland for the family labor potential.

Forms of farming diversity
   - monoculture   = one-sided production program
   - specialized farming = moderately varied production program
   - mixed farming   = highly varied production program
   - diversification   = increasing production variety
   - specialization (farming simplification) = reduction of production variety.

Efficiency concepts (see Fig. 3)
   - *gross return* = Farm income plus value of normal withdrawals for personal needs, retirement payments, rent in kind, and perquisites, plus value of inventory additions in livestock and farm-produced inputs.
   - *reimbursement share* = Gross cash returns less variable special costs (cost of seed, livestock purchases, veterinary services, feed concentrates, chemical fertilizers, plant protection agents, drying, cleaning, etc.; partly also fuel, machine, and hand labor costs; also interest payments on working capital).
   - *farm income* = Gross returns minus expenditures, farm taxes and encumbrances = Income from production factors of land, labor, capital, and management.

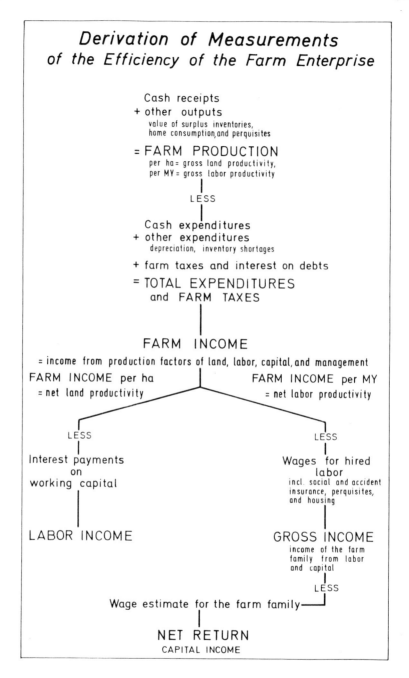

Figure 3

Glossary

- *gross income* = Farm income minus hired labor expenditures = Income of the farm family from land, labor, capital, and management.
- *labor income of the farm family* = Gross income minus interest payments on working capital.
- *net return* = Gross income minus wage claims of family labor = Interest payments on working capital plus operator's profit.
- *gross land productivity* = Gross output per ha FL
- *gross labor productivity* = Gross output per MY
- *net land productivity* = Farm income per ha FL
- *net labor productivity* = Farm income per MY

Livestock production based on grain feed can be:
hog fattening,
poultry raising, and
egg production.

Abbrevations and concepts for the agriculture of the East Bloc

— *Agricultural-Industrial Complex (AIC)*	= Totality of the economic domains which contributes to agricultural production or shares in the manufacture of foodstuffs (GDR – German Democratic Republic).
— *Agricultural Producers' Cooperative (APC)*	= Legally autonomous collective farm; differentiated into types I to III according to the amount of production factors used (GDR).
— *Agro-Chemical Center (ACC)*	= Specialized service enterprise that stores, transports, and provides fertilizers of all kinds for participating farms (GDR).
— *Brigade*	= Collective of farmers that fulfills production goals set by the cooperatives (GDR).
— *Cooperative Organization (COO)*	= New farm type, developed through the collaboration of the APC and PE (*People's Estate*) organi-

zations, service organizations, as well as units of the Food and Trade Economies (GDR).

— *Kolkhoz* = Farm type in the Soviet Union; cooperative formed by the collectivization of former peasant farms (USSR).

— *Kombinat for Industrial Fattening (KIF)* = Large, state-owned industrial farm specializing in the production of a specific animal product, e.g. cattle, hogs, poultry (GDR).

— *Machine Tractor Station (MTS)* = A former machinery-lending station in the Soviet agricultural system, with repair shops. Now replaced by the *Farm Machinery Association (FMA)*.

— *People's Estate (PE)* = State farm created after 1945 through expropriation and consolidation of farmland (GDR).

— *People's Commune (PC)* = Type of agricultural organization in China; embraces entire districts; organized into production bridges. The commune is a multi-purpose institution and includes, besides agricultural production functions, those of a political, administrative, and communal nature. (PRC — People's Republic of China).

— *Sovkhoz* = Farm type in the Soviet Union; state farm created through expropriation of large estates or land reclamation (USSR).

Glossary of Tropical and Subtropical Crops

English	Botanical	German
Luxury Plants		
cocoa, cacao	*Theobroma cacao* L.	Kakaobaum
coffee	*Coffea* spec.	Kaffee
tea	*Thea (Camellia) sinensis* L.	Tee
tobacco	*Nicotiana* spec.	Tabak
Rubber and Gum Plants		
Para rubber, caoutchouc tree	*Hevea brasiliensis* Muell. Arg.	Parakautschukbaum
gutta-percha	*Sapotaceae*	Guttaperchaliefernde Pflanzen
gum arabic tree	*Acacia* spec.	Gummiakazien
Oil and Fat-Producing Plants		
oil palm	*Elaeis guineensis* Jacq.	Ölpalme
coconut palm	*Cocos nucifera* L.	Kokospalme
olive	*Olea europaea* L.	Ölbaum
soybean	*Glycine max* L.	Soja
peanut, groundnut	*Arachis hypogaea* L.	Erdnuß
sesame	*Sesamum indicum* DC.	Sesam
castor bean, castor oil plant	*Rincinus communis* L.	Rizinus
Roots and Tubers		
manioc, cassava	*Manihot esculenta* Crantz	Maniok
yam	*Dioscorea* spec.	Yam
sweet potato	*Ipomoea batatas* Poir.	Batate or Süßkartoffel
Fruit Crops		
banana, plantain	*Musa* spec.	Banane
pineapple	*Ananas comosus* Merr.	Ananas
citrus fruits	*Citrus* spec.	Zitrus
date palm	*Phoenix dactylifera* L.	Dattelpalme
fig	*Ficus carica* L.	Feigenbaum
mango	*Mangifera indica* L.	Mangobaum
Sugar-Producing Plants		
sugarcane	*Saccharum officinarum* L.	Zuckerrohr
Grains		
rice	*Oryza sativa* L.	Reis
corn, maize	*Zea mays* L.	Mais
sorghum	*Andropogoneae*	Hirsen, große
millet	*Paniceae*	Hirsen, kleine

English	Botanical	German
Fiber Plants		
cotton	*Gossypium* spec.	Baumwolle
kapok, silk cotton tree	*Ceiba* spec.	Kapok
jute	*Corchorus* spec.	Jute
deccan hemp, kenaf	*Hibiscus cannabinus* L.	Kenaf
roselle hemp, sorrel	*Hibiscus sabdariffa* L.	Roselle
ramie, China grass, rhea	*Boehmeria* spec.	Ramie
sisal	*Agave sisalana Perrine*	Sisal
abaca, Manila hemp	*Musa textilis Nees*	Faserbanane

Introduction: Origins and Evolution of Agriculture – A Three-Stage Theory

The development of agriculture can be divided roughly into three epochs. These three stages, all of which have been experienced by the industrial countries, still exist side by side in the developing areas.

1. The representatives of *economies that simply occupy, or appropriate space* (gatherers, hunters, fishermen, and herders) struggle against the pressure for expanding their radius of action. Entire races, such as the Bushmen, pygmies, aborigines, or Patagonians, have starved or languished to the extent of their depletion of food sources. Their problem has been a growing lack of space. The primitive exploitive stage and early phase of gathering takes up the overwhelming part of the historical period of the economic development of man: 98–99 % (Manshard 1968, p. 20).

The nomadic population of the six Sahel nations has grown in the last three decades at an annual rate of about 1.7 %. Overgrazing has been the result. When, then, a severe drought also set in during the last few years, a catastrophic famine ensued, its severity shocking the world. Nature had cruelly corrected the biological balance, which economies at this primitive level and with growing populations can hardly maintain. The spatial problem had become a survival problem.

2. Limited food reserves and a narrow radius of action for man and animals thus set limits to expansion, and sooner or later force human groups to turn to *exploitive economies*. The stage of hoe culture, later of plow culture, is now attained. Clearing and cultivation supplement harvesting. Agri*culture* has now evolved from the simple collections of wild plants. Only now can one speak of farmers. Their history takes in only about the last 10,000 years.

Steppe-shifting, bog-burning, and forest-burning systems of shifting cultivation are examples of such soil-exploiting economies. Man draws deeply from nature's budget without having as yet the means for reimbursement. As a result, fertility-depleting agriculture must be followed in a few years by many years of grass, bush, or forest fallow, so that the fertility and ecological balance which the deficient hand of man has damaged or destroyed may be restored.

These events are not historical reminiscences. Shifting cultivation is still predominantly practiced by more than 200 million people over an area of more than 30 million km². Nor does the cause of this system lie in the economic irrationality of a population that is practically at the cultural stage of the Neolithic. So long as settlement remains dispersed, this economy can even be carried on with great efficiency. By using lavishly extensive and easily available land supplies, labor expenditures are minimized and almost all capital investment is dispensed with, and thus a minimum cost combination in the prevailing factor cost relationships is attained.

Growing population nevertheless leads to a fateful vicious circle: more cropland is needed and for that reason the fallowing period is shortened. A shorter fallowing period leads to incomplete regeneration of soil fertility. Decreasing crop yields stimulate further expansion at the expense of the fallowed land – and so on. With such a self-destructive system the Mayan culture shifted more and more from the center of the Yucatan Peninsula to the periphery until, reaching the ocean, was extinguished. (P. Gourou: The Tropical World. New York, 1961).

3. With increasing population density, therefore, exploitive economies too are finally no longer able to provide sufficient food. More *forms of land use emphasizing cultivation* are now required.

Once again the spatial problem becomes more discriminating. Plow culture requires plowable soil and thus cleared land. According to the principle of minimum expenditures, therefore, all agricultural peoples have first selected the treeless steppes and prairies, the natural grassland. The coniferous forests, which set down shallow roots and allowed clearing by burning, followed. Only after this did permanent cropping penetrate the regions of the deep-rooted forests.

When Slavs came from the east and settled Mecklenburg, they simply passed through the beech and oak forests which were closer to them in the east. They settled first in the more distant western part of the country because there in the coniferous forests the work required for clearing was modest. Also, they could use their iron hook plow better in the light soils.

Thus the available agricultural technology determines the choice of area. Malawi is an overpopulated agricultural country and yet possesses extensive unutilized land resources. But the soils are so heavy that they still cannot be worked at the stage of hoe culture that prevails there. Only if more powerful energy sources are made available in the form of draft animals or tractors can these areas be converted to cropland.

Up to now only the subsistence economies have been considered. Since cities and markets develop earlier and more vigorously than infra-structures, a new spatial, or locational problem confronts market-oriented agriculture in its supply and demand situation. The ancient cultural centers of mankind did not lie on the Nile, in Mesopotamia, on the Indus

Introduction

and Ganges, on the Mekong Delta, or on the Yangtze simply because of the food-production capacity of the fertile alluvial soils and the possibilities for irrigation; here waterways guaranteed what at the time was, even more than today, the cheapest form of transport. To the same extent that the growing population had to push the settlement area farther and farther into the interior, the problem of transport access worsened. When Thünen's "Der isolierte Staat" appeared for the first time in 1826, the steamship (from 1807) and the railroad (from 1825) were only in their infancy, and sixty years would have to pass before the first motor vehicle (1886) appeared. Otherwise Thünen's space picture would have looked different. The concentric rings would have retreated to a more radial arrangement of farming types. In the less accessible areas within the developing countries agricultural systems are often less a function of market distance than they are of distance from the major transport arteries.

Only when the transport network thickens and transport charges lower as national economic development progresses, does the farmer again free himself of the locative disadvantage of the more outlying areas. Then he can adapt much better to the natural locative restrictions. With the greatly increased number of production methods this is also now far more necessary and far more possible.

Eventually, further economic growth and a wealth of technical advances lead to such a complete control of natural forces by means of irrigation, drainage, fertilization, plant protection, adaption of the genetic potentials of plants and animals, and the like, that the farmer can carry on his business much more independently of the natural spatial or areal factors. Consequently in highly developed industrial countries the *personality of the farm operator* dominates to a degree heretofore unknown. Emotional, and not spatial or locational problems now come to the fore.

Thus the evolution of agriculture might be viewed as a variation on the well-supported three-stage theories of Richard Krzymowski (Geschichte der deutschen Landwirtschaft [History of German Agriculture]. Stuttgart, 1951) and Eduard Hahn (Von der Hacke zum Pflug [From the Hoe to the Plow]. Leipzig, 1914), one that would normally proceed as follows:

Occupation	→ *Exploitation*	→ *Cultivation*
(Appropriation)	(Depletion)	(Husbandry)

I. Agricultural Geography as a Science

To keep the focus of this book on specific agricultural-geographic subjects as much as possible, this chapter was made very brief and fragmentary. In spite of its methodological significance, the author believed allowances could be made since major statements already are available on this subject. Besides noting the cited sources, the reader should refer to the bibliography and especially the pertinent works on economic geography listed there.

1. Definitions

Agricultural geography is defined by Otremba (1976, p. 62) as the science of that part of the earth surface transformed by agriculture, not only as a whole but in its parts, its physiognomy, its inner structure, and its interrelationships. *Agriculture* is understood to be the management of the land for purposes of producing plant and animal products to satisfy human needs. Agriculture certainly is also a way of life, but primarily it is an economy. Thus agricultural geography is obviously a branch of cultural geography, but even more a component of economic geography.

The following basic agrospatial concepts, according to Manshard (1968, p. 10 f.), are of help in understandig agricultural-geographic relationships:

Agricultural space is a term applied to any land surface that is used agriculturally in any form whatsoever. In contrast, an *agricultural area* is defined as a spatial unit characterized by a clear dominance of farming. Structurally, it is a part of a larger heterogeneous spatial unit such as a state or physical area.

Agricultural zones are agricultural areas that can be associated with different climatic and vegetation belts and exhibit a definite structural and physiognomic uniformity.

The *agricultural region* is a part of the earth surface which, on the basis of defined geofactors and social and historic realities, possesses specific field and settlement forms, technical resources, and crop and livestock economies, and whose *agrarian structure* bears definite uniform characteristics. Further definitions can be found in the glossary on page 15.

I. Agricultural Geography as a Science

2. Objectives and Significance of Agricultural Geography

A major objective of agricultural geography is the analysis of the agriculturally structured areas and their natural, economic, and social relationships and organizations as reflected spatially. The primary goal of agricultural geography, according to Manshard (1968, p. 9), consists of investigating the spatial differentiation of the various manifestations of agriculture. The results of such agricultural-geographic studies are necessary for any transforming activity of man, insofar as they are spatially oriented.

Agricultural geography provides help for decision-makers: the demographic planner, who plans public services and utilities; the transportation engineer, who has to determine the routing of auto expressways or the siting of new ports; the regional planner, who is looking for the most favorable location for recreation areas; the agricultural specialist, who wishes to improve the agricultural structure; the hydraulic engineer, who is planning new dams; the food economist, who wishes to optimize the production and distribution of foodstuffs; and numerous other specialists. The smaller the planning area, the more the agricultural sciences can help; the larger the planning area, however, the more agricultural geography must provide the necessary aid (e.g. for the UNO, FAO, UNESCO). Thus it is no surprise that agricultural geography is especially well developed in the two large countries of the U.S. and U.S.S.R.

Further objectives of agricultural geography and their connection with agricultural economics were already sketched in the foreword of this volume.

3. Work Methods of Agricultural Geography

As a subdiscipline of economic geography, agricultural geography combines the elements and work methods of geography with those of the economic and agricultural sciences (Manshard 1968, p. 9). Thus work methods are as diverse as problems raised. In principle, one frequently takes the following steps in agricultural-geographic investigations:

1. Precise formulation of the problem.
2. Determination of the research design.
3. Data collection
    a) Collection and organization of primary data (building types, crop rotation systems, etc.),
    b) Collection and processing of secondary data (global and especially regional statistics; in part, even bookkeeping information and farm surveys).

4. Areal organization of data by agrospatial units, if possible by mapping; in other cases, graphs, diagrams, and tables may help.
5. Spatial analysis of the processed data with the help of methods, information, and theories from geography, the natural sciences, and economics.
6. Derivation of basic conclusions through elimination of exceptional cases.
7. Answering of the questions first raised with the principal results of the investigation.

II. The Climatic Zones of World Agricultural Space and their Significant Features for the Agricultural Economy

Whoever wishes to study and to understand the spatial arrangement of world agriculture must keep in mind the climatic zones. They are the most important differentiating force on a world scale. It is on them, therefore, that our first and foremost interest must be focused. Fig. 4 indicates the most important climatic zones of Africa, and it is supplemented by Fig. 5.

1. Tropical Rainy Climates

The tropical rainy climates are identical with the collective notions of "humid tropics" or "inner tropics." From the agricultural-geographic standpoint one must be concerned with at least a threefold subdivision:

a) Rainforest Climate

This climate is found in the equatorial lowlands. The Congo Basin and the Guinea Coast stand out in Fig. 4. The Amazon Basin and large parts of Indonesia also have a rainforest climate. This climate is hot and moist throughout the year. Two rainy seasons totaling at least 1,500 mm of precipitation, no month with less than 60 mm rainfall, an average temperature varying only between 25° and 28°C during the entire year, and a relative humidity seldom falling below 90% (6:00 a.m.) produce a moist climatic type that brings forth an evergreen, ombrophylic rainforest. This is the proverbial luxuriant tropical vegetation that the layman usually identifies with all tropical vegetation, though it occurs only in a relatively small part of the tropics, and just in the rainforest climate. The rainforest climate has eight-and-a-half to twelve wet months and thus must be characterized as humid to perhumid.

Since the natural vegetation of this climate is forest, it is no surprise that bush and tree crops are favored among the cultivated plants: cacao, rubber, oil palms, coco palms, and coffee (*Coffea robusta*). With the exception of the palms, these crops provide no basic foodstuffs. That partly explains the sparse settlement of the Amazon and Congo basins. Bananas, sugarcane, manioc, yams, corn, and under certain conditions, rice, are also cultivated. Conditions for livestock raising in this climate are extraordinarily unfavorable, as they also are for man, and particularly for whites, who cannot tolerate well this constantly moist and hot climate.

II. The Climatic Zones of World Agricultural Space

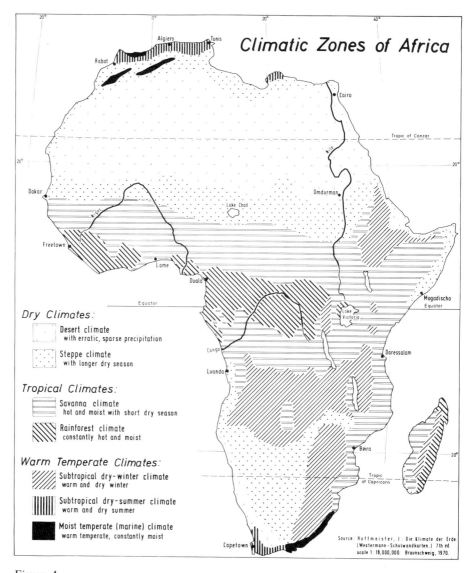

Figure 4

b) Humid Savanna Climate

The humid savanna bounds the rainforest belt on both northern and southern margins. Its climate, with its six to eight-and-a-half wet months and 600 to 1,500 mm of rain, is subhumid. Rainfall is concentrated in one long rainy period in the summer, followed by a short dry period in the winter. Grasslands occur as luxurious grass savannas with gallery forests, and woodlands take the form of monsoon forests.

1. Tropical Rainy Climates

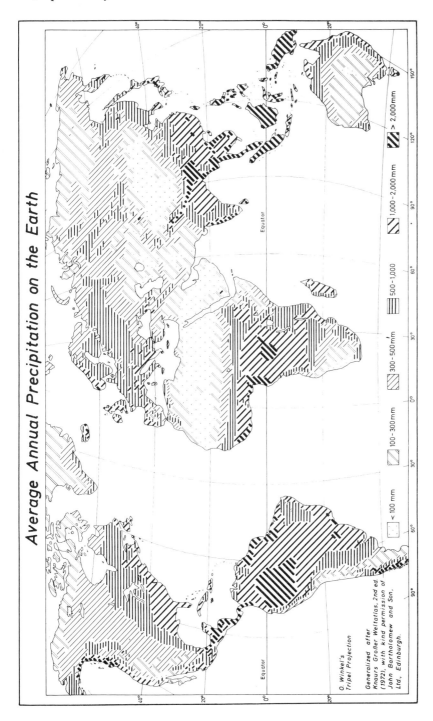

Figure 5

The raising of ruminants is now possible since the danger of disease, in contrast to the situation in the rainforest climate, is less. On the demand side, livestock raising is of course still restricted because of the low purchasing power of the population. The contact zone between the humid savanna and rainforest climates is designated as the wet boundary of grazing. The crops of the rainforest climate that demand the most water, such as cacao, rubber, and coffee, are little cultivated in the humid savanna climate, and the cultivation of oil palms and manioc also diminishes. Even yams become somewhat uncertain. In contrast, new types enter the cultivated plant society, crops that require a dry period for ripening: bush beans (Phaseolous) and peanuts.

c) Tropical Highland Climates

One cannot speak of a tropical highland climate in the singular, for here, depending on the altitude, slope, exposure, and conditions of moisture, light, and heat, can be found more diverse climatic variants than anywhere else. The association with the tropical rainforest climates is also problematical, because dry climates also occur in tropical highlands, and temperate or even Arctic climatic elements appear in very high regions of the Himalayas and Andes.

Tropical highlands begin, by general agreement, at about 100 m above sea level. The average annual temperature in Madras, at sea level, comes to 27.8°C, but at the 2,280 m-high hill station of Ootocamund, in the Nilgiri Mountains, reaches only 13.8°C. This is a difference of ± 0.6°C per m. As a result, in the tropical semiarid lowland millet, sorghum, and peanuts dominate the landscape and in the moist monsoon climate of the Indian southwest coast rubber, pepper, bananas, and manioc are grown. In the high mountains, however, which are still frost-free, tea, coffee (*Coffea arabica*), potatoes, and vegetables are cultivated (Piekenbrock 1958, p. 20). Thus somewhat the same succession of agricultural landscapes that we encounter on a trip from the Congo to the Mediterranean or from Amazonas to the La Plata can be experienced again with an ascent of a few hundred kilometers in high mountains along the equator.

There are specific altitudinal levels, physiologically determined and economically relevant, for our cultivated plants. In the equatorial zone (10° N.L. to 10° S.L.) many crops have an

– *altitudinal minimum.* Economic cultivation of coffee, e.g., is only possible above 950 m, tea above 1,300 m, potatoes and passion fruit above 1,600 m, and wheat only above 2,000 m. Also diverse is the

– *altitudinal range.* Coffee (*Coffea arabica*) extends from 950 to 2,000 m, corn from 0 to 2,800 m. All crops, however, have an

– *altitudinal boundary*. In Costa Rica, e.g., cultivation of rice stops at 1000 m, coffee at 1,450 m, sugarcane at 1,500 m, cooking bananas (plantain) at 1,700 m, and criollo grasses at 2,000 m, whereas imported grasses, red clover, corn, potatoes, and European vegetable types ascend to 2,800 m (Spielmann 1969, pp. 42–51).

2. Dry Climates

The dry climates are set off from the tropical rainy climates by the climatic dry boundary, where the number of humid months in the year now comes to, at most, six. While the tropical rainy climate is usually characterized agriculturally by a water surplus, all tropical dry climates, without exception, suffer from a water deficit. This situation also applies to what are called the outer tropics.

Dry climates are not only found in the tropics, but in the subtropics and in the temperate and even cooler latitudes as well. Common agricultural characteristics nevertheless appear to allow their treatment as a group.

a) Dry Savanna Climate

This type clearly belongs to the tropics, as the vegetation formations of the savannas occur almost exclusively there. The dry savannas usually insert themselves as a very small belt between the humid savannas and shrub savannas. Larger areas of this climatic type are found throughout almost the entire Zambezi Basin, in the southern part of the Sahel Zone, in western Malagasy, as a belt through India, in northernmost Australia, and in parts of Mexico. Three-and-a-half to six humid months permit rainfed farming, i.e. cropping without irrigation. Grasslands take the form of short grass savannas, and woodlands appear as deciduous dry forest (e.g. miombo). In contrast to the humid savanna, the rainy season is much shorter and less favorable for production.

With only 300 to 600 mm of rain a year, the climate is definitely semi-arid. It is an example of the typical alternately wet-and-dry tropics.

Cropping now can be supported only by the most drought-resistant crops, mostly millet, sorghum, peanuts, and bush beans. Even the cultivation of yams is no longer possible because of the long dry season. Shifting cultivation must often be resorted to in order to save water. In contrast to the temperate climate, however, livestock raising becomes more important relative to crop farming because of the increasing competitive advantage resulting from the dryness. Thus dryland cropping and extensive grazing are competitive in this climatic zone from the standpoints of both farming system and regional physical environment. Their

b) Shrub Savanna Climate

The shrub savannas are set off from the dry savannas agricultural-geographically by the so-called dry boundary, the boundary of rainfed farming. Thus there is practically nothing but extensive grazing in the shrub savannas. Indeed, irrigation can gain little additional farmland because of the low water table and the fact that even large rivers have no steady flows because of the very long dry period. Only an exceedingly one-sided agricultural landscape can develop in this semiarid climate, with its short rainy season of up to four humid months at the most and just 100 to 300 mm precipitation. Natural pastures are used with little-demanding animals, more with cattle in the direction of the equator but more with sheep in the poleward direction. This type of livestock grazing is practiced in the New World on cattle farms (ranches) and in the Old World still frequently by nomads.

The shrub savanna climate includes the largest part of Southwest Africa and the Kalahari, and the northern portion of the Sahel Zone. Further

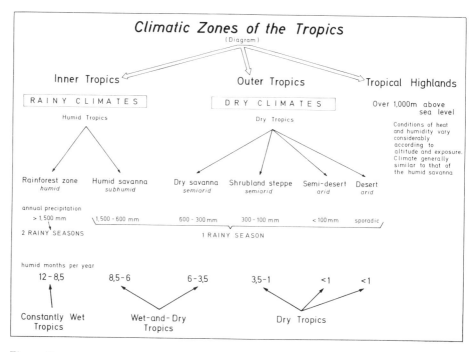

Figure 6

2. Dry Climates

examples are southern Somalia and South Ethiopia and a broad zone of northern Australia.

Fig. 6 may facilitate the overview of the tropical climates with its schematic presentation (p. 36).

c) Steppe Climate

Steppes have an agricultural-geographical character that is similar to that of the dry and shrub savannas. They are often not suitable for rainfed farming, though they are pasture areas par excellence. Overall, they suffer from a lack of water for man, plants, and animals.

But the steppes lie beyond the tropics of Cancer and Capricorn, in the subtropical and middle latitude climates. This leads to a modification of the economic characteristics that apply to the dry and shrub savannas:

- Precipitation in the shrub and dry savannas declines — as everywhere in the tropics — in the summer, but in the steppes usually in the winter. The result is that animals in the steppes are exposed to greater stress, since calves and lambs must be nursed while the vegetation is dormant — and all this in the greater summer heat.
- The network of watering places must be denser because the forage-poor season coincides with the heat period.
- Under certain conditions there is a need to keep feed in reserve for the dry season.
- Under certain conditions there is also the need in the continental cold-winter climate to protect the animals from inclement weather with primitive stalls.
- Under certain conditions less productive animal breeds or varieties must be given priority because of their greater cold resistance, such as the yak in interior Asia or the llama in the Andes.
- Cultivated plants are now completely different from those in the dry tropics because of the varying light and heat conditions from place to place.

Large steppe areas are found in the Intermountain states of the United States, in North Africa, in the Spanish Meseta, in the Don area, and in large parts of interior Asia.

d) Semidesert Climate

The arid semidesert with less than 100 mm precipitation per year form the transition from the shrub savannas and steppes to the dry-hot deserts with their still scantier and only sporadic rainfall. Deserts usually lie beyond the dry boundary of habitation, though they are partly utilized by desert nomads, and industries of sorts develop in the vicinities of oases and oil wells.

It is well established climatologically that almost all large deserts of the world lie on or near the tropics of Cancer and Capricorn. This is pretty much the situation for the Sahara and the Arabian Desert, the Namib and all of interior Australia, and the Iranian, Pakistani, and interior Asian deserts (in Kazakhstan, Uzbekistan, eastern Turkestan, etc.). Thus the semideserts surrounding the deserts also lie near the two tropics and therefore cannot be easily differentiated as to whether they belong to tropical or subtropical areas.

Only the somewhat more humid zones of the semideserts can still be used for sedentary grassland farming. The remaining portions can be used only episodically by nomads, hunters, and gatherers. The dry boundary of livestock grazing thus extends through the middle of the semideserts.

3. Humid Warm-Temperate Climates

The three variants of this climatic group are distinguished from each other, from the agricultural-geographic viewpoint, not only in temperature conditions but in the amount of precipitation and especially its distribution.

a) Subtropical Dry-Summer Climate

Representatives of this climatic type are the Mediterranean area, the south coast of the Black Sea, a wide zone from the Caucasus to almost the Persian Gulf, the hinterland of Capetown, the largest part of California, central Chile, and parts of the south coast of Australia.

Agriculture bears the stamp of warm, moist winters and hot, dry summers. This climate becomes more extreme as we approach the dry climates. In North Africa, only the typical winter rainy season allows rainfed farming, whereas in the summer drought and heat preclude all cropping without irrigation.

Of the four regions in Table 1, all but the Po Plain belong to the subtropical dry-summer climate. One can see that the total yearly precipitation, though in itself not insignificant, is unfavorably distributed. Thus in the June-to-August quarter only 104 mm fall on the Campanian coast, no more than 86 mm fall in northern Apulia, and just 71 mm fall on the coast of the Ionian Sea. These amounts are respectively 8.3, 13.3, and 8.9% of the yearly total, and their use effectiveness, moreover, with a mean temperature of 23°C, has to be very low. Under such extreme ecological restrictions agricultural land use is a compromise decision between

3. Humid Warm-Temperate Climates

Table 1: Climatic Data for Italy (Ten-Year Averages)

Quarter	Lower Po Plain (Bassa Padania)	Tyrrhenian Coast (Campania)	Northern Apulia	Coast of the Ionia Sea
I. Precipitation in mm				
Dec. – Feb.	186	440	213	324
March – May	200	259	161	169
June – Aug.	*160*	*104*	*86*	*71*
Sept. – Nov.	230	378	187	231
Annual Total	776	1,181	647	796
II. Mean Temperatures in °C				
Dec. – Feb.	2.5	7.9	6.5	7.9
March – May	12.3	13.0	12.4	13.0
June – Aug.	22.4	23.4	22.7	23.4
Sept. – Nov.	13.7	16.9	15.8	16.9
Annual Mean	12.7	15.3	14.4	15.3

Source: Annuario di Statistica Agraria 1948, Rome, 1948. Cited by M. Rolfes, Die betriebswirtschaftlichen Grundlagen des Zuckerrübenbaues in den Ländern der Europäischen Wirtschaftsgemeinschaft, Teil II. Manuscript reproduced by the Forschungsgesellschaft für Agrarpolitik und Agrarsoziologie e.V., Bonn, 1961, pp. 47 and 56.

– the restriction of cropping to the fall, winter, and spring months;
– the shift to bush and tree crops, which through their deep root systems can better endure the summer dryness (grapes, olives, almonds)/ and
– irrigated agriculture.

Usually, in the interest of maximizing income, it is not a question of selecting one of these three options but of deciding on the appropriate combination of two or all three.

b) Subtropical Warm-Summer Climate

For this climatic type, Table 1 offers some data for the Po Plain. It is apparent that temperatures are lower and precipitation is more evenly distributed. Here the summer heat can be used for agriculture without irrigation. Important world farming areas are associated with this climate: a large part of China, southern Japan, parts of the Australian east coast, large parts of northern India, southeastern United States, Brazil south of the tropic, and the La Plata countries.

If summer temperatures are sufficient, then heat-loving crops like rice, peanuts, soybeans, or cotton can be cultivated. If the winters are still mild enough, year-around crop production with two or even three harvests per year is possible. Then crops which like a cooler climate, such as grapes, winter grain, or forage crops, are planted in the winter, and those which require a warm climate, such as rice, soybeans, sweet potatoes, or vegetables, are grown in the summer. In this way, crop rotations can be developed within a calendar year, as the following examples show.

Crop Rotations in Southern Japan (Tsuzuki 1963, p. 210)

Example A:	*Example B:*
Winter: renge grass	Winter: w. grain
Summer: rice	Spring: vegetables
Summer: rice	Summer: vegetables

c) Marine Cool-Summer Climate

This climatic type, also designated as oceanic, has its widest distribution in the European Community area, which except for Italy is almost totally encompassed here.

Mild winters and cool summers are characteristic: a climate which vigorously promotes fodder growth and consequently is also favorable for cattle raising. This is still more the situation when longer and even year-around pasturing is possible (Cornwall, parts of Normandy, areas around the Bay of Biscay). Such climatic conditions reduce feeding and building costs and thus increase profitability. Root crops (sugar beets, potatoes) and grains dominate the fields. Among the grains, winter-grown types rank first because the long fall easily permits their cultivation, even after the root crops.

All other areas with marine cool-summer climate are very much smaller and, with the exception of a northerly coastal strip in western North America, lie in the southern hemisphere: the tip of southeast Australia, New Zealand, the southern tip of South America, and large parts of the Republic of South Africa.

4. Humid Cool-Temperate Climates

Under this designation are subsumed climates in which the most important agricultural criterion is not the amount of precipitation, but the minimum temperature.

4. Humid Cool-Temperate Climates

a) Continental Warm-Summer Climate

For the continental warm-summer climate, this restriction still does not apply to the summer, to be sure, but to the winter whose lengthening leads to a longer rest period for vegetation. Significant grain corn areas are found in this climatic area, thus the American Corn Belt south and west (up to 100° W.L.) of Chicago and the greatest part of the Balkan area. China north of Peking, Korea, and large portions of northern Japan also belong here. Besides corn, soybeans also grow well to some extent.

b) Continental Cool-Summer Climate

Now the winter becomes still longer and colder and the summer also cooler. Precipitation falls predominantly in the summer.

Agriculture reacts with an emphasis on spring-grain and fodder cropping. The latter already assumes well-established forms in such areas as the Dairy Belt of northeastern United States and southeastern Canada, central Sweden, and the Baltic Provinces.

Only three large areas display these climatic features: one, the area north of the Sea of Japan; another, a broad belt in North America which extends from New York and Halifax through the Great Lakes and well beyond Winnipeg; and a third, the great block of central and eastern Europe, delimited approximately by the cities of Oslo – Szczecin (Stettin) – Vienna – Magnitogorsk – Leningrad.

c) Subarctic Climate

Finally, poleward of the continental cool-summer climate, and naturally only in the northern hemisphere, is the subarctic climatic belt which extends northward beyond the Arctic Circle. Characteristic features are a moderately warm summer, a very long and cold winter, a concentrated summer precipitation, and a natural coniferous forest flora. The region includes almost all of Scandinavia north of the three capitals, the U.S.S.R. approximately north of a line through Leningrad – Tomsk – Irkutsk, as well as a broad belt of northern Canada and Alaska extending from the regions around southern Hudson Bay westward to the Bering Sea. The Arctic Circle passes through all these large areas. The same climate also appears at the higher medium and high mountain elevations of more southerly latitudes, such as the Alps, Carpathians, and Andes.

For agriculture, the most significant of all climatic features is the short growing season. It allows fodder cropping to dominate even more than in the continental cool-summer climate, and of course always in well-established forms. Fodder cropping is superior here despite the long stall-feeding period because it completely utilizes the effective plant growth period, from the first day to the last, whereas with crops which have to be replanted annually, a part of the costly and all too short growing season is lost in planting, cultivating, and harvesting.

To this handicap should be added the rapid impoverishment of the domesticated plant community toward the north, as the specific polar boundaries are exceeded. The cultivation of sugar beets finally stops at 61° N.L., that of wheat at 63°, and that of spring barley and potatoes at 70°. Thus crop rotations in the far north are only possible in ley farming systems, in which four to six or more years of grassland cropping are followed by one or two years of spring barley and potatoes.

III. The Delimitation of World Agricultural Space

1. The Expansion of World Agricultural Space as a Current Problem

By 1976, world population had exceeded the 4-billion mark, or sixteen times the number, quarter of a billion, living in the time of Christ. By 2007, that number is expected by the UNO to reach 7.7 billion, and by 2050 to come to as much as 14 to 15 billion. According to projections by the FAO, world food demands by the year 2050 will be seventeen to eighteen times that of 1960.

Already more than two thirds of the world population lives in the developing countries. Nearly 400 million people suffer from hunger. By about 1985, the number of afflicted may well rise to 750 million. During this period, to be sure, sizable food surpluses in the developed countries must also be taken into account; but in the developing countries an increase in demand of around 70% will only be matched by an increase in production of less than 50% (Matzke 1975, p. 362). What can be done so that food production will not just keep pace with population growth, but will surpass it in providing the adequate diet so urgently needed?

Basically, agricultural productivity can be increased three ways:

1. through *an increase in operational intensity* by means of augmented investment in irrigation, fertilization, pest control, plants and animals of high genetic quality, etc;
2. through *an increase in organizational intensity,* that is, through successive substitution of intensive enterprises for extensive ones, thus e.g. roots and tubers in place of grains or dairying instead of cattle fattening; and
3. through *an expansion of agricultural space* beyond its present boundaries.

As Table 2 shows, only about a third of the world land surface is now used agriculturally, and that with a 2:1 ratio of permanent grassland to cropland. The agricultural share of land in the agricultural countries is clearly lower than in the industrial countries. Countries with extremely small agricultural shares are almost without exception developing countries, whereas those with extremely high agricultural proportions are to be found among the industrial countries. Thus agricultural proportions are the lowest where most people are undernourished or even starving. In contrast, where agricultural surpluses are accumulating, as in Western Europe, agricultural shares are the highest. There must be reasons for this.

Table 2: Proportion of Land in Agriculture

Geographic or Political Unit	Permanent Grassland	Cropland	Bush and Tree Crops	Total Agricultural Area
	% of Total Area			
A. World	22.4	11.0		33.4
B. Continents				
South America	21.6	4.9		26.5
Africa	26.3	6.9		33.2
Asia[1]	19.5	17.5		37.0
Europe[1]	18.4	29.3		47.7
C. National Groups by Stage of Economic Development[2]				
Agricultural Countries	20.6	10.1		30.7
Industrial Countries	27.6	12.1		39.7
D. Countries with Extremely Small Proportion of Land in Agriculture				
Bahamas	0.1	0.1 0.1	0.9	1.1
Egypt	—[3]	2.7	0.1	2.8
Oman	4.7	0.1	0.1	4.9
Libya	4.0	1.3	—	5.3
E. Countries with Extremely High Proportion of Land in Agriculture				
Nigeria	27.1	23.6		50.7
West Germany	21.7	30.5	2.2	54.4
Australia	59.1	5.8	—	65.0
Denmark	6.7	61.6	0.4	68.7

[1] Excluding U.S.S.R. — [2] Excluding countries with centrally controlled economies. Source: FAO, Production Yearbook, Vol. 27 (1973), Rome, 1974, pp. 3 ff.
[3] — = none, or in negligible quantity.

The boundaries of agricultural space can be drawn on the bases of cold, drought, wetness, soil salinization, market isolation, and other agents. Accordingly, one speaks of polar boundaries, altitudinal boundaries, dry boundaries, wet boundaries, settlement boundaries, and so forth. It is by no means as simple to delineate these boundaries of world agricultural space. This is because they are not lines, but zones, and within these zones three different boundary lines have to be defined:

(1) the *effective boundary,* the actual limit of agricultural production according to land use statistics;
(2) the *profitability boundary,* the boundary where returns equal zero; and
(3) the *technological boundary,* the boundary up to which land could be farmed according to the current stage of technology, if economic considerations were waived.

Only the first, the actual boundary, will be discussed. It is not fixed, but on the contrary fluctuates with the successes and failures of economic life. After the last war, in Siegerland, the unproductive and finally extinct *Hauberg* farming system was revived. By 1976, a total of 310,00 ha in West Germany was no longer being used agriculturally, though it was still arable (social fallow). The fluctuation of the actual boundaries of agricultural space is almost identical with that of the limits of the inhabited area. That the correlation is not complete is shown by examples such as oil drilling in the desert. With greater population density, minor income demands, and food shortage, the actual boundary shifts toward the profitability boundary or even coincides with it, though the profitability boundary also advances under such conditions because of higher prices for agricultural products. Thus under these conditions the three boundaries are close to one another, i.e. the boundary zone is narrow.

The situation is quite different with low prices for agricultural products and high income demands of the rural population. The actual boundary then retreats from the marginal locations, while the technological boundary is characterized by far-flung outposts. The boundary zone is therefore wide and within it occurs what we today, for example, label as highland exodus in the western European mountains of moderate altitude or in the Apennines. Here the highland boundary of agricultural space is lowering. An analogous phenomenon is found today in the semidesert areas around the Sahara, where the meager pastures can no longer be used. The dry boundary of agricultural space is retreating here, the effective dry boundary moving away from the technological.

It is obvious that at present the reverse movement, the advance of the farming boundaries, is stronger and more frequent. Thus, for example, in the tropical rainy climates with trifling population densities, agricultural boundaries are being extended outward with every new road and railroad. A broad band of pioneer farms originates to the right and left of these transportation arteries because, although not shortening the spatial distance from market, they do reduce the economic distance so that the effective transportation boundary can expand around the market site in eccentric rings. The chains of pioneer farms that are pushing forward in this way into the Amazon or Congo basins in the tropical forest region soon result in country-town or city-like settlements, so that the spatial distance to market is also reduced and additional settlements become viable on the periphery. From all this it follows that the boundaries of the farming area can be conditioned not only ecologically but economically.

2. Ecological Boundaries of Farming

In the following we shall first look at the characteristics of the individual ecologically-determined boundaries of the farming area and their approximate courses.

a) Polar Boundaries

The polar boundaries of cultivated plants are more important for the industrial countries than for the developing ones. As indicated in Fig. 7, potatoes and spring barley push farthest into the high north, up to about 70° N.L. This is possible only through ley cropping and (or) fallowing, as Overview 1 makes clear.

Overview 1: Crop Rotation Types in Lapland

A Arctic Circle ←	B	C	D → North Cape
1.–5. Grassland 6. Potatoes, oats 7. Fallow 8. S. barley	1.– 8. Grassland 9. S. barley 10. Potatoes 11. S. barley	1.–10. Grassland 11. Potatoes 12. S. barley	1. Fallow 2. Potatoes 3. Fallow 4. S. barley
		% CL		
62.5 8.0 17.0 12.5	72.7 9.1 18.2 –	Grassland Potatoes S. barley, oats Fallow	83.4 8.3 8.3 –	– 25.0 25.0 50.0

The growing season is so short that one must use as much of it as possible for the growth of the cultivated crops without losing time for so much as plowing a furrow. Thus in crop rotation A, land preparation begins in the last grass cropping year with a prompt plowing and in the fallow year. In crop rotation D, land preparation, potatoes, and spring barley alternate with one another so that for crops, cultivation can directly follow the drying of the soil in conjunction with the snow melt and harvesting can beat the outbreak of the next winter. Thus a maximum growing season is attained, one closely corresponding to physical circumstances. Ley cropping and fallowing are methods of extending the growing season for food crops and in general are prerequisites for their cultivation. This also holds true for other locations with extremely short growing seasons (Overview 2).

2. Ecological Boundaries of Farming

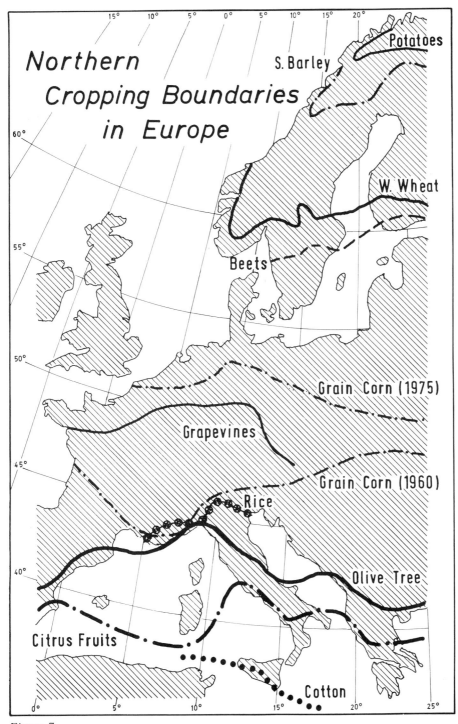

Figure 7

Overview 2: Crop Rotations with an Extremely Short Growing Season

A	B	C	D
Siberian Coniferous Forest Zone	Graubünden 1,500 m Altitude	New York State Harsh Altitudes	
1. Complete fallow	1.–10. *Naturegart*	1.–5. Grass-clover	1.–8. Grass-clover
2. S. grain	11. S. barley	6. Corn silage	9. Green oats
3. Complete fallow		7. Oats	
4.–5. S. grain			

The Siberian fallow-crop rotation resembles the northern Swedish rotation D. In Graubünden, the cultivation of spring barley as a bread grain in spite of a short growing season is forced. In the case of crop rotation D, oats can no longer ripen. They are, however, still raised green to rejuvenate the grass-clover turf. Without the oats, a conversion from the more productive rotation grassland to the less productive permanent grassland would be necessary.

Wheat is found only up to about 63° N.L., and only in western Norway can it be extended farther poleward, through the influence of the Gulf Stream. The northern boundary of beet cultivation corresponds approximately to the northern boundary of the Swedish East Göta Plain (61° N.L.). Grain corn offers a good example of how advances in plant breeding can push back the polar boundary; in the last two decades it has made a truly victorious march through central Europe.

The map (Fig. 7) shows that the polar boundary running through southern Europe restricts the distribution of some useful plants that are extremely important for developing countries: rice, olive trees, citrus fruits, and cotton.

If we now direct our observations to Fig. 8 and Overview 3, beginning at the equator and moving gradually into the middle and higher latitudes, then the following can be said:

– Coconut and oil palms favor locations particularly near the equator (polar boundary of 15° or 16° N.L.).
– Sisal, cacao, coffee, bananas, manioc, and rubber have very low polar boundaries of 19° to 25° N.L. because they find their physiological optimum under humid tropical conditions.
– Sweet potatoes, cotton, sugarcane, peanuts, tea, and citrus fruit push farther north, up to 35° to 42° N.L., thus into the subtropics. Some of them, however, such as sweet potatoes or sugarcane, also find suitable locations directly on the equator.

2. Ecological Boundaries of Farming

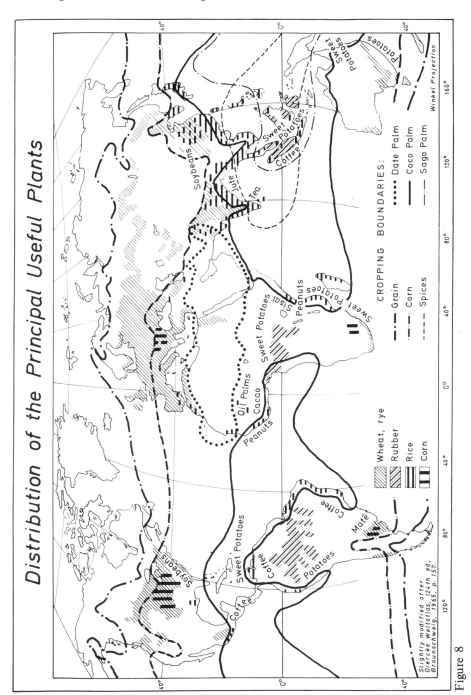

Figure 8

— Finally there are the cultivated plants of the tropics, such as soybeans, rice, or grain corn, which extend well into the mid-latitude climates (45° to 54° N.L.). They are, on the basis of their great distributional range, world food crops par excellence.

Overview 3: Polar Boundaries of Some Useful Plants, in Degrees N.L. (Approximate Values)

Useful Tropical Plants		Useful Tropical and Subtropical Plants	
Coconut palm	15°	Sweet potatoes	35°
Oil palm	16°	Cotton	38°
Sisal	19°	Sugarcane	39°
Cacao, *arabica* coffee	22°	Peanuts, tea	41°
Bananas, manioc	23°	Citrus fruit	42°
Rubber	25°	Soybeans	45°
Useful Tropical, Subtropical, and Mid-Latitude Plants		**Useful Subtropical and Mid-Latitude Plants**	
Millet, sorghum	45°	Olive tree	45°
Rice	52°	Grapevines	51°
Tobacco	53°	Beets	51°
Grain corn	54°	Wheat	63°
Beans (Phaseolus)	...	Barley, potatoes	70°

Sources: Franke, G. et al.: Nutzpflanzen der Tropen und Subtropen. Vol. I (2nd ed.) and Vol. II. Leipzig, 1975 and 1967. — Schütt, P.: Weltwirtschaftspflanzen. Berlin and Hamburg, 1972.

b) Altitudinal Boundaries

„The altitudinal boundary of cultivation, compared to the polar and dry boundaries, excludes only small islands from the world's arable land. It is like a cold boundary in its characteristics and as a result is similar in structure to the polar boundary, into which it blends imperceptibly in the high latitudes. ..." (Otremba 1976, p. 101). Relief also plays a decisive role.

The distributional picture of useful plants, as affected by both polar and altitudinal boundaries, is extremely distorted (see Overview 3 and Table 3) For some cultivated plants, it is one of polar and altitudinal boundaries that are both closely confining, such as for coconut palms, oil palms, cacao, and kapok. For other useful plants, it is one of crops advancing widely both poleward and altitudinally, as with spring barley, potatoes, and wheat. Then there is the pattern of plants like sisal, teff, pyrethrum,

2. Ecological Boundaries of Farming 53

tea, and passion fruit, whose polar boundaries are narrowly drawn but whose altitudinal limits are widely spread. Finally, there is the distribution of plants that thrust to the pole but avoid highlands. Beets, soybeans, and peanuts are among this group.

Table 3: Altitudinal Boundaries of Some Useful Plants in Meters
(Approximate Values)

Crop Type or Useful Plant	a) Tropical Rainy Climates	b) Dry Climates	c) Humid Temperate Climates	Observation Area; a), b), and c) = Climatic Zone
Pasture	2,800	5,210	4,700	a) Costa Rica; b) Andes; c) Nepal
Cropland	4,300	4,300	4,600	a), b) S. America S.L.; c) Central Asia
Cacao	1,300	–[1]	–	a) Western Colombia
Bananas, sugarcane	1,600	·[2]	·	a) Costa Rica
Citrus fruit	2,000	·	700	a) SE Asia; b) California
Dry rice	2,000	800	3,000	a) Western Colombia; b) Himalayas
Manioc, sweet potatoes	2,000	2,000	·	a), b) Equatorial areas
Arabica coffee	2,280	–	200	a) Western Colombia
Tea	2,400	–	2,300	a) Sri Lanka
Millet, sorghum	2,500	2,500	·	
Vegetables	2,800	·	3,600	a) Costa Rica; b) Tibet (Lhasa)
Sisal	3,200	1,800	–	a) Coastal areas
Wheat	3,300	3,300	3,600	a) Tanzania; c) Tibet (Lhasa)
Grain corn	3,900	2,800	1,300	a) Peru, Mexico; c) Inner Alps
S. barley	4,100	4,100	4,600	a) Andes; c) Central Asia 29° N.L.
Potatoes	4,300	4,300	4,400	a) S. America 15° S.L.

[1] – = None, or negligible quantity; [2] · = data not available.
Sources: Franke, G., et al.: Nutzpflanzen der Tropen und Subtropen, Vol. I (2d ed.) and Vol. II. Leipzig 1975 and 1967. – Otremba, E.: Die Güterproduktion im Weltwirtschaftsraum, Stuttgart, 1976, pp. 102f. – Schütt, P.: Weltwirtschaftspflanzen. Berlin and Hamburg, 1972. – Spielmann, O.: Viehwirtschaft in Costa Rica. Dissertation, Hamburg, 1969, pp. 42 ff.

A definite fortuitousness is associated with the values of Table 3, for well-known altitudinal limits are the ones recorded although they are probably exceeded by less-known sites. Table 3 also leaves open the matter of whether cultivated plants are actually cultivable at sea level or whether the altitudinal boundaries include an altitudinal threshold.

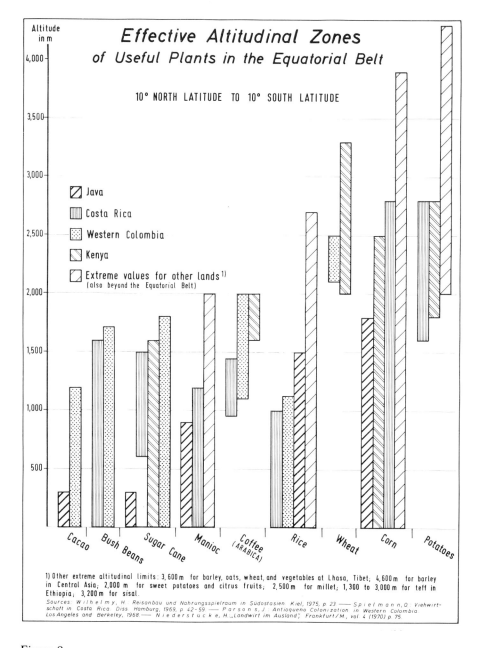

Figure 9

2. Ecological Boundaries of Farming

Figure 9 sketches the altitudinal zones of some cultivated plants in four small countries of comparable geographic location, namely the equatorial belt. Three interesting facts are revealed:

1. Of the nine listed cultivated plants, only three have an *altitudinal minimum:* coffee ends at 950 m, potatoes at 1,600 m, and wheat at 2,000 m. There are, however, other outstanding examples of highland crops which in the equatorial belt are cultivated only above the following minimum elevations: tea and cotton, 500 m; European vegetables, 700 m; chinarinde, 1,000 m; teff and pole beans, 1,300 m; red clover, 1,500 m; passion fruit, 1,600 m; and pyrethrum, 2,100 m.
2. Despite similar geographic locations, the altitudinal boundaries exhibit in part notable *place-specific differences* that are predominantly conditioned by relief. Thus grain corn in Java ascends only to 1,800 m, whereas in Kenya it is still cultivated up to 2,500 m and in Costa Rica even as high as 2,800 m.
3. The *altitudinal range of cultivation* for the individual cultivated plants is extremely varied. *Arabica* coffee has only a small altitudinal tolerance, being cultivated from only 950 to 2,000 m above sea level, thus within an altitudinal range of only 1,050 m. In contrast, grain corn is cultivated from sea level up to 2,800 m.

Fig. 10 schematizes the distribution of terrace farming in Java and compares the cultivated vegetation with the natural vegetation of the

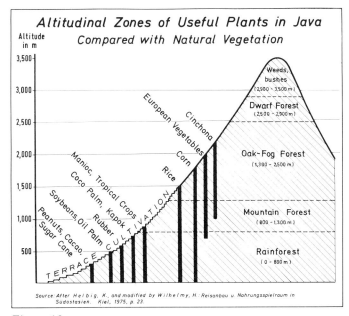

Figure 10

various altitudinal zones. An idea of how strongly the altitudinal zonation of economic plants of world importance can mark the landscape is shown by the cultivation of tea in Sri Lanka, as indicated by the following (Sarfalvi 1970, p. 301):

Up to 500/600 m: Tea planted locally, in part in single bushes as a cash crop on small farms. Poorer grades for the domestic market and for export to poorer countries.

500/600 to 1,000/1,200 m: Transitional zone between peasant farm and plantation economies. Altitudinal zone of the dual economy. Rice cropping in the valleys, tea farms and plantations on the slopes.

1,000/1,200 to 2,300 m: Tea landscape, tea monoculture. The presence of man is evidenced only by one-sided tea cultivation.

In all cases, however, it has been confirmed that with increasing altitude the cultivated plant community of any one level, with its particular climatic and exposure conditions, undergoes a gradual impoverishment. This accentuates an unfavorable seasonal distribution of labor all the more as the growing season becomes shorter, so that sowing or planting on the one hand and harvesting on the other are pushed still further together. Pure crop farming would thus mean an unbearable labor stress in the summer and a lack of opportunity for productive work in the winter. For this reason, cropping can ascend to increasingly greater heights the more it is combined with livestock raising, which requires winter work.

Fortunately it so happens that the altitudinal boundary of ley farming lies largely beyond that of cropping. In the Andes, for example, north of 15° S.L., the potato thrives up to 4,300 m elevation, whereas usable grassland ends only at 5,210 m.

Overview 4: Adaption of Crop Rotations to Altitudinal Zones of the Southern Alps

Altitudinal Zones	Meters Altitude	Cropping Sequence
Very warm low altitudes	250	Grain corn – w. barley
Warm low altitudes	400	Grain corn – w. rye
Warm medium altitudes	700	Grain corn – w. wheat
Mild medium altitudes	900	Potatoes – w. rye
Cold medium altitudes	1,100	Barley – w. rye
Mild high altitudes	1,300	W. rye – w. rye
Cold high altitudes	1,600	Pasture – w. rye – oats
Very cold high altitudes	1,800	Pasture – s. rye – oats

Source: Löhr, L.: Bergbauernwirtschaft im Alpenraum. Graz and Stuttgart, 1971, p. 36.

2. Ecological Boundaries of Farming

Special difficulties are caused by the insertion of winter grain into the crop rotation with increasing altitude, i.e. with a shortening growing season. It is evident from Overview 4 that earlier-ripening preparatory crops must be chosen in increasing amounts, until finally at very raw altitudes winter activity is completely excluded from the cropping enterprise.

c) Dry Boundaries

Much more important than the altitudinal boundaries from a worldwide view are the dry boundaries of the agricultural area. They occur as broad bands or zones, in which farming types with low productivity but drought adaptability compete with those that buy even greater drought tolerance with still lower productivity, until man finally surrenders to drought and aridity.

Drought-tolerant farming types group around an axis that is designated as the agronomic dry boundary, the dry boundary of rainfed farming. In the tropics it runs between the dry and shrub savannas with about eight-and-a-half dry months and in the subtropics between the deciduous hardwood forest and the shrub steppe with about eight dry months.

Overview 5: Precipitation on Agronomic Dry Boundaries in mm/Year (Examples).

Tropics		Subtropics		Mid-latitudes	
Large parts of Africa	250–400	Khuzistan (Iran)	250	Saskatchewan	230
Mauritania	350	Algeria, Tunisia	300	Kazakhstan	250
Sudan	400	Iran, South Australia	300–350	Columbia Plateau (USA)	250–300
Namibia	250–500	Orange Free State	300–400	South Dakota (USA)	300
West Africa	500	Libya, Cape Province	350	Kansas (USA)	350
Parts of East Africa	up to 600	Arabia	350–400	Central Asia	350–400
Kenya (2 rainy seasons)	750	Texas	375	Patagonia (Argentina)	500–550
Tropical margins,	up to 1000	Chaco (Arg.)	500–550		

Overview 5 shows that the annual precipitation for the agronomic dry boundary, or better, the dry boundary zone, ranges from 250 to 500 mm in Africa, 300 to 350 mm in Iran, and 350 to 400 mm in Arabia and Central Asia. With a poor seasonal distribution of precipitation, these critical boundaries lie notably higher, such as 600 mm in

Overview 6: Dry Boundaries of Some Useful Plants and Farming Systems. Approximate Values in mm Annual Precipitation

\multicolumn{8}{c}{A. Useful Plants by Climatic Demands}							
Tropics		Tropics and Subtropics		Tropics, Subtropics, and Mid-Latitudes		Subtropics and Mid-Latitudes	
Bananas	2,000	Tea, yams	1,500	Corn	800	Sugar beets	450
Rubber, oil palm	1,500	Sugarcane	1,400	Grain corn	760	Potatoes	400
Cacao, coconut palm	1,500	Citrus fruit	1,000	Beans (Phaseolus) ·		Dried peas (Pisum)	300
Arabica coffee	900	Cotton, sweet potato	500	Tobacco	500	Wheat	300
Manioc, kenaf	500	Sesame	400	Millet, sorghum	250	Barley	250
Sisal	250	Peanut	300			Olive tree	200
\multicolumn{8}{c}{B. Farming Systems}							
Dryland Cropping				Grassland Farming			
Dry farming with 33% fallow			400	Dairy farming			500
Peanut-millet or peanut-sorghum rotation			350	Cattle raising and feeding			350
Dry farming with 50% fallow			300	Cattle feeding with purchased calves			250
Steppe shifting cultivation			200	Sheep raising for wool or hides			150

Sources: Andreae, B.: Die Farmwirtschaft an den agronomischen Trockengrenzen. Erdkundliches Wissen, No. 38. Wiesbaden, 1974, pp. 22 ff. – Franke, G. et al.: Nutzpflanzen der Tropen und Subtropen. Vol. I (2d ed.) and Vol. II. Leipzig, 1975 and 1967. – Schütt, P.: Weltwirtschaftspflanzen. Berlin and Hamburg, 1972.

parts of East Africa and even 1,000 mm in many zones on the tropical margins. Finally, if one excludes these examples from consideration and simply compares normal precipitation amounts with the moisture demands of useful plants, as done in Overview 6, then it becomes celar how impoverished the cultivated plant society is in the vicinity of the agronomic dry boundary. This situation is shown even more emphatically by the fact that practically nothing but millet, sorghums, and peanuts are possible between the tropics of Cancer and Capricorn and that poleward of them, about only wheat and barley survive.

Fig. 11 to 13 provide an impression of the course of the agronomic dry boundaries under various ecological and economic conditions.

The drought-tolerant farming systems presented in Overview 6 reflect the following basic principles:

A. Dryland Cropping (humid side of the dry boundary):
  1. Favoring of short-lived field crops, particularly crops like peanuts, millet, sorghum, and spring barley.
  2. Insertion of bare fallow into the crop rotation, since the bare ground evaporates less water than a plant cover and thus allows the part of the rain that penetrates the fallow to be conserved for the following crop.

2. Ecological Boundaries of Farming

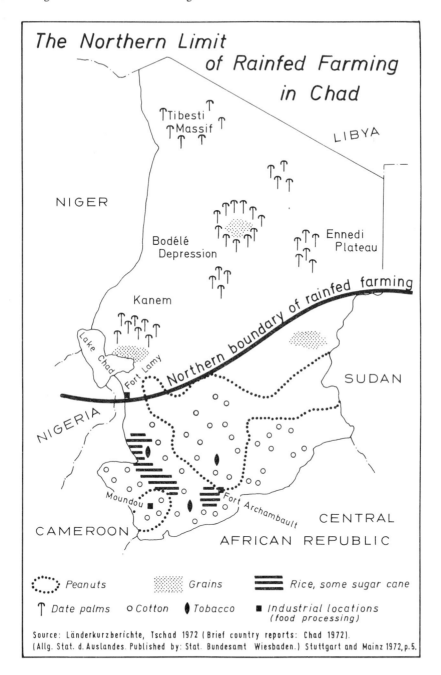

Figure 11

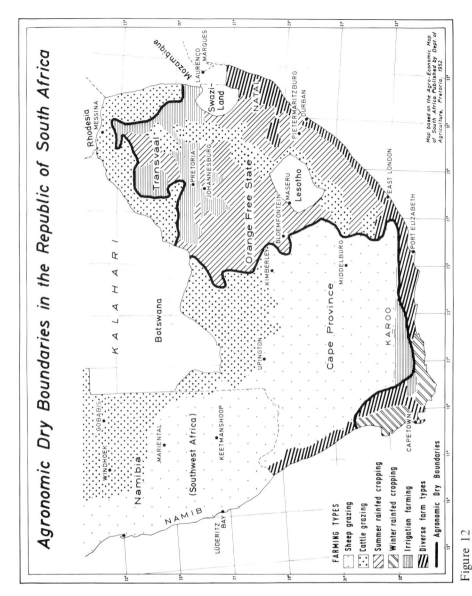

Figure 12

3. Introduction of grain cropping into the savanna or steppe for the utilization of the accumulated soil fertility, thus permitting shifting cultivation.

B. Grassland farming (dry side of the dry boundary):

1. Raising of calves instead of dairying so that the cow needs only to nurse its calf and can terminate lactation in the advanced dry seasons.

2. Ecological Boundaries of Farming 61

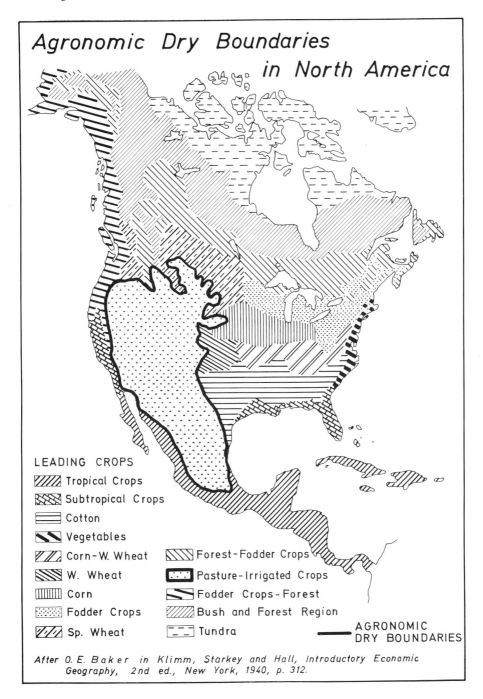

Figure 13

2. Eventually only the feeding of cattle, an enterprise less endangered by drought than is the raising of calves.
3. Raising of sheep for wool or hides, particularly drought-tolerant systems which make it possible to graze very selectively and still utilize feed that is rich in fiber. Raising of karakul sheep has the further advantage that market lambs are slaughtered shortly after birth, so that no further demands are made on the maternal organism through nursing. Such types of sheep raising are practiced with as little as 100 mm and in some cases as low as 75 mm annual precipitation.

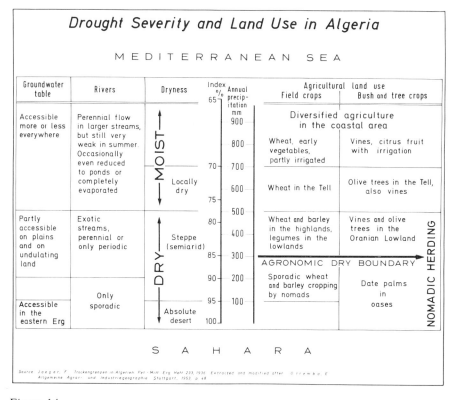

Figure 14

Fig. 14 shows how the agriculture of Algeria changes its form from the Mediterranean coast to the Sahara. The most important stimulus is the pressure for adaption to the increasingly drier climate.

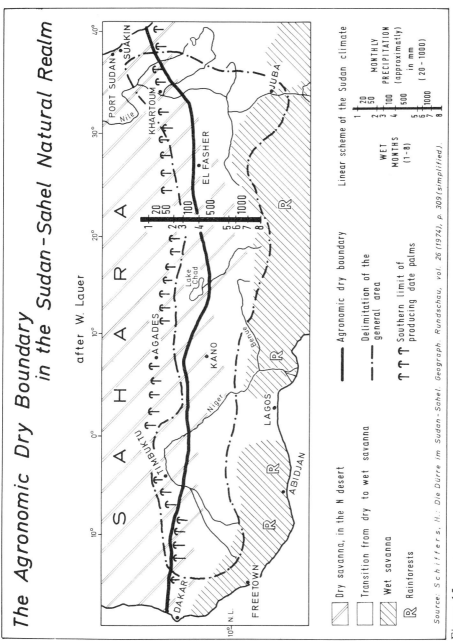

Figure 15

Finally, Fig. 15 allows us to recognize the agronomic dry boundary as it runs directly through the Sahel Zone, which has been thrust into the public eye so tragically the last few years. Many places in the tropics, in spite of governmental measures, remain closed to crop farming because the *seasonal distribution of precipitation* is too unfavorable. While the agronomic dry boundary in the tropics usually runs between 350 and 500 mm annual rainfall, it moves up in Kenya to 750 mm because this amount is divided into two rainy seasons as a result of the equatorward location; thus neither of the wet periods is sufficiently long and productive enough for cultivation. As a result, wherever annual precipitation falls below 750 mm only ranching can be practiced, which not only has a much smaller food production capacity than cropping but because of its capital intensity and cost structure is little suited for farm settlement. The unfavorable precipitation regime thus leads to important socio-economic consequences. In the border areas of the tropics there are even zones with 1000 mm annual rainfall that are hostile to cropping because of great precipitation risks. On the other hand one can find agricultural regions which, because of greater stability in precipitation amount and seasonal distribution, can support an agriculture under sizable population pressure, though they have only two-and-a-half humid months per year (four is usually the minimum).

Bare fallowing is again one of the most effective tools in the struggle with aridity when irrigation is impossible. The grain-fallow economy (or *dry farming system*) makes use of the fact that bare ground gives up less water than plant cover. Part of the precipitation falling in the fallow year is thus stored in the soil and adds to the precipitation of the next year to increase the yield of the following crop. The less precipitation falls or the farther cropping is projected into the natural pasture of ever drier areas, the larger must be the share of fallowed land and the smaller will be the share of grain land in the total crop area. Overview 7 shows a schematic stage-sequence.

Overview 7: Stage-Sequence of the Dry Farming System near the Agronomic Dry Boundary

Annual Precipitation			
500 mm	400 mm	300 mm	250 mm
1. *Fallow*	1. *Fallow*	1. *Fallow*	1. *Fallow*
2. Wheat	2. Wheat	2. Wheat	2. *Fallow*
3. *Blue lupine*	3. Wheat		3. Wheat
4. Wheat			
25	33 Fallow % CL 50		67

2. Ecological Boundaries of Farming

Farmers in the course of human history have developed a multitude of additional operational techniques in order to wrest one more harvest with minimal precipitation and to push back the agronomic dry boundary. Two practices may be described here. An especially interesting and simple agricultural type that is adapted to a low stage of technical development and is carried on in spite of severe drought is the "*molapo farming*" of the Shorobe area of Botswana. Maun receives an annual precipitation of 476 mm. Of that, 423 mm or 89% falls from November to March, whereas the period from June to September is almost rain-free, with an average of but 5 mm. Under such conditions cropping dependent solely on rain involves almost unbearable risk, all the more so since precipitation variations from year to year are sizable in all zones near the tropics of Cancer and Capricorn. However, farmers in the Shorobe area have succeeded in supplementing rainfall with the water of the Okavango Swamp for their field crops.

Because of the retarding effect of the swamp, the water in the lower part of the delta rises only after the rainy season and floods the adjoining *molapos* (moist fields). Toward the end of the winter the higher-lying *molapos* are the first to become dry and can then be cultivated. The moisture in the soil makes it possible for the seeds to germinate before the rainy season sets in. Seeding generally is from December to February and harvesting is from May to June, before the *molapos* are again inundated. Yet large fluctuations in climatic and hydrographic conditions in Ngamiland can cause considerable deviations from this normal cycle, which again cause stress situations.

A somewhat later onset of the rainy season is of course seldom a problem since the soil still has sufficient water. An early flooding of the *molapos*, in contrast, has a disastrous effect. If the water is already rising so quickly in May and June that the fields are being flooded, then the crops in at least the lowest-lying fields can no longer ripen because here inundation takes place first where seeding was done last. The growing season then is no longer adequate. Cropland per farm in *molapo* farming fluctuates from 2 to 20 ha, with 5 ha as the most frequent figure (E. Klimm, Ngamiland. Cologne, 1974, pp. 142 ff.).

In Israel, an experiment which concentrates rainwater falling on large areas onto smaller surfaces has produced significant results. Barren mountain slopes are sprayed with a plastic film to increase runoff. Thus with 30 to 50 ha of drainage area, it has become possible to develop a hectare of fertile cultivated land of the highest productivity.

Below the agronomic dry boundary (Overview 5) *extensive grassland farming* can still be practiced, which, above all in various types of ranching, possesses a high capacity for adapting to diminishing annual precipitation and a shortening rainy season. To be sure, market milk production in central Namibia requires about 500 mm annual rainfall and as a result projects into the boundary areas of crop farming. With 450 down to 350 mm of rain, however, the self-sustaining cattle-fattening farm becomes possible; here the operator has his own breeding herd, raises his own calves, and sells feeders. With 350 down to 250 mm annual precipitation, calf production is no longer possible. Calves must then be purchased from areas growing more feed, which are then raised and sold as feeder stock.

Finally, with only 200 down to 100 mm of rain, cattle raising is replaced by the raising of sheep, which need less feed, can range more widely, and can graze more selectively. In the zones of the two tropics a seasonal drought stress combines with one of heat.

In the marginal zones of grassland farming herds must overcome their seasonal feed shortages through extensive *periodic migrations.* Sedentary farmers and ranchers then give way to nomads; place-bound settlements are replaced by camp sites; farms based on civil law are supplanted by tribal land; animals that are highly demanding and can range only narrowly but are quite productive (especially cattle) cede to types that are little demanding and can range widely though they are less productive (sheep, goats, donkeys, camels); market-oriented production changes into predominantly home-oriented activity.

Often the migratory movements of nomads are not primarily motivated by the striving for feed balance, but are dictated by the search for water. Many grazing areas remain unused because the distance to the nearest water hole exceeds the action radius of the animals. In such cases the cattle and feed capacity of the nomadic grazing areas can be significantly raised with tube wells. Overcoming the water problem in the developing countries with wells naturally means more problems with labor and capital.

d) Wet Boundaries

Eighty percent of the cultivated area in Africa is afflicted by either a periodic or permanent water deficiency (Tropics of Cancer and Capricorn zones) or too much moisture (equatorial area). Thus there are both dry and wet boundaries of agriculture.

The wet boundary of grazing in the tropics applies to the marginal zone between the humid savanna and the tropical rainforest. In the central Mozambique lowlands there are wet areas that can only support jute. In extensive flooded areas of the river deltas of Bangladesh, rice and jute are an indispensable condition for any agriculture. On the Bangkok Plain of Thailand, which is regularly flooded in the rainy season, rice is the only useful plant and all attempts to form a crop rotation have failed.

2. Ecological Boundaries of Farming

Why have the Amazon and Congo basins up to now been so little developed agriculturally? They lie in the equatorial lowland with the maximum precipitation of their continents. More than 2,000 mm, and indeed in some cases more than 4,000 mm of rain falls during the year in the equatorial belt (5° N.L. to 5° S.L.). This belt, with a zenithal rainfall that is frequently reinforced in coastal vicinities by the monsoon influence, is the home of the evergreen tropical rainforest. Two rainy seasons a year are characteristic, with no month having less than 60 mm of rain. Temperatures are high and uniform and average about 26°C, while mean relative humidities register at about 95% at 6 : 00 a.m. and 74% at 12 : 00 noon.

The sum of these moisture factors leads to the following stress effects in agriculture:

1. extremely vigorous weed growth, which cannot be effectively controlled in the developing countries with the herbicides available;
2. formation of sorption-weak argillaceous minerals;
3. rapid degradation of organic material. A cleared virgin forest soil loses 95% of its organic material down to a depth of 40–50 cm in little more than a year, thus leading to
4. the destruction of the soil structure after clearance of the forest cover; and
5. severe soil erosion.

e) Soil Boundaries

Soils can also set limits to agricultural activity. In severely overpopulated Malawi extensive areas lie fallow because soils are too heavy to be worked by the hoe, which is still widely used in that country. A large part of the northern Sacramento Valley in California is, because of its heavy and wet soil, suitable only for rice cultivation and would again become wasteland if rice farming were no longer economic. Although date groves irrigated by sea water are found in the vicinity of Basra and Abadan, certainly no other cultivated plants could be raised there with good economic reasoning. The date palm is the most salt-tolerant of the useful plants. Winter barley is the lead crop in the first years after diking of the new polders on the Baltic Sea coast, because under central European conditions it tolerates the highest salt concentrations.

In dry areas, the proportion of the water in soluble salts is often high because of insufficient leaching by precipitation. The result can be salt efflorescence on the soil surface along irrigation canals and salt damage to cultivated plants. Salt carbonates in the irrigation water are often a problem. They not only raise the salt content of the soil, but initiate alkalinization and increase the pH-values of the soil to 8 or 9. This leads especially to the immobilization of the trace elements of iron, manganese, and zinc (H. Marschner).

In India and Pakistan, most notably in the irrigated areas of the floodplains, the area of salt-damaged soils already amounts to 10 million ha. Of that, 600,000 ha are no longer cultivable because the salt content it too high. In Egypt, the fertility of the old croplands of the Nile Delta has been impaired by the building of the Aswan Dam. Before its construction the annual floods not only supplied water, organic material, and nutrients, but also flushed superfluous salts away either superficially or with the groundwater. The ecosystem was in balance. After the dam was completed, flooding ceased. Irrigation now had to be instituted, which brought with it only minor humus and nutrient additions and a heightened danger of salinization. Areas on the lower course of the Nile that have been irrigated for only a few years are already no longer cultivable because of salinization (H. Marschner).

The olive tree in the Mediterranean area is content with thoroughly sterile, shallow, rock-infested, and steep slopes, which indeed are still agriculturally useful only because of it.

f) Slope Boundaries

The *angle of slope* can also describe limits to agriculture, though they sharply fluctuate with farming technology. According to Löhr (1971, p. 33) the following zonation of the correlation between slope and farming type can be found in the Alps:

- with 40% slope, *Kunstegart* with 4 cropping years;
- with 50% slope, *Kunstegart* with 3 cropping years;
- with 60% slope, *Naturegart* with 2 cropping years;
- with 70% slope, *Naturegart* with 1 cropping year;
- with 80% slope, permanent meadow with periodic grazing; and
- with a 100% slope, permanent meadow without grazing.

Thus, just as with increasing altitude, steepening slope brings about the increasing displacement of money crops by fodder cropping. One reason for this shift was already described in connection with the polar boundary of cropping, which as a cold boundary is naturally related to the altitude boundary: with a shortening growing season, fodder cropping becomes increasingly superior to cash cropping competitively because it completely utilizes the growing season, from the first to the last day. Photoperiodism also plays a role here. In addition, however, the difficulty of working on a steep slope must be considered. Tractors can be used to cultivate slopes of up to only 25%, and in grassland farming they can negotiate gradients not exceeding 40%. In the industrial countries, farming on the contour by means of expensive animal adaption stops

2. Ecological Boundaries of Farming   69

completely at 35 to 40% slope. At that point it must yield to the use of
lines with winch or block and tackle, a technique whose upper limit lies
at almost 100% slope — but only technically, for it is questionable economically.

Generally with exaggeration of slope one can observe:
− the transition from tractor through draft animal to tow rope;
− the displacement of food crop cultivation by fodder cropping;
− the sequential succession of *Kunstegart* to *Naturegart* to permament
  meadows; and
− the progressive impoverishment of land use, with first the cropping of
  winter wheat, corn for silage, then potatoes, then winter grain, and
  finally, only spring grain.

In countries with insufficient land and low wages *terrace farming* can
still be intensively practiced on very steep slopes. Sixty percent of the
cultivated land on Taiwan is in rice terraces. Terracing of steeply sloping
land allows for a more thorough soaking of the soil and reduces erosion.
Terraces conserve soil as did the former natural forest cover. Economically they represent an artificially distributed valley soil. Flooded rice
terraces in Java (*sawahs*) go up to 1,200 m, on the flanks of volcanoes
up to a hight of 1,500 m. *Sawahs* are found in central Sumatra at
1,500 m elevation, in northern Luzon even up to 1,800 m. Also long-
famed are the rice terraces of Bali, Sri Lanka, Assam, Japan, and South
China (Wilhelmy 1975, pp. 23 and 37).

Terrace farming as a means of overcoming or at least lessening the difficulty of slope has of course the great disadvantage that it lacks adaptability to economic growth. What was produced with an extremely low
wage level through incredible labor inputs, turns out to be more and more
difficult with the increasingly mechanized operations associated with
economic growth, until finally the use of terraces must be discontinued.
Labor investments in terracing can then no longer be recovered and
must be written off as losses. The result is symbolized by the many
steeply sloping sites in the vineyard areas of southwestern Germany,
where terraces lie deserted because of the flight from the slopes in the
technological era.

g) Unfavorable Exposure

Finally, as the last natural stress factor, exposure must be mentioned,
which through this or that orientation can produce impediments to
growth and farming operations. A schematic cross-section of one of the
great Alpine longitudinal valleys, such as the Inn, Enns, or Drau, shows
a somewhat typical pattern. On the cold, moist northern slope the forest
extends almost to the valley bottom. On the south-facing slope, in con-

trast, cropping often rises to well over 1,000 m above sea level. A pasture belt follows, until finally at 1,400 to 1,600 m altitude the forest sets in, which in turn is terminated at timberline by a region of Alpine pastures.

The value of a particular exposure naturally depends on the kind of crops being grown, the geographic latitude, the altitude, and other conditions. Shade-loving plants will under certain circumstances grow better on northern than on southern slopes, especially in lower latitudes.

Löhr (1971) determined that Alpine farms on grades averaging 39% have, at noon in the summer, an angle of incidence of solar radiation of almost 90% on the south-facing slope, whereas in winter it drops to 0° on the north-facing slope. Further, at an altitude of 1,500 m the duration of the snow cover on the south-facing slope was measured at 130 days, but on the north-facing slope at 169 days. It is clear that exposure-determined climatic differences of such kinds can not be a matter of indifference to agriculture, which is so unusually place-specific in its characteristics.

3. Economic Boundaries of Farming

Besides the ecological boundaries of farming that are qualified by cold, drought, wetness, and the like, there are also the economically generated marginal zones of agricultural space.

a) Settlement and Industrial Boundaries

Quite obvious boundaries are posed for the farmer by settlement and industrial areas and by paths, streets, and railroads. Where settlement is dispersed in the form of individual farms or hamlets, agricultural and settlement space are closely meshed. Where man is established in large clustered villages, as in parts of West Africa, agricultural and settlement space are sharply set off from each other, and for the farmer the transportation problem becomes one of internal rather than external management. Finally, where a strong urbanization tendency exists, as in South America, settlement boundaries of agricultural space are sharply pronounced. Then usually a horticultural belt is interjected between the settlement and agricultural areas to form a transition.

b) Transport Boundaries

As an *"involuntary transport business,"* agriculture is subject to still another economic stress factor, or at least so long as there is commercial

3. Economic Boundaries of Farming

production: the distance to market. Normally all transport costs incurred in supplying the market with products from the farm and purchasing manufactured inputs from the market work to the disadvantage of the farmer. Thus the price-cost relationship for the farmer becomes more unfavorable with increasing market distance. The profit margin of almost all farming enterprises diminishes, and naturally increasingly so the more transport costs are involved.

All boundaries of agricultural space that are characterized by a peripheral transport location that is too unfavorable, thus too great an economic distance to market, should be designated as transport boundaries. Thünen already recognized such boundaries in his "isolated state" (1826). The more sensitive a production enterprise is to transport, the smaller is the radius of its transport boundaries around the market. Accordingly, Thünen found in his empirical model study that on a plain that was uniformly fertile and uniformly accessible by transport, and surrounded by an impenetrable wilderness, farming systems with differing transport sensitivities had to be grouped in concentric circles around the sole and central market site. There is still an almost classical picture of such a Thünian spatial arrangement in northern Sweden (see Overview 8).

Overview 8: The Thünen Spatial Pattern in the Storsjön Area (63° N.L.)

Inner Circle (more intensive)	Middle Circle (normal intensity)	Outer Circle (more extensive)

1.–3. Grass-clover
4. Winter rye,
   s. barley, oats
5. S. barley
   spring peas,
   potatoes
6. S. barley
   underseed
	1.–4. Grass-clover
	5. S. barley,
	   spring peas,
	   potatoes
	6. S. barley
	   underseed
		1.–5. Grass-clover
		6. S. barley
		   underseed

Östersund, a large city located in the Storsjön area of Sweden, is the only noteworthy market for an agricultural region whose service area ultimately disappears in the unending forests of the taiga. The isolation of the market area has allowed Thünen circles to develop around the city, and they become increasingly more extensive with the pushing back of grain cropping by clover-grass cropping. The progressive reduction of expenditures from the inner, through the middle, and to the outer circle is brought about by not working the soil in the clover-grass years and by fodder cropping for livestock raising instead of milk production. The transport hindrance is removed by marketing ever more calves, butter, cheese, and pork instead of milk and grain.

As indicated in the schematic Fig. 16, every commercial production enterprise must ultimately cease at the market distance at which its profitability reaches zero value. This situation occurs the earliest with milk production. Cream is much more capable of being transported. Feeder stock can even go to market on the hoof.

In the outermost circle, one must concentrate on enterprises that produce for every 100 ha of pasture land products that have very low market weight and can be easily preserved and stored, such as wool, hides, and karakul fleece. Many areas of interior Australia can sell only wool or hides because of the great market distance, although they could still produce many other products. Getting a good price for an agricultural commodity is often even more difficult than it is producing it.

Fig. 17 does not show transport boundaries of agriculture, but rather agronomic transport limits. In northeastern Zimbabwe 700 to 800 mm of summer precipitation makes grain corn cultivation ecologically possible everywhere. The increasing distance from the Salisbury market, however, is an inducement to reduce gradually the place of grain-corn cropping in the farming system and eventually to abandon it altogether in favor of extensive livestock grazing.

In the service area of Moscow a Thünian agricultural spatial arrangement was produced by the centrally planned economy, as will be shown in Chapter VIII, 3-f.

There are a whole series of partial, *product-specific transport boundaries*. Export bananas can be cultivated no farther than 100 to 150 km from the port because they combine a high weight per value unit with high perishability.

Sugarcane and sisal leaves are extremely transport sensitive. Consequently these raw materials can only be produced near the processing center. Monocultural landscapes of sugarcane, sisal, and the like thus originate around these centers. The radius of these landscapes, though, is circumscribed by transport costs, which thus also limits the capacity of the processing plant. Processing costs therefore cannot be reduced simply by expanding operations. This handicap is to be viewed as another transport-conditioned stress factor.

In a similar connection, Aereboe (1923) describes how the populations of isolated Balkan valleys formerly raised rose petals as a cash crop because their oil content obtained such a high price per unit weight, that the great distances from the market and over difficult roads could be overcome by bicycle and knapsack.

3. Economic Boundaries of Farming

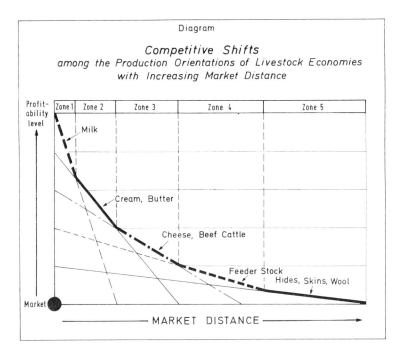

Figure 16

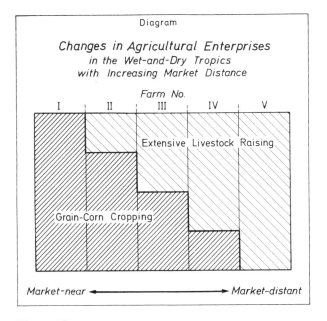

Figure 17

c) Commercialization Boundaries

A qualification must now be noted, in that not all farm sizes, or more exactly, commercialization stages, are subject to the same transport boundaries. The Indian farmer who cultivates sugarcane for his own consumption is in this respect independent of market location and thus knows no transport boundaries. The same is true of the Ethiopian and Costa Rican farmers raising their cooking bananas, the farmer in the African equatorial zone cultivating manioc and yams, or the Southeast Asian farmer cultivating sweet potatoes, all for their own subsistence. The work forces for the ranchers are able to grow corn, millet, or sorghum for themselves far beyond the agronomic commercial boundaries of the market-oriented ranching system so long as precipitation is sufficient. This can be widely observed in Argentina, Colombia, Zimbabwe, and other places.

Where the amount of precipitation permits cropping and only the distance to market prevents it from becoming an income source, improvisations are still possible. Large haciendas in Argentina formerly leased small areas to subsistence settlers and distributed these fields among the pasture lands. After three to four years the settler was required to sow his land with alfalfa and to return it. The *estanciero* thus gained alfalfa land for the dry season, which he himself was not able to establish because of the price-cost ratio, especially since alfalfa cannot be sown with the first plowing of virgin land and can only do well after several years of prior cultivation. The alfalfa pastures then provided the feed for fattening cattle, since the alfalfa with its meter-long tap roots was able to utilize the groundwater much better than natural grass was able to. The settler, however, also profited from the shifting of this subsistence land over the pasture lands in that he obtained high yields with a regeneration of soil fertility that cost him nothing.

Such a symbiosis between large and small farms still exists today in Colombia. Uhlig (1965, p. 20) has reported on a Colombian coastal lowland that has only four wet months and receives barely 400 mm precipitation, so that the natural vegetation consists mostly of shrub savanna with succulents and cactus, but also partly shrub forests. Thus the extensive grazing economy in this area, with haciendas of 1,000 to 50,000 ha, is highly vulnerable to the danger of shrub invasion. As a result, the *hacendados* relinquish a part of their land to small tenants for eighteen months without cost. The farmers clear and burn the shrub-infested pastures and then cultivate them once or twice for corn, yucca (fiber plant), and other subsistence crops. The compensation made by the tenant for the abandonment of land use becomes manifest when, after a year-and-a half, he returns the land as fresh pasture, seeded with guinea grass.

3. Economic Boundaries of Farming

The reason for this symbiotic relationship between large and small farming systems, besides the impossibility of supplying highly remote markets, is the difference in the scarcity ratios among the production factors for the two farm size classes. The large operator has much land but little labor at his disposal. For him, therefore, it is economically practical to substitute land for labor, in which the work of clearing is paid for by a moratorium on land use. The small capital-poor tenant, in contrast, has none of his own land but has at his disposal a relatively plentiful family labor supply. For him, it is economically advantageous not to have to pay cash for the use for the land but rather to make compensation with the labor expended in clearing and sowing grass.

Agriculture has still many more possibilities for avoiding or at least ameliorating the locative stress caused by great distance to market. This involves the conversion of primary crop products into more transportable form through *livestock raising* or more *technical auxiliary enterprises*. Heavy harvest volume and weight, as well as high perishability of the harvested product, may not mean in every case that particular land use types can be carried on only close to the market. For many, transformation of the harvested products can reduce the transport barrier so much that they are still worth cultivating in market-distant locations. There are numerous transformation stages, for example, in the utilization of the potato. Many light soils in Farther Pomerania would hardly have been cultivated before 1945 had it not been for the potato. This presupposed, however, the transformation of the crop into pork (peasant farms) or alcohol (large farms).

In many overpopulated countries of the world, transformation of farm products through *cottage industry* increases the income base for the same amount of land. Examples such as the processing of jute, kenaf, cotton, wool, or silk come readily to mind. Since it is a question of mostly hand processing of small quantities, high labor productivity can rarely be achieved. Embroidery with cashmere or the hand weaving of rugs by oriental nomads would, at our income level, be unthinkable. Also, the further processing of flax into linen, still a component of productive winter work in the spinning and weaving rooms of many German peasants 150 years ago, has long fallen victim to industrial mass production. Today the viticultural industry is increasingly shifting from the peasant wine cellar to the technologically, economically, and managerially superior wine growers' cooperatives.

All of these examples of rural home industry produce not only a profit, but also a saving in transport costs, for the transformed product is basically more capable of transport than is the raw material. Under the most extreme conditions (very small farms without draft animals, nomadism, isolated high mountain settlements), such forms of cottage industry generally offer only the possibility of market delivery, even if the raw material is already quite transportable as are wool and silk. Cottage industry is thus capable of pushing the commercialization boundaries outward.

The commercialization of farms, however, is not something absolute but relative. It varies by degrees. Four stages can be coarsely differentiated:
1. *subsistence farms,* with operators generally producing only for home consumption and with sales amounting to less than 25% of the gross output;
2. *little commercialized farms,* with operators already pursuing a program of systematic commercial production that accounts for 25 to 50% of the gross output;
3. *strongly commercialized farms,* with operators selling 50 to 75% of the gross output; and
4. *fully commercialized farms,* with operators using less than 25% of the gross output to cover home needs.

Transport boundaries will contract from 1 to 4. The greater the marketing share of production, the greater will be the dependence on location with respect to the market. Also playing a role in this gradation is the fact that the farmer in grade 2 can select for his only partial market production enterprises that encounter only small transport resistance. A fully commercialized farmer cannot be similarly flexible in production planning, both for economic reasons associated with labor and fertilization and matters regarding rotation and risk.

4. Boundary Shifts with Economic Growth

From all that has now been said, one could get the false impression that the boundaries of farming are firmly given for all time. This is in no way the case. Rather, the marginal zones of the agricultural area are not stable, but fluid. They shift quite easily as farmers adapt to the progress of technological and economic development.

a) Mechanical Technological Advances as a Cause

Mechanical technological advances can noticeably displace the commercialization boundaries just noted. Every *extension of transportation access* lessens the economic isolation of the farmers and produces more favorable purchasing power relationships. Every new rail line, every new road, and every reduction of transport charges must necessarily reduce the differences between the various farms in transport costs and moderate the burden of transport costs for all of them, and with that, increase prices at the farm gate for agricultural products and reduce prices of industrially-produced farm inputs. The farms which up to now had been marginal operations now begin to provide a differential rent. Beyond these farms a new, additional ring of pioneer farms develops, extending outward until the differential rent, in reacting to the transport situation, finds its zero value. Here lies a new transport boundary. The market area now has widened, the agricultural area has expanded.

Advances in *harvesting technology* can also decisively influence the spatial organization of agricultural production. Where farming with large machine aggregates enters the picture, as in the dry farming of wheat and barley, production costs are notably lowered so that a competitive advantage over the ranching operation is obtained. Before full mechanization had been achieved in the Columbia Basin of northwestern United States, wheat yields of 6.0 q/ha were necessary to ensure the competitiveness of wheat farming with the drought-tolerant ranch; today only 4.5 q/ha are required. This has meant that the precipitation minimum for dry farming has been lowered by about 50 mm, from 300–350 mm/year to 250–300 mm/year. The agronomic dry boundary has thus shifted outward.

In Central Europe, the earlier stress on labor by the grain harvest was greatly alleviated by the combine, that of beet cultivation by thinning with herbicides, and that of the sugar beet harvest by the beet harvester. The technological development of beet farming shows how strong the desire was to reduce these labor pressures and how eagerly the considerable capital investments were made in order to accomplish this. In England, labor expenditures for harvesting and loading sugar beets decreased from 180 MH/ha in 1954 to 65 MH/ha in 1965. In sugar beet farming in the United States, 35.79 MH were still being expended per ton of sugar (unrefined) in 1953/55, whereas in 1971/73 only 16.66 MH were being spent. These technical advances have allowed present sugar beet areas to expand and have opened up new, additional locations for planting.

How much labor has been saved by *draining* the Central Schwerin Marsh! Many marsh soils on the Baltic Sea coast and along the lower courses of the Elbe, Oste, Weser, and Ems, which earlier supported land fit only for permanent grassland, have been made cultivable only through drainage. Thus the soil boundaries have shifted outward.

There are heavy soils that, at the hoe culture stage, still cannot be worked. They can only be added to the cropland at the stage of *plow cultivation, using oxen.* Other soils are still heavier and can only be made usable with the *tractor.* Again, advanced technological methods expand the agricultural area.

In the zones near the tropics of Cancer and Capricorn there are areas with such a short rainy season that practically all of the land is needed if millet, sorghum, or barley are to bear seed. So long as the energy source of the farmer is human muscle power or animal draft, cultivation and seeding will take too long to leave a sufficient growth period for the grain, and just so long will these sites remain absolute natural patureland. Only the much greater power of the tractor will permit the agronomic dry boundary to be extended into these regions.

A hundred years ago, large areas of central Southwest Africa were used only periodically by the nomadic Hereros because of the difficulty in *providing drinking water.* Only the tube wells of the white settlers made the sedentary ranching system possible. Large areas on the margins of the Sahara remain unused by nomads because of lack of drinking water. Tube wells can push back the dry boundary of livestock raising and contribute to the urgently necessary expansion of living space.

b) Biological Technological Advances as a Cause

Biological technological advances also can expand the boundaries of agricultural space. *Advances in veterinary hygiene* have been able to advance the limits of livestock raising by reducing the distribution area of the tsetse fly, the carrier of nagana, a serious animal disease. It is also precisely in the tropics that crop losses from pests, diseases, and weeds are great: in Africa, 39% of the wheat, 40% of the peanuts, and 45% of the sorghums and millets are lost in potential harvests. If new developments in phytomedicine succeed in halving these damages, the competitiveness of the favored useful plants will be strengthened; they will advance beyond their present boundary locations.

In the industrial countries of the middle latitudes, the *invention of caustic, phytohormone, and soil herbicides* as labor-saving means of killing weeds in combination with other technical advances has significantly contributed to the fact that pure grain crop rotations are now possible. In the developing countries of the humid tropics, the effects of these herbicides on cropping will probably be revolutionary as soon as the evolution in price-cost relationships allows their greater use. Among ecological restrictions, weed growth is also much more formidable than in the mid-latitudes and agriculture there is in good part an eternal battle against it.

Plant breeding has accomplished much in the expansion of the agricultural area. It has, for example, through the development of quick-growing spring barley varieties, gradually advanced dryland cropping in the southern Mediterranean countries more and more toward the Sahara and at the same time made possible the increasingly poleward thrust of spring barley cropping into the high north.

The poleward advance of hybrid corn in Central Europe during the last two decades is a further example. Before the last war, corn was cultivated as the main crop on the Po Plain and in southwestern France. Today the Paris Basin is the biggest cropping region of the European Community. In the Federal Republic of Germany grain corn cultivation has won at least a firm place in the valley and basin regions as far up as the Rhine-Main area. Even in Schleswig-Holstein about 289 ha of grain corn were already being cultivated in 1976.

More important still, in its time, was the advance of new spring wheat strains into Canada with the step-by-step development of earlier-ripening varieties (Hohnholz 1975, pp. 48 ff.):

- In 1763, wheat cropping was possible only along the St. Lawrence River.
- From 1880 to 1890, the Red Fife variety advanced from the East to the West.
- Beginning in 1909, Marquis wheat, which ripened safely in 120 days and had a growing season 6 to 10 days shorter than that of Red Fife, helped to push the polar boundary of wheat cultivation some 300 km to the north.
- By 1926, Garnet wheat, which ripened 5 to 7 days still earlier, was available, and it allowed the polar boundary to shift some 200 km still farther north.
- Today there are varieties that ripen in less than 100 days.

When the polar boundary of spring wheat was able to be pushed northward in Finland from about 63° to 66° N.L. between 1922 and 1946, and in Canada by about 500 km between 1890 and 1926, these displacements wrought by plant breeding and selection were primarily shifts in the profitability boundary. Secondarily, however, this development also reduced the stress on spring wheat along the original polar boundary farther south.

The benefits are also twofold where the high yielding varieties (HYVs) of rice and wheat double and triple yields. Firstly, the stress of farm size is reduced, in that more food can be produced from the same area and a more desirable production combination can be achieved. Secondly, rice and wheat are made more competitive with other land use types, so that more of these grain types can be planted in the present rice and wheat areas and new lands for the two crops can be opened up.

c) General Economic Growth as a Cause

General economic growth is characterized by the following features:
1. increasingly more technological advances and with that, a transition to ever newer production functions with greater efficiencies;
2. increasing scarcity and cost of cultivable land with the pressure for higher land productivity;

3. reduction of costs of all capital goods through technological advances, labor division, and industrialization; and with that,
4. growing labor productivity as a result of rising labor incomes.

All these growth phenomena influence the marginal zones of agricultural space decisively. This is seen, for example, in the outer tropics, the dry savannas, as well as, with minor variations, in certain steppe areas of the subtropics and middle latitudes. The displacement of the agronomic dry boundary in the course of economic development is discussed below.

For generalizing statements like the above, economic theory is more reliable than empiricism. If one takes as a base a location between the agronomic and climatic dry boundaries having an annual precipitation of about 400 mm, thus a site technically capable of cultivation, then the succession of farming systems with economic growth will have to take place in approximately the following stages (Andreae 1974, pp. 61 ff.):

Stage 1: Extensive grassland farming

At the beginning of development, with sparse settlement and an almost complete lack of division of labor, *extensive grassland farming* dominates the scene almost completely. This is Stage 1. There are still no markets. The scarcely differentiated society consists of various autonomous domestic economies that produce everything that is consumed and consume everything that is produced. Because of the wealth of land this purely agrarian society can, despite the food loss involved in livestock raising, base its diet on animal products, which are supplemented by the gathering of wild plants, tubers, and the like. The animal products, which also include hides, hair, bones, fuel material, horns, and so forth, are first obtained from wild animals (hunting), later by migratory herding (nomadism), and finally by sedentary livestock raising (ranching). In the latter forms, animals are first kept together by bulls and stallions, then by herds, and later by fences. Several types of animals are raised so as to be able to satisfy human needs as completely and diversely as possible. They include, at the very least, cattle, sheep, or goats, or combinations thereof, and in the drier areas camels and donkeys as well. All animal types are milked because milk is available daily in small amounts and therefore is especially suitable as a basic food item.

Stage 2: Steppe shifting cultivation

But since extensive livestock raising has only a minor food-producing capacity, a growing population must sooner or later force the adoption of some cropping and thus a shift to Stage 2. This takes place chiefly in the form of *steppe shifting cultivation.* Just as *nomadic herding* marks the beginning of livestock raising, so does shifting cultivation characterize the initial production of useful plants. The reason for this lies in the fact that the yields of millet, sorghum, spring barley, and the like sharply decline after only a few years because of deficient cultivation, absence of mineral fertilization, and heavy losses from diseases and pests, while labor expenditures increase because of worsening problems like weed growth and damage to soil structure. The cost-production ratio becomes increasingly unfavorable so that after two to four years it becomes economically sensible to transfer the small farming plot to another natural pasture.

Stage 3: Dry (grain-fallow) farming

If, then, cropland increases with the further growing population, then sooner or later it will gain the largest part of the arable area of the farm. Then shifting of the cropland will stop. The functions of shifting are now replaced, at this third stage, by fallowing, thus by practicing *dry*, or *grain-fallow* farming side by side with extensive grassland farming. In the fallow years, water is stored in the soil for the next crop.

Stage 4: Integrated farming systems

Later, national economic development stimulates further production intensification, either through a reduction of the food-producing area as a result of population growth or through an incentive to take up commercial cropping so as to be able to trade for industrially-produced inputs. An important means of production intensification then is the *integration of cropping and livestock raising* through the substitution of fodder cropping for fallowing. With that, Stage 4 is attained.

Fodder cropping serves purposes that are preeminently pedological and not economic. It is first introduced primarily in the interests of grain cropping and not for the support of livestock raising. This does not keep it from providing grazing possibilities during the dry season, however, or from being called upon for hay production. In this way it forms the binding element between the farming enterprises of cattle fattening and grain cropping, which are still isolated from each other in Stage 3. Fodder cropping provides grain cropping with an improved crop rotation and cattle fattening with a broader feed base for the dry season. It offers root humus for crop farming and winter feed for livestock on pasture. Livestock transform this feed, in part, into manure, which again is to the benefit of grain cropping. In this way the two farming enterprises of slaughter cattle production and grain cropping, types previously operating parallel to but independent of each other, now give rise to an association, an integrated whole, a diversified operation, a farming system.

Stage 5: Semi-intensive grassland farming

Finally, with advanced stages of national economic development the degree of industrialization encourages the rapid expansion of job opportunities and individual income, whereas the population growth rate becomes retrogressive. Then dryland cropping is again dislodged from its marginal locations and more space is returned to livestock raising, but now in larger and more capital-intensive forms than in Stage 1. The competitive superiority of ranching over dryland cropping at this highest stage, Stage 5, rests on the fact that even with modern production techniques field crop yields are still too small to satisfy the now very high income demands. The surplus of rural population, however, which originates with the abandonment of cropping, is drawn off by new industrial job openings. The possibility of a decline in food production because of the restriction of dryland cropping need scarcely be considered because the yield-increasing capital inputs notably expand the land productivity of the areas in the nation that are wetter and more capable of intensification. Finally, of course, irrigation farming also spreads very rapidly at this stage.

The succession of stages that has been sketched shows that the dry boundaries of cropping cannot be established for all time. This could be done if they were determined only ecologically and economic motives were absent. It would therefore really be better if the popular term, agronomic dry boundaries, were not used. The dry boundaries of cultivation are extremely unstable, quick to react to the progress of national economic development. There are actually only economic dry boundaries of cropping, and they shift according to the particular national economic situation of the time.

5. Contraction of World Agricultural Space as a Future Prospect

If our view is expanded to include the tropics as a whole with its four outstanding groups of farming types, then the rural space picture of the lower latitudes, as affected by competitive shifts and locational displacements, may be expected to develop as follows (Andreae 1972, pp. 183 f.):

A. Expansion will occur in:
   1. the bush and tree crop farming zones, because
      a) world markets for most of their products are expanding,
      b) the farming enterprises are intensive, and
      c) the inner tropics urgently need this kind of farming because their soil fertility is being endangered.
   2. the irrigation farming zones, because
      a) national economies are increasingly exploiting their water resources through large-scale projects such as the impounding of water by dams,
      b) irrigation increases both land and labor productivity, and
      c) flooding preserves soil fertility much as natural vegetation does.

5. Contraction of World Agricultural Space as a Future Prospect 83

B. Contraction will occur in:
   1. the rainfed farming zones because they are being pushed back from three sides:
      a) in the rainforest belt by farming types with bush and tree crops,
      b) in the humid savanna by irrigation farming, and
      c) near the present agronomic dry boundaries by the penetration of extensive grassland farming.
   2. the extensive grassland farming zones because
      a) although they are gaining land in the present border areas of dryland farming, which are losing their value for cultivation because of low yields and a rise in wage levels,
      b) this gain is being more than offset by losses in the semideserts and shrub savannas, where extensive grassland farming will have to be increasingly abandoned because of its extremely low productivity.

The zones of regular cropping will thus be compressed on two sides, on the borders of the tropics by extensive livestock raising and on the equator by bush and tree crops. Increasingly larger shares of the tropics will thereby acquire a vegetation cover approaching the natural state. The outer tropics will have more grassland and the equatorial belt more trees, even though the natural landscape has now been transformed into an agricultural landscape.

Overall, agricultural space in the tropics will suffer losses in the zones of the tropics of Cancer and Capricorn but will still gain areas in the vicinity of the equator. In the more immediate future the second tendency might be the strongest.

However, the defusing of the place-specific stress situations for world agriculture as a whole now requires evaluation and judgement from a quite different and less conclusive aspect. Agricultural production will in time, under the pressure of wage and income levels and with increasingly cheaper capital inputs, diminish in the less productive locations and expand all the more in the areas capable of intensification. This thesis must naturally be demonstrated from the perspective of the world food problem.

The agricultural capacity of mankind can be expanded basically in two ways:

1. *by expansion,* i.e. by land reclamation, through which agricultural space is enlarged; or
2. *by intensification,* i.e. by increasing hectare-yields, through which agricultural space, circumstances permitting, can be reduced.

For the present, both methods will have to be employed in order to master population growth and food shortages. But in the long view much

speaks for the development of intensification into the stronger tendency because:
- plant breeding and reduction of costs of all yield-increasing farm inputs will raise yields by as yet a still undetermined amount;
- ecologically restricted peripheral locations of today will for lack of sufficient income drop out of the agricultural area in countries with higher development stages (mountain areas, dry areas, etc.). In Finland the polar boundary of spring wheat cultivation has already retreated, as has the agronomic dry boundary, in part, in the United States and the Soviet Union. In the High Alps, Apennines, and higher places in the mountains of central Germany the phenonemon of rural flight from high altitudes has been observed for two decades;
- social boundaries of agriculture (social fallow) are beginning to appear. From 1960 to 1970 the Federal Republic of Germany (FRG) lost 4.5% FL, and Belgium as much as 6.8%. It is estimated that the loss for the 1970–1980 decade will amount to about 3.5% for the FRG and 6.0% for Belgium. In 1973, fallowed land took up 30% of the arable land in the district (*Kreis*) of Siegen and perhaps as much as 40% in the Dill district;
- more and more cropland will be needed for settlement, industrial transport, and social purposes. In the FRG, from 1935/38 to 1974, the agriculturally-used area declined from 14.6 to 13.3 million ha, and in the U.S., for the 1950–1969 period, the fall was from 453 to 435 million ha.

Thus the higher income expectations and wage levels rise, the higher yield levels will have to be. Hence as development proceeds, areas that up to now have been high quality farmland will become marginal and those that now are marginal in quality will drop out of production entirely. Agriculture is retreating more and more to the lands that are the most productive or at least the most receptive to intensification measures. It is producing an increasing amount of foodstuffs on ever smaller areas. This concentration process in agricultural space is already far advanced in such places as the High Alps, Apennines, southern France, and parts of Rhineland-Pfalz.

It is certainly difficult to believe that the world agricultural area will shrink in the near future despite the pressure of the world food problem, and in the face of information from the FAO that during the period from 1956 to 1973 the proportion of the world land surface in permanent grassland rose from 18.3 to 22.4%, that of cultivated land from 10.1 to 11.0%, and that of total agricultural land from 28.4 to 33.4%. Still 17 years are not much in the process of world economic growth.

Just as very intensive handicraft industries like hand weaving of rugs or hand carving of ivory are possible only in countries with low wages, so

5. Contraction of World Agricultural Space as a Future Prospect

also are there agricultural enterprises that lose their ability to survive above a certain wage level. Thus unproductive cheese making in the high Alpine pastures gave way to livestock raising, until finally it too was abandoned. It was more economic to compensate for the loss in production by intensifying operations around the farmstead in the lowlands. The altitude boundary of livestock raising was consequently lowered.

How many of the once far more numerous flocks of sheep that grazed the North German moors still remain? Most can no longer provide the herder with a living wage. The limits of the grazing economy are becoming ever more confining. In Italy wages are lower to be sure, yet even the summer grazing of the High Apennines by roving flocks is more and more being discontinued because the year-around intensive production of the fertile coastal plains has removed a link from the system responsible for supplementing the feed supplies for the migrating shepherds. The economic systems of the coastal plains thus are forcing an exodus from the High Apennines. Areal expansion is being limited by yield intensification.

Whoever restricts the living space of the nomads and finally forces them to become sedentary, must fully realize that their present grazing grounds will in most cases become a wasteland; result: retreat of the dry boundaries of livestock raising.

If income expectations in agriculture require a labor productivity that only the tractor can ensure, then the maximum angle of usable slope becomes much more shallow than that negotiated by horses, sinking to 40% with pasturing and 25% with cropping (Löhr 1971, p. 75 f.); result: retreat of the slope boundaries with use of the tractor.

In sum, development policy strives for increase in income, and economic growth makes this possible by increasing labor productivity. High labor productivity, in turn, causes the limits of agricultural space to contract. Production losses can and must not only be compensated, but more than compensated by capital investments promoting yield increases in the areas capable of intensification.

The task, then, is one of vigorously expanding food production on a shrinking agricultural area. The necessary technical resources are already widely available. Their application depends on the development of the overall economic structure of the individual countries.

6. Summary

1. The agricultural area amounts at present to only 33.4% of the total world land surface and more specifically, 30.7% of the total area of the developing countries and 39.7% of that of the industrial countries.

2. The geographic boundaries of agricultural activity are determined by place-specific deficiencies such as cold, dryness, or wetness; obstacles stemming from soil type and relief (great difficulty in working the soil or steep slopes); and economic deficiencies of location such as too great a distance from market.

3. Hoe and plow cultivation are still possible in the humid tropical rainforest belt, but grassland farming is not (wet boundary). In all other marginal areas the boundary of grassland farming goes beyond the boundary of cropping. Thus cropping (spring barley, potatoes) in the high north finds its limit at 70° N.L., whereas the reindeer of the Lapps go still farther into the tundra pastures (polar boundary). In the Andes north of 15° S.L. the potato thrives up to 4,300 m altitude, while the pasture zone ends only at 5,210 m (altitude boundary). Barley and millet on the margins of the Sahara penetrate the dry zones with 250 mm annual precipitation, while sheep grazing is still possible with 100, indeed even 75 mm (dry boundary).

4. Three types of boundaries are distinguishable in all marginal zones: effective boundaries (currently operating; profitability boundaries (profit = 0); technological boundaries (technologically attainable).

5. All boundaries are influenced by technological advances and price-cost shifts. They vary, therefore, with the progress of economic development. They are dynamic rather than stable.

6. At present the world food situation is stimulating a maximum effort at expanding the boundaries. The battle to conquer the marginal areas is being waged with various weapons. On the polar boundary the breeding of more cold-resistant and quicker-growing plants is being emphasized. On the dry boundary all farming practices are being subordinated to the principal goal of effective management of the scarce water supplies. In the sparsely settled developed countries much cropland has been gained through improvement of the infrastructure, since the transport boundaries are shifting into the zones more distant from the market.

7. For the more distant future, however, a contraction of agricultural space must be forecast, for all agricultural boundary zones have production disadvantages that stand in the way of income demands. The volume of food production is marked by a powerful growth rate, but it will be achieved on a shrinking agricultural area. The present intensification zones of agriculture will become still more intensive. The currently most extensive zones, however, will drop out of production.

IV. Farms as Building Stones of the Agricultural Region

The smallest units of agricultural space, agricultural areas, agricultural zones, and agricultural regions are in every case the farms. Thus every analysis of these geographical units must end with an analysis of the farms, their structure, their locational orientation, and the possibilities for their development; whereas every rural planning investigation must begin with an evaluation of the farm management situation. This chapter therefore is devoted to the forces that shape the farm.

A farm is understood to be an economic enterprise in the realm of primary production, consisting of both the residence of the operator and the agriculturally utilized area. Usually, especially in cooler climates, the farm has a variety of buildings in addition to the house, including a service area, which collectively are termed a farmstead. Special forms of the farm are, e.g.

- the *ranch*, which is an extensive grazing operation;
- the *bush* or *tree crop farm*, with a specialization, often exclusive, in crops that are planted rather than sown; and
- the *plantation*, commonly with a specialization in bush and tree crops, but which also has transportation facilities or installations for additional processing of the products (sugarcane mills, oil mills, sisal factories, tea factories, etc.).

The economic goal on all farms is primarily the satisfaction of the needs of the farm operator, his coworkers, and his family. Under conditions of little agricultural development, farms assume a domestic, subsistence character. They are closed natural systems, in which everything that is produced is consumed and everything that is consumed is produced (food, fiber, wood for fuel and construction, luxury products, etc.). With further economic development, integration with the market grows, so that finally with the incorporation of the farms into the national economy, operators now specialize in production that is the most suitable for the area, sell everything not required for their basic needs, and buy consumer and production goods with their sales proceeds.

The number of farm types is simply immeasurable, for there is no branch of the economy that is so dependent on location as agriculture. Actually, there are hardly two farms on our globe that are completely alike. What follows in this chapter, therefore, is a short overview of all the forces and groups of forces that lend shape and form to the agricultural operation.

Aereboe (1923) and Brinkmann (1922) have already strikingly demonstrated that there are two completely different groups of forces that determine operational organization in agriculture. One group is of an intrinsic operational nature. It is reflected in the striving for the greatest possible reduction in production costs, which is accomplished by combining in one farming operation all the enterprises that complement each other in their demands on labor, fertilizers, feed, and soil fertility. This group of forces thus impels one to diversify the farming operation as much as possible.

On the other side are the forces that make it profitable to favor some farm enterprises more in one location and others more in another. They are the forces that usually affect the farm from without, in particular the natural and economic locational factors.

Thus every farming system is the result of a compromise between the forces pressing for diversification on the one hand and those inducing one-sidedness on the other. It is in this way, with first one group gaining the advantage and then the other achieving superiority, that the development of the various farming systems takes place.

If one also wants to understand the nature of transformation and development of farming systems, the moving forces of economic life, then he must look for the explanation in the dynamism of technological development and in price-cost shifts, hence a third group of forces. The first two groups of forces explain for us the agricultural-geographic variations of farming types at the same time; the third group of forces gives information on the agricultural-historical variations in the same place, or on the evolution of farms.

1. Reasons for Diversified Farm Production

One of the most outstanding characteristics of agriculture is the predominance of farms with diversified production. The reasons for a more or less sizeable diversity are the following:

a) Work Spacing

The greatest share of farming costs is, in almost all cases, contributed by expenses for farm labor and draft, as well as machinery (release of labor). Hence the farmer must strive to get by with as little labor and labor supplementation as possible and to employ what is available as productively as he can throughout the year, i.e. spacing the work.

Such a degree of work spacing in agriculture, however, is only rarely obtained with one or two farming enterprises. To be sure, livestock rais-

1. Reasons for Diversified Farm Production

ing (if one disregards fodder cropping) provides uniformly distributed labor demands, and stall feeding of hogs or the modern, factory type of poultry raising offer by their very nature a steady level of employment. But in most land use types, planting, cultivation, and harvesting are sharply fixed in time and have different labor needs. Periods of labor slack alternate with pronounced labor peaks. Spacing work to minimize labor costs then can be achieved only by combining different enterprises in the farm operation, and of course not those whose principal labor culmination points touch upon one another, but rather those that complement each other in time so that the labor valleys of some enterprises are filled with the labor peaks of others, and vice versa. There are only a few farming types that by themselves provide adequate work spacing. Extensive grassland farming without winter feeding and modern grassland farming with mowing of pasture belong to this group, as do viticulture and citrus farming. As a rule, though, spacing of labor requirements sufficient to lower costs can only be attained through a significant diversification of farm operations.

A qualification of this restriction is that not all operations, naturally, are dependent on work spacing in the same way. If large operators employ migratory labor, they only need to space work through the summer and winter, and not in the intervening seasons. So long as extreme under-employment ruled in southern Italy and cheap seasonal labor was available for any short time period, the proprietors of the latifundios were able to restrict themselves to a very small permanent work force for tending animals and operating machinery. They were also able to do away with work spacing completely and to practice pure grain farming. One-, two-, and if possible, three-crop operations are the rule in California fruit and vegetable farming, since Mexican and Filipino migratory workers are available at any time, thus making it even advantageous to concentrate work in a few short labor peaks. Where mechanized labor is cheaper than hand labor, one can rely on enterprises whose labor peaks are readily amenable to mechanized operations. Thus grains, rice, or cotton can be raised in the United States in part as a single-crop operation. Finally, where a particular farming enterprise has such a marked economic superiority that it can provide more income with just a half-year utilization of labor than a more diversified farm using labor throughout the year, work spacing has to be abandoned and the "suitcase farming" of the American Plains states is the result.

Yet as a rule, agriculture remains dependent on work spacing and that can only be attained, at least in the mid-latitude climatic zones, with di diversified production. The degree of requisite diversification naturally varies with the stage of national economic development.

Spacing of agricultural work at the lower and middle stages of technological development is achieved only with highly diversified economic practices. In contrast, at the highest stage of technological development with only a minor investment in hand labor, striving to minimize labor costs induces specialization because agricultural machines are designed for only a few work procedures. Large areas must therefore be included in the same enterprise so as to spread costs, although the same amount of land can also be cultivated within a diversified operation (overmechanization). As the farm enterprise expands, high labor costs also force a reduction in the time lost to field preparation and farm travel and in the length of exposure to possible crop damage. Full motorization also permits specialization earlier than animal draft because the tractor, by its very nature, forms a smaller share of the fixed costs than that of draft animals. The principle of work spacing will thus lose its force as time passes.

b) Crop Rotations

Also responsible for diversified production in agriculture is the principle of plant alternation. To utilize fully the productive capacities of the soil, crop rotations are employed so as to alternate plants setting down deep roots with those establishing shallow ones, plants consuming humus with those producing it, and plants destroying friability with those improving it — to name only a few of the divergent soil-degrading and soil-building processes of plants. In principle, crops with heavy soil demands are allowed to follow those that greatly enrich the soil, and crops with only small soil demands are made to succeed those that add to soil in only a minor way. The real supporters of the crop rotation are members that combine the rich legacies of a preparatory crop with moderate soil demands (potatoes, alfalfa, broad beans, rapeseed), whereas the principal beneficiaries of the rotation are those that make great demands on their preparatory crop but offer only minor value in that role themselves (wheat, barley). What nature attains through the spatial diversity of plant societies, man strives for with temporal diversity in the usual single-crop fields of agriculture. The yields of almost all land uses increasingly decline the more the same crop is returned to one and the same field.

The pressure for diversified production resulting from the requirements of the crop rotation is naturally not everywhere the same. This applies even to the requisite cropping interval for the same crop. In a dry climate, red clover can be repeated only after six years without a decline in yields setting in, but in the Rhineland it can be done already after five years, and in England after four. Considerations for nematodes require a four-year cropping interval for sugar beets on light soils, but only one of three years on loess soils. In Emilia and the Paris Basin, sugar beets are planted every other year without a second thought. Above all, however, not all crops are dependent on a crop rotation in the same way. The

1. Reasons for Diversified Farm Production

examples of continual rye cultivation or the overseas monocultural system based on rice, corn, cotton, or sugarcane show that there are crop types that are so self-compatible that they allow permanent cropping.

Besides particular place-specific and crop characteristics, technological advances are also capable of modifying the significance of plant alternation and thus the diversity of the farming operation. When the soil is more thoroughly prepared by motorized traction power, crops that loosen the soil can be abandoned. When cheap spraying and dusting equipment are made available for combatting plant diseases and pests, crop rotations do not have to be instituted simply for plant protection. When weeds can be controlled with hormone weed killers, crop rotations can become more flexible. Thus the shackles of the crop rotation loosen more and more in the course of the national economic integration of agriculture.

c) Fertilizer Balance

The necessity of a fertilizer balance that takes into account the needs for adequate nutrients and sufficient material for the formation of humus is also a motivating force for diversified production. So long as chemical fertilizers are unavailable or still very expensive, the nutritive capital of the soil must be fully utilized through cultivation of different crops with different requirements. Then one must depend on deep-rooting plants, fallowing and precise soil preparation measures for increased nutritive availability. There are other methods at higher developmental stages. Fertilization now enables industrial states to achieve the nutrient balance between crops producing nitrogen and those consuming it, between crops hungering for potash and those not needing it, and between crops thriving on calcium and those repelled by it, compromises which up to as late as fifty to eighty years ago were being achieved in Central Europe mostly through crop rotations. Chemical fertilizers have become so cheap in the last few years that a specific nutritive effect can be more conveniently obtained with an appropriate amount of application than with a more or less highly complex crop rotation.

On the other hand, the production of humus in many agricultural zones gains importance with increasing intensification. If no humus-consuming crops are planted, humus production is of no concern. With minor hoe crop cultivation but intensive fodder cropping, the root humus of the forage crops may itself suffice. Manure is still a waste product. If the forage crop area is reduced at the same time that hoe crop cultivation is expanded, then a purposeful manuring economy must be instituted. The farmer must then produce artificially a balance between the humus-producing and humus-consuming members of the crop rotation. If finally fodder cropping disappears altogether and hoe crop cultivation increases still further, then crop farming will often become economically dependent on other sources for its humus supply; humus is then obtained from permanent grassland, heath, peat bogs, or through purchase of harvested humus-rich crops, and converted into a more effective form via a manuring economy and applied to the fields. Finally, bush and tree crops are completely dependent on humus-producing crop types.

As evidence of the need for humus, the American Dairy Belt farmer produces corn silage in order to use the surplus of humus from ley farming. Alternatively, the vineyard operator emphasizers grain and fodder cropping on the other fields so as to satisfy the great humus demand of the vines.

d) Feed Balance

Since the most important means of maintaining the feed balance in midlatitude climates is manuring and this presupposes livestock raising, forage needs provide a further motive for diversified production. Only rarely can a single enterprise in fodder cropping guarantee sufficient amounts of high-quality feed throughout the year, adequate nutritive concentrates, and the synchronization of supplies of succulent and dry feed and of pasture and barn feed. Usually several must supplement each other.

Of course, livestock raising and fodder cropping promise to be more simplified and rationalized than they are now. As our road network continues to improve, truck transport will become still cheaper. This development will make it possible for feed and hay to be exchanged among farms and agricultural areas. Then it will no longer be necessary for every farmer to produce all the hay and feed he needs, and he may also be able to have a surplus to sell. A division of labor between different farms can then be established, with the result that the individual operator can restrict his livestock raising to one or two enterprises or even abandon it altogether. The feed balance will thus be facilitated and diversification curtailed. Technological advances also are permitting a simplification of fodder cropping. Today a farmer can get forage, hay, succulent feed (silage), and even feed concentrates (dry-green) from the same pasture, and thus with only one fodder cropping enterprise produce the same feed balance that in earlier times could only have been achieved through the combined but independent enterprises of livestock grazing, raising of root forage crops, cultivating of crops for roughage, and grain farming. When the cubing or wafering of hay becomes cost-favorable one day, the same pasture will be able to produce still another commercial product (as is already the case for alfalfa in the U.S.).

1. Reasons for Diversified Farm Production 93

e) Self-Sufficiency

Improvements in modes of transport and new achievements in technology also sooner or later make farming for one's own food needs superfluous, indeed even uneconomic, for they facilitate the purchase of food. The thickening density of transport networks, the improvement of customer services, the automobile, the refrigerator, and the telephone all encourage the farm family to buy its food supplies, so that the unmechanizable garden work required of the wife is also being constantly reduced. Earlier, the diversification of farming operations was based in good part on the desire of the farmer to produce all the food and raw materials that he needed from his own soil. Today, subsistence still plays a noteworthy role as a cause of diversified farming only in areas well off transport routes, as in high mountain areas, undeveloped steppe zones, or the tropical belt.

f) Spreading Risk

Every farmer must nautrally be alert to the threat of losses. He will always strive to distribute his production and marketing risks over several enterprises. The degree of farming diversity necessary for spreading risk, however, varies according to space and time, quite exclusive of the fact that the kind of risk also varies considerably according to the nature of the farming operations.

The oceanic climate offers a smaller production risk than that of the steppe. Irrigation farming can stabilize returns much better than dryland crop farming. Irrigation thus allows a higher degree of specialization. Where a market with fairly firm prices is guaranteed, farming need not be as diversified as it is where market risks and large price fluctuations are the order of the day. In West Germany, the first situation applies to milk and the second to vegetables, and that is why we are well acquainted with pure grassland farms, whereas production diversity in vegetable farming is great.

On the whole, the pressure for more mixed farming in order to spread risk is diminishing. When more powerful machinery is employed, the optimal planting, cultivation, and harvesting periods can be better observed and thus risk is reduced. Technological advances in plant protection reduce the risks of diseases and pests. Also, various dangers for the farmer are being increasingly reduced by society. Insurance companies cover harvest risk. Market risks, though, are being assumed more and more by the state, be it through marketing orders or guarantees of minimum prices, or through development of even better political measures for reducing cyclical and seasonal price fluctuations.

2. Reasons for Spatial Differentiation of Farms

For all the reasons mentioned above, diversified agricultural production in principle is advantageous almost everywhere and anytime, even with its variations. But what enterprises are combined into a farming system at any one time is a question of the locational peculiarities of the moment. The spatial differentiation of farming types is thus a question of the locational orientation of agricultural production. The influence of some of the locational forces is indicated in what follows.

a) The Physical Location of Production

Even the most fleeting observation of the smallest area demonstrates that differences in soil, climate, and terrain evoke extraordinary modifications in farming types. Our domesticated plants place very different demands on the natural conditions of their locations. Thus every location must be endowed with a quite specific fertility for one crop or the same crop group. Varying yields of a plant in different locations may, however, under otherwise similar conditions, lead to different hectarages for the crop at the different locations.

Nevertheless, it is not the absolute amount of return, but the return per input that is decisive. A specific expenditure of labor and capital in a climate that is cool and moist and on a soil that is extremely heavy will be best utilized with an emphasis on fodder cropping and extensive cattle raising. In a cool and dry climate and on easily worked soils with gentle terrain, this expenditure will be best turned to account with more intensive hoe crop cultivation; in the warm and dry Mediterranean climate, with the cultivation of specialty crops; in the shrub savanna, with extensive grassland farming; and in the constantly wet and hot tropical rainforest belt, with the raising of particular bush and tree crops. The overall production of foodstuffs has the tendency to be distributed over the physical production sites in such a way that the minimum amount of inputs is achieved. Adaption to physical conditions is one of adaption to production costs.

In the Europe of the past, this selective, differentiating influence of the advantages of physical location could take effect only partially because of national tariff restrictions. The strivings by the states for autarky in the agricultural economy stimulated the production in most countries of products that were not areally suitable and could have been produced more cheaply in other countries. This contributed to the diversity of German farms.

The European Common Market changed this situation. The individual physical areas of the EC can specialize more intensively than before in particular enterprises since they are united in a large, single economic area, thus one based on free trade and the absence of tariff barriers. By this means, a stronger interregional competition of production orientations is developing, with the result that each crop is produced where it is least encumbered by production and marketing costs. This specialization tendency of agriculture in the Common Market area is further strengthened as all energy supplies become relatively cheaper over the long run, so that falling freight rates make commodity exchanges possible over wider areas. Thus fruit and vegetables now stream from Holland to Germany and early potato production has shifted to climatically favored sites in France and Italy, to give but two evidences of this increased areal specialization.

b) Population Density, Educational Level, and Personality of the Farm Operator

The fluctuating input-output ratio in one and the same enterprise from region to region and the resulting differences in the relative efficiency of the various enterprises from farm to farm are not, however, simply the expression of the influences of nature; they also reflect the density and educational level of the agricultural population and above all, the differing capabilities of the farm operator. These subjective stimuli are also responsible for one farmer favoring one type of production and another farmer emphasizing another.

When, after the last war, West German farmers suddenly had at their disposal a host of refugees who had to offer their labor at any price, production procedures that had for decades been considered as antiquated because of high labor expenditures were in part successfully reinstated. In central Africa the shifting cultivation system still dominates immense areas because its food-producing capacity is sufficient only with sparse settlement. In the largest parts of the overpopulated Southeastern Asia areas, in contrast, farmers have to concentrate on wet rice cultivation with its two to three harvests a year and a carrying capacity many times that of shifting cultivation, and even this is not adequate. In the almost completely unpopulated Amazon Basin wild rubber is still gathered. In the already densely settled Congo Basin rubber is obtained on small tree crop farms that are widely distributed. In over-populated Java, however, only intensive plantations at the highest technological level are still possible.

The influence of the educational level of the agricultural population confronts us at every turn. In the subtropical dry belt sugarcane is still frequently grown at locations where the winter cultivation of sugar beets would be more economic, this because sugarcane cultivation requires less "know-how." In many dry areas sprinkler irrigation still cannot displace furrow irrigation because it demands more technical skill than is available. The largely agricultural country of Malawi is still one of hoe farmers without cattle raising and with a high quota of illiteracy. In itself, the time would now be ripe for yoking oxen. Yet it might be asked whether this stage ought not to be skipped so as to proceed immediately to the use of tractors. It is obviously easier to find a few thousand capable tractor drivers than to be able to teach every farmer how to work draft oxen. The backwardness of southern Italy in contrast to the Po Plain is, among other things, also an education question.

The personality of the farm operator is becoming a stronger factor in farm organization with the progress of economic development, since the requirements in knowledge and ability on the part of the farmer in every enterprise are becoming ever greater. Already just the purely technical information that is required has increased tremendously and continues to do so. As late as fifty years ago, there still was not much known in Central Europe about mineral supplements for animals, methods of ensilaging, plant protection, and other measures. Today these practices already require an extensive branch of science. Even a good knowledge of the techniques of agronomy and animal husbandry has long become insufficient in itself. Now the farmer must also be a machine specialist, a skillful manager, and an able business man. All of this applies even more to West German agriculture as it has come under the sharpened competitive pressure of other parts of the European economic area which because of locational advantages can produce more cheaply. Moreover, since supply is stronger than demand for most agricultural products in the EC, competition for the market will force agriculture to concentrate much more on quality production than it has up to now. Quality production, though, places an especially high premium on technical knowledge and ability. Although vocational education is improving, it is becoming impossible for the individual farmers to acquire the necessary specialized information and experience for all the enterprises that are now commonly carried on in a single-farm operation. The practical farmer, as in all other occupational groups, will definitely have to become a specialist if he wishes to mange his farm in the future successfully.

c) Size of Farm, Ranch, and Plantation

As to the spatial differentiations of the farm that can be traced back to farm size, the predominant cause is the varying relationship of manpower to usable land area. Figure 18 illustrates this point schematically. The smaller the farm, the higher its labor inputs will be per 100 ha FL and the less the farmer will be able to substitute mechanized labor for hand labor in his striving for a higher return in DM per ha FL, in short, the more intensively the farm will have to be run.

2. Reasons for Spatial Differentiation of Farms

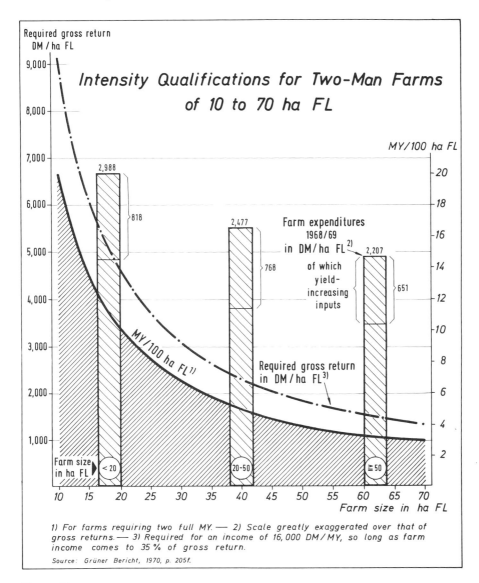

Figure 18

The effect of farm size variation on the nature and intensity of the mechanization of the operating unit is shown in figures 19 and 20. Investment in machinery on small farms is restricted on two sides:

— on one, by a relatively abundant labor supply whose opportunity costs (cost = sacrificed opportunities) are small because the work capacity released by mechanization can hardly be productively employed in any other way; and

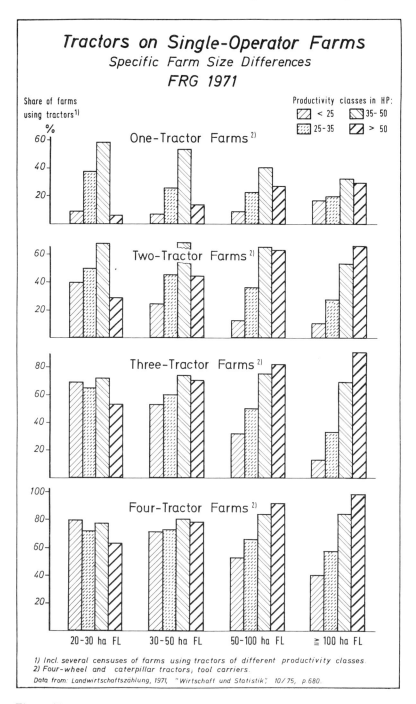

Figure 19

2. Reasons for Spatial Differentiation of Farms

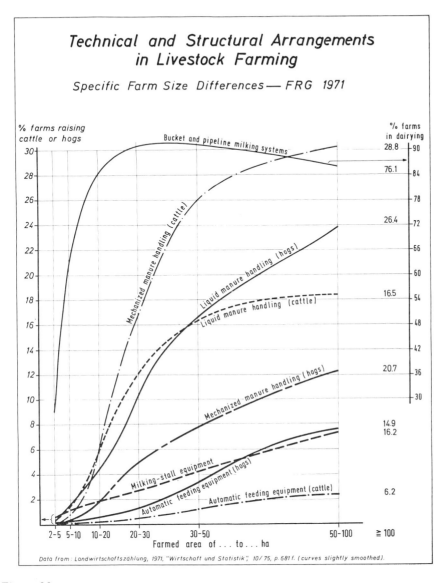

Figure 20

— on the other, by relatively high machinery costs, when insufficient work area and livestock inventories encumber the labor unit with high fixed costs.

The latter disadvantage can be eliminated or at least reduced through the cooperative use of machinery (machinery cooperatives, machinery pools, custom hiring, etc.). Fig. 21, however, shows that apparently

IV. Farms as Building Stones of the Agricultural Region

the degree of availability of the various machines for interfarm use is highly variable.

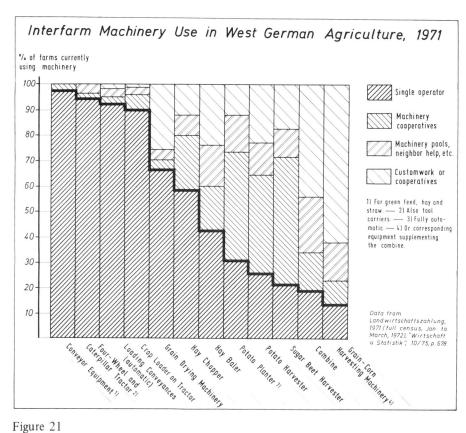

Figure 21

                     Size of Two-man Farms

30 ha FL		90 ha FL
6.67	MY/100 ha FL	2.22
1. *Sugar beets*	Crop Rotations	1. *Sugar beets*
2. W. wheat		2. Oats
3. W. barley		3. W. wheat
4. *Sugar beets*		4. W. barley
5. W. wheat		5. *W. rape*
6. W. rye, oats		6. W. wheat
		7. W. barley
		8. W. rye
	% CL	
33.3	Hoe crops	12.5
66.7	Combine-harvested crops	87.5

2. Reasons for Spatial Differentiation of Farms

In all cases, the smaller farmer must strive to maximize his returns in DM/ha FL by increasing labor intensity. If he has a location that allows production elasticity, then he can adapt his production program to his goals. Thus, for example, the crop rotations shown on page 98 are frequently encountered in the Central European loess strip. More difficult is the accommodation of farming intensity to a location allowing little production elasticity, such as heavy marshland (Table 4). Here production orientation can only be modified in a limited way. Adaption of intensity to the labor inventory of different farm sizes must therefore be accomplished through specialization intensity: higher intensity of fertilization, greater fodder yields, more careful preservation of feed, higher livestock inventories, and shift of the ratio between dairy cattle and feeders in favor of the former.

Table 4: Operational Objectives on the Pure Grassland Farm in the North German Marsh; Required Gross Income = 25,000 to 30,000 DM for 2 MY

Farm Size ha FL	up to 25	25–30	30–50	over 50
Fodder yield, kStU/ha	over 5,000	5,000	4,500	4,000
Livestock inventory, LLU/ha	3.0	2.5	2.2	2.0
Farm type	Dairy	Dairy	Dairy-Feeder	Dairy-Feeder
Dairy cows: young stock = 1:	0.4	0.4–0.9	1.5–3.0	3–6

Source: Blohm, G.: Die Ökonomik der Grünlandnutzung. In: Schriftenreihe der landwirtschaftlichen Fakultät der Universität Kiel, Vol. 42 (1968), pp. 105 ff.

Table 5: Ranches of Different Sizes in Southwest Africa, 1965/66

Ranch No.	Size ha	Annual Precipitation, mm	Number of Drinking Places	Number of Enclosures	Sizes of Enclosures ha	Pasture ha/CU[1]	Gross Return Rand/CU[1]
I	7,000	280	6	18	389	10	20.14
II	9,000	350	7	13	692	12	14.69
III	11,000	395	8	12	917	8	14.18
IV	11,500	350	5	10	1,150	8	13.95
V	12,500	350	10	13	962	7	11.72
VI	17,500	320	6	24	729	13	10.37
VII	18,000	350	11	26	692	9	9.24

[1] 1 CU (Cattle Unit) = 1 head of cattle or 6.2 sheep or 5 goats or 2 donkeys or 0.83 horse or mule.

Source: Andreae (1966), p. 61.

Adaption of farm organization to different farm sizes becomes even more difficult, then, when it is a matter of simple appropriative economies, where the range of specilization intensity is also narrowly restricted. Table 5, however, shows that even here there are still some possibilities. Although nature and this kind of economy allow no improvement of pasture through fertilization, irrigation, cultivation, and other measures, gross returns per cattle unit nevertheless can be steadily increased and overall more than doubled with a dimishment of farm size from 18,000 to 7,000 ha.

Finally, there are still even more extreme locations in which farm adaption for all practical purposes is no longer possible. Even a small sugarcane mill, with a daily capacity of about 800 tons of cane and an output of 75 tons/ha needs 2,000 ha FL in sugarcane monoculture and 4,000 to 5,000 ha FL with a rotation (Gnielinski 1968, p. 285). If the plantation is smaller, then this may be partially compensated by fertilization. But this measure cannot ameliorate the cost progressions that must be put up with, especially in the mill.

d) Location of the Farm for Transport

There is still one more fact responsible for the differentiation of production orientation of the various farms, the distinct falling off of farm gate prices for various agricultural products and inputs from place to place. This in turn leads to areal variation in the competitive positions of enterprises with each other, so that in one place one branch of production enjoys economic superiority and in another place another production type holds sway.

The adaption of farming types to the transport situation is one of accommodation to transport costs. The reasons for this are the following (see Fig. 22): Inasmuch as the transport of farm products and inputs from farm to market and back is at the expense of the farmer, growing market distance changes the relevant price-cost relationship for a farm. Augmenting market distance lowers the price at the farm gate of agricultural products to the same degree that transport costs climb. The farmer of course realizes only the difference between market price and transport costs. Inputs of agricultural origin generally cannot behave differently from agricultural products in their price gradations. Inputs of industrial origin, in contrast, must increase in farm gate price with increasing market distance because the farmer has to pay his transport costs as well as the market price.

Cash wages also increase in industrial states – equivalent purchasing power assumed – with increasing market distance since the farm worker consumes more industrial than agricultural resources. They increase, though,

2. Reasons for Spatial Differentiation of Farms

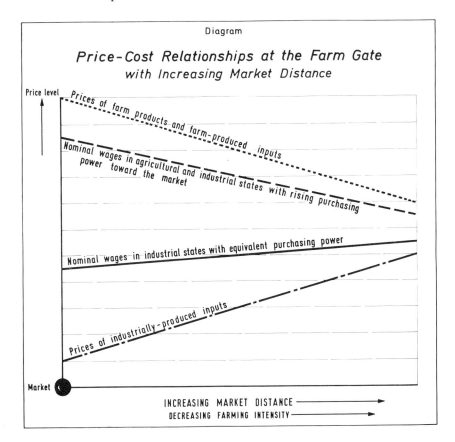

Figure 22

less strongly than prices of industrially-produced inputs because the increase in prices of consumer goods of industrial origin is partially compensated by the cheapening of consumer goods of agricultural origin. As a rule, we even find a falling off of cash wages with growing market distance, since the above premise, equivalent purchasing power, is usually not given; on the contrary, purchasing power increases on approach to the market because of the competition of industries for labor. In agricultural states the fall of cash wages with increasing distance from market is even given in principle, and of course because here the "agricultural share" of cash wages is greater than the "market share." Hence the increase in prices of market goods is more than compensated by the cheapening of agricultural goods.

Generally differences between prices of agricultural products and those of wages and inputs increase with approach to the market; thus increasing proximity to the market center signifies an improvement of the transport situation so that:

1. operational and organization intensityal can increase;
2. specifically intensive enterprises gain priority;
3. periods of idling land are shortened; and
4. farm expenditures shift more and more from hand labor to industrially-produced capital goods.

Being closer to market also means that farming can be more diverse because areas with this location, under the pressures of the integrating forces for diversified production, can carry on extensive as well as intensive farming enterprises, whereas areas more distant are dependent entirely on the extensive types. Farmers closer to market can also pursue, besides hoe cropping and dairying, grain cropping and cattle raising. Farmers farther away, though, must restrict their diversity because transport-sensitive products like seed potatoes and milk, though certainly produceable, cannot be marketed. Fig. 23 now needs no further explanation.

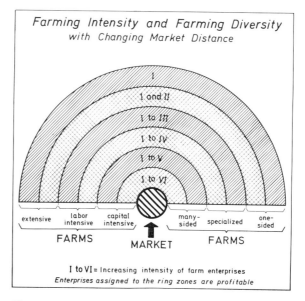

Figure 23

If it is asked whether transport situation influences farms and agricultural regions more in industrial states or more in agricultural states, then the answer must clearly be: in the latter. In many developing countries the transport system today is still less developed than it was in Germany 150 years ago, when Johann Heinrich von Thünen's famous work, *The Isolated State ...* (1826) appeared. Thünen's conceptual model has a much greater relevance for the developing countries than for the highly industrialized states, which have been so prominently developed through transport.

2. Reasons for Spatial Differentiation of Farms

Over a century later Laur (1930, pp. 61 ff.) attempted to systematize and to map the farming types of the world. As delimiting criteria he used the transforming forces of transport situation, population density, and certain climatic characteristics. The result was the following systematic representation:

    a) farming types of the caravan zone   Market-distant
    b) farming types of the pasture zone
    c) farming types of the agricultural zone
    d) farming types of the plantation zone
    e) farming types of the industrial zone
    f) farming types of the local zone
    g) farming types of the residential zone   Market-near

Laur spoke of these areas as "economic zones of world transport." In this formulation the dominating influence of market location is to be seen just as much as it is in the above designations of farming types or agricultural regions.

e) The Diversified Farm in the Tension Field of Force Groups

Let us look back. Two groups of forces determine the stage of production diversity of the farming operation, one pressuring for one-sidedness and the other for diversity. If there were only the differentiating group of forces, the locational forces, then only farms with monoculture could exist. The division of labor that would naturally follow would confer on every individual farm a very one-sided character. Diversity would apply only to the synoptic view of all farms. If, on the other hand, only the group of forces pressuring for diversity were effective, then all conceivable enterprises would have to be combined on every farm. Uniformity would apply only to the totality of operations. Only because of this fact, that the force groups work against each other, has there originated the great variety of farming types that confront us in real life. (Brinkmann 1922, p. 65).

The preceding discussions, however, have now shown that at least some of the forces encouraging one-sidedness in the course of development are gaining strength, and that all of the forces pressing for diversity, in contrast, are losing significance. The force field is therefore shifting. One pole, one-sidedness, is becoming stronger and the other, diversity, is becoming weaker, so that farming systems are shifting more from the second path to the first. Here lies the real root of the farming simplification and specialization that is being strived for in industrial states.

In many developing countries the numerous crop and livestock enterprises of the *many-sided farms* ensure strong self-suffiency, full employment of the permanently available hand labor and draft power, utilization of wastes, and great stability; that the operators of these farms can neither fully master nor mechanize the many enterprises is the serious disadvantage.

Contrarily, the limited enterprises of the *one-sided farms* in many steppe areas guarantee the exploitation of expensive single-purpose machines (e.g. tractors with detachable implements, sprinkler systems, combines) and capitalize on special regional conditions or particular capabilities of the farm manager. But they also pose, in their failure to provide self-sufficiency and their threat to soil fertility, high production and market risks.

On *diversified farms,* concentration is on only a few important crop and livestock enterprises, those that are adapted to the locale, are more reliable producers, and are the most profitable. Thus the aim is to combine the advantages of the economic extremes: the durability and stability of the many-sided farm operation and the cost-favorable mechanization capability of the one-sided operation. Operations of this type are indispensable wherever the investment of labor in land has greatly diminished and the diversity of farming operations has up to now hindered a vigorous mechanization. In most European agricultural regions today the diversified farm is the appropriate form.

3. Reasons for Temporal Changes in Farms

The distribution of farming types in world agricultural space, as illustrated in the following chapters, is still not fully explained by the forces named so far. Still to be considered are:
– price-cost differences and
– differences in the state of technology.

These two forces are even more important for the understanding of the adaption processes of farms to technological developments and to economic growth, for they are the impulses that provoke new developments and evolutions, the motivating forces of economic life. They are the forces that explain for us the metamorphosis of farming types in the course of development.

a) Price-Cost Development

It was established that the same market prices affect farms with different transport situations (market situations) differently, since prices at the farm gate differ from market prices. But now even market prices have been subjected to change in the course of national economic development. The effects of this on the farming operation are the object of this section.

3. Reasons for Temporal Changes in Farms

Absolute prices and costs cannot be decisive for the farmer. Rather it is always a matter of relations of prices to one another, of relations of costs to one another, and of the relations of prices to costs. Essentially three price-cost relationships are relevant for the farmer:

aa) Price Relations between Agricultural Products

This price relationship decisively determines the combination of enterprises because it substantially influences the terms of competition for the enterprises. Where animal products are low in price in comparison to plant products, agriculture will be of an essentially *primary*-production character (e.g. China). In contrast, where prices of animal products surpass those of crops, *secondary* production will come to the force (e.g. Denmark). Where hoe crops surpass overall agricultural price levels, intensive hoe cropping will occur (e.g. Germany before the last war). Where grain crops are clearly favored in price, cropping systems with an emphasis on grain will predominate. In a country with high milk prices but low prices for slaughter cattle, milk production will be promoted quite differently (e.g. Sweden in the 1950s) than in a country where the reverse is the case.

Many other examples of this kind could be cited in which the prices of agricultural products relative to each other determine the production orientation. Yet even a few examples show that one also cannot simply express the converse; the production program is not completely dominated by the price relations among enterprises. From 1966–67 to 1973–74 the price ratio between beef and milk in the FRG shifted only a little. The selling price for feeder cattle rose on the average by 26%, that for milk by 20%. Yet beef production increased by 18% during this period, while milk production stagnated (Statistisches Jahrbuch 1975, pp. 119 and 234). This was due to rapidly rising wage levels, by 75%, which favored the wage-tolerant beef production over the wage-sensitive milk production. Swedish farms without cattle developed at a time when animal products exhibited a quite favorable price level in comparison to that of crops. Here increasing wage and building costs restricted livestock farming in its competition with crop farming. In West Germany, from 1961–63 to 1973–74, the price ratio between grain and hoe crops shifted in favor of the latter because the selling price for grain sank by 5%, whereas that for hoe crops rose by 20%. Yet from 1957–61 to 1974 grain farming expanded its share of the cropland from 61.3% to 70.2%, during which hoe cropping fell from 23.2% to 14.7% CL. Here, also, wage levels were the driving force that decisively shifted the competitive ratio between grain and hoe cropping in spite of the inverse development of the product price.

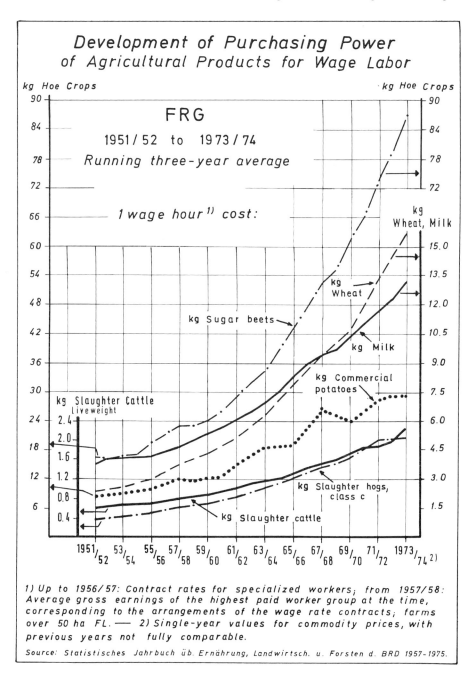

Figure 24

3. Reasons for Temporal Changes in Farms

bb) Cost Relations between Agricultural Inputs

The second group of price-cost relationships important to the farmer is that of the relations of input costs to one another. It determines the combination of inputs or production factors. The optimal productive combination of inputs is always attained when the marginal productivity of every production factor corresponds to its cost. But since the application of every production factor is subject to the law of diminishing returns, each individual production factor can be applied more heavily in relation to the others the lower its costs turn out to be. On the other hand, the expensive production factors may be used sparingly. Where chemical fertilizer ranks lower than human labor in prices, efforts to maintain the nutritive balance will emphasize chemical fertilizers (e.g. United States). In contrast, where the situation is the reverse more emphasis will be put on manuring (e.g. Germany at the beginning of the 20th century). Where supplemental feeds can be purchased relatively cheaply (e.g. Denmark and Holland), operators can satisfy their special feed needs quite differently from those in areas where feed is more expensive (e.g. developing countries). Where rural wage levels still are low but farm machinery is expensive, an agriculture with intensive manual

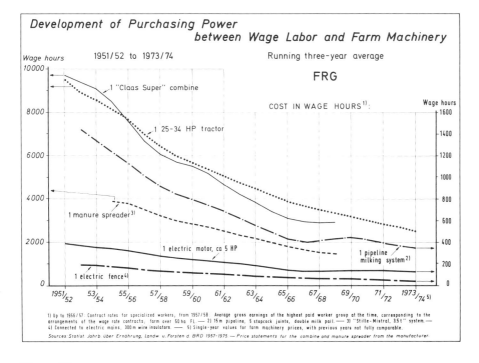

Figure 25

labor will emerge (e.g. Southeastern Asia). Where extremely high wages and income expectations prevail among farm populations but where machines are cheap, a high degree of mechanization gives agriculture a capital-intensive style (e.g. Unites States, Canada, England, Sweden). Fig. 24 clearly shows how swiftly and drastically the purchasing power of agricultural products for wage labor can sink in only two decades in a country with uninterrupted growth. Fig. 25, though, shows that agricultural machinery, as measured in wage hours, is becoming cheaper at the same time. Mechanization, then, is the prescribed path. An especially clear development from labor- to capital-intensive operations has taken place in German sugar beet farming, which around 1900 still required 1,400 MH/ha and in 1945 at least 500 MH/ha, but today, without thinning and with fully mechanized harvesting, manages with only 55 to 60 MH/ha.

cc) Price-Cost Relations between Agricultural Products and Inputs

These relations determine the intensity of agriculture. Where price levels for agriculture are still low but cost levels high, efforts to wrest a harvest from the soil must be made with the most frugal use of labor and capital. This can only be done with extensive farming types like grain-fallow farming or extensive grassland farming without stabling (e.g. the large Sahel area). On the other hand, where a dense population with high purchasing power brings an increase in agricultural prices and where at the same time a highly developed industry can provide agriculture with cheap inputs of industrial origin, crop yields can be powerfully increased with greater expenditures per hectare (e.g. Europe, especially Germany, Belgium, Holland, etc.).

The price-cost ratio also determines the degree of intensity of each individual farm enterprise. Up to 300 kg of pure nitrogen per hectare are applied to the Dutch grassland marsh because nitrogen prices are low in relation to the price of milk. Contrarily, in New Zealand, one of the most important milk producing areas of the world with 94% FL in permanent pasture, nitrogen is applied at a trifling rate of 2.4 kg/ha because the price ratio between nitrogen fertilizers and milk is much more unfavorable. The rice yields of the United States are higher than those of East Asia, though the U.S. expends only 20 to 30 MH/ha versus the 600 to 1,200 for East Asia; the reason quite simply is that the price-cost ratio permits heavy fertilizer applications in the U.S. and not in East Asia. Countries with high milk production per cow are also those in which prices for feed concentrates are lower than milk prices. Countries with low milk output per cow, conversely, are those in which low milk and high feed prices hardly allow the use of concentrates.

In the course of economic development the purchasing power of agricultural products for hand labor will ordinarily become less favorable, but for all capital goods more favorable. Thus, for example, in the Federal Republic of Germany,

1 MH in 1950/51	cost	15 kg of sugar beets, but in
1974	already cost	83 kg; whereas
1 kg N in 1913/14	cost	50 kg of sugar beets, but in
1974	cost only	17 kg.

b) Technological Advances

Technological advances make up the strongest force for economic development. Two major groups of technological advances can be distinguished:

aa) Organic Technological Advances

These apply to the breeding of more productive cultivated plants and domesticated animals, the development of more efficient feeds, fertilizers, insecticides and herbicides, and the other advances of a biological nature. Only when the breeding of hybrids was successful did it become possible to lower the costs of egg production significantly. Development of hybrid corn has opened up new cropping areas for grain corn in large parts of the world. The breeding of hard winter wheat varieties has carried wheat cultivation in Scandinavia, Russia, and North America far into the northern latitudes. The breeding of short-season varieties of spring barley has made possible the cropping of the grain, on the one hand, in regions with long winters, like Lapland, and on the other hand, in regions with a short rainy season, such as North Africa. Organic technological advances enable agriculture to intensify and thus promote economizing on land. As a result, they are especially expedited in countries that are dependent on land-productive economies because only a little and very expensive land is available in relation to the population. Central Europe and East Asia are examples.

bb) Mechanical Technological Advances

Mechanical technological advances, in contrast, bring about a diminishment of agricultural intensity. They contrast with the land-economizing organic technological advances in their land-opening character. We need only note the effect of the invention of the combine, which carried the fully mechanized grain farming of semiarid climates far into the areas of extensive livestock raising. Also to be cited are the great regional development projects in Siberia, which became possible through mechanical technological advances.

Sugar beet farming in West Germany has continued to increase in the last decade despite increasing wage levels because the development of the sugar beet harvester made harvesting less onerous. Lessening drainage costs also allow us to assume that cropping of heavy marsh soils is still possible in the technological era. The electric fence has allowed strip grazing, which saves on labor, to be substituted for indoor feeding, which is also land-productive but costs more labor. The effects of hay cubing appear to be unlimited.

Mechanical technological advances thus are especially important for countries and farms where high labor productivity must be eliminated, for mechanical technological advances are labor-saving, whereas organic technological advances are land-saving. That even here, though, boundaries are nowhere to be drawn sharply is shown by the example of the more recent technological developments in sugar beet cultivation. In order to reduce the critical labor peak, efforts had long centered on the mechanical pulverization of the beet top and on thinning with machines, thus on providing help with mechanical technological improvements. But these technological aids proved to be beneficial only for countries that could, in opposition to the situation in German agriculture, dispense with high land productivity. We now know that in densely populated countries it is not the mechanical but the organic technological advances, the breeding of single-germ beet seeds and single-grain seeds and the row-spraying of herbicides, that make it possible to crop without weeding and thus save the most on labor.

Two other examples of the labor-saving effect of mechanical technological progress deserve special mention. The vegetable canning industry as an agricultural enterprise has been basically transformed in the last few years by harvesting machinery. This development has helped to move vegetable cropping on large farms into the supply areas of the canneries (pea viners, bean pickers, carrot harvesters, spinach harvesters). Technological improvements in sprinkler irrigation, such as lighter pipes, better pipe couplings, and better dispersal of water jets, have opened up new and of course labor-saving possibilities for irrigation farming. They are also significant to the extent that many fragmented plots in the desert, heretofore incapable of irrigation and therefore any farming, can now be suddenly transformed into green and thriving fields.

This short overview of the forces responsible for the shaping of the farm may now be concluded, particularly since the latter ones will again have to be discussed in the interpretation of the spatial distribution of farming types, which now follows.

V. The Principal Farming Systems of World Agriculture

The terms, "farming system" and "agricultural system," are commonly used to designate the form of agricultural operations. In both cases, the emphasis is on systems as a purposeful intermeshing of various forces. It indicates that agricultural production is generally one of diversified farm production. For the characterization of agricultural systems, therefore, the identification of degrees of diversification is of significance. Further, the factor combination for each economic unit ought to be considered if one wishes to delimit agricultural systems. Finally, and above all, the production program serves as a proven classification principle for farms, ranches, and plantations.

If one wishes to characterize an agricultural operation concretely, exactly, and in detail, then all the above mentioned criteria must be used: degree of diversification, factor combination, and production program. The result is of course such a wealth of combinations that the taxonomy becomes too bulky and eventually unintelligible. A world view, as intended here, can be attained only through a simplified and coarse classification system which at times is restricted to only one of the three criteria.

In this chapter the geographic juxtaposition of current agricultural systems is treated in a brief overview, the locational orientation of agricultural production. The classification of outward manifestations is made according to the most important agricultural-geographic criterion, the production program, and here again primarily in conformity with the land use structure. Chapters VI to VIII will proceed in the same way, though they are more location-oriented and dwell more on regional details and functions.

The other two distinguishing criteria, operational diversity and factor combination, are more suitable for an agrohistorical inspection. They will therefore be treated in the concluding Chapter IX, which has as its content the structural changes in world agricultural space that accompany economic growth.

A simplified world map serves as a partial guide to the following farming classification (see colored map at the end of the book, Fig. 26)*.

---

* See note on page 114.

1. Grassland (Grazing) Systems

Although the following discussion of the grassland, annual-cropping, and perennial-cropping systems is based on their production programs, it must be remembered that they do not always occur in pure form. Transitional and mixed systems frequently exist in forms such as that of a grassland farm in the Elbe Marsh on which 30% FL is cultivated; a grain-fallow operation in the state of Kansas or a coffee plantation in the state of Sao Paulo, where a not insignificant amount of land is left in natural pasture; or a grape farm in Calabria or cacao farm in Ghana, where some cropland is set aside for self-subsistence and work spacing. Yet, if one wishes to recognize the essential and gain an overview, then only the systems representing the economically most important production programs can be highlighted. This applies not only to the three major categories of grassland, annual-cropping, and perennial-cropping systems, but to their subtypes as well.

a) Nomadic Grazing

The nomadic grazing economy is found today only in extremely dry areas with unfavorable transport situations. The nomadism of the population has its origin in the pressure for feed balance. Livestock herds are not actively driven by man; rather, man moves passively behind his animals which seek their seasonal feed balance through extensive migratory movements. Hence this type of economy is favored where highland chains, mountains, and valleys make for regional differences in the annual vegetative rhythm. Grzimek, in his book, *Serengeti Shall Not Die*, shows that African wild animals are for reasons of feed balance dependent on great migrations, and that therefore the Serengeti National Park ought not to be reduced. Domesticated animals must also migrate if no feed comes from human hands and they wish to achieve a seasonal feed balance. Man, who is dependent on these domesticated animals (little demanding types in every case), then follows his herds.

We find a modified form of pastoral nomadism in our migrating sheep farms in Germany. An essential difference, though, is that here man actively intervenes in the operation, whereas in nomadic livestock grazing he participates in the migratory movements more passively.

---

* Tr. note: This map was extracted from another publication and the English translations were made by another translator. For the discussion in this book, the following translations of the German terms are more appropriate: *Agrarzonen der Erde* = World Agricultural Zones; *Ackerbausysteme* = Annual-Cropping Systems; *Dauerkultursysteme* = Perennial-Cropping Systems; *Hackfruchtbauwirtschaften* = Hoe (Root) Crop Farming; *Pflanzungen* = Bush and Tree Crop Farms.

b) Sedentary Extensive Grassland Farming (Ranching)

Here also it is a question of agriculture in the marginal zone of human habitation. In the American desert state of Arizona extensive sheep raising is carried on with as little as 150 mm annual precipitation. In the southwest African desert, the Namib, we even find sheep raising practiced with an annual precipitation of 100 mm. The upper precipitation limits of extensive grazing are located where grain-fallow farming becomes possible. Besides the amount of precipitation, a great many other factors influence the mutual competitiveness of these two farming systems that are so important for the dry areas, extensive grassland farming and grain-fallow farming. Thus a firm precipitation boundary cannot be drawn between them.

Organizational modifications of extensive grassland farming are also clearly correlated with the amount of precipitation. In addition, the transport situation is decisive for the types of extensive grassland farming because it determines whether milk can be sold or at least cream can be delivered weekly, or whether one must be restricted to cattle raising or fattening.

In the great gamut of farming systems extensive grassland farming constitutes a pronounced extreme. It is, above all, the most extensive farming system. Draft animals are almost completely lacking and chemical fertilizers are not employed. As a result, land productivity is extremely low. Net labor productivity, in contrast, attains relatively high values. Capital investments consist essentially of livestock and land. The value of land per hectare is naturally extremely low; yet land capital carries great weight since extensive grassland farming is only possible with large areas. To that must also be added the sizable investments that must be made for fencing and watering.

Extensive grassland farming is possible only in large areas because productivity per hectare can only be minor and because of the cost structure. This cost structure makes it clear that the preponderant part of farming costs consists of expenditures for labor, farm vehicles, and engines. All of these costs, however, are heavily dependent on the size of the herd. The investment in farm vehicles, engines, and labor scarcely varies if the livestock inventory rises from 300 to 800 cattle or from 800 to 2,000 sheep. But since the livestock investment can be expanded only by increasing hectarage and not operational intensity, fixed costs become less encumbering as the farm, and with it the livestock herds, become larger.

c) Sedentary Intensive Grassland Farming

While the extensive grassland systems are characterized by the farmer who passively adapts to times of feed deficiency through production orientation; animal selection; timing of gestation, sheering, and selling periods; and the like; intensive grassland systems are distinguished by the active overcoming of periods of feed deficiency by means of feed production and feed purchases. This is possible only with a favorable price-cost ratio, which allows high labor and capital expenditures per hectare. Intensive grassland systems are therefore to be found only in strongly industrialized states.

There are many economically and ecologically determined variants. Here only three climatically-conditioned types may be noted:

1. The *polar fodder cropping systems* that ring the globe: Scandinavia, northern Russia, Alaska, and northern Canada. The reasons for the growing prominence of fodder cropping with increasing poleward location are the shortening of the growing season and the radiation and temperature conditions in the domain of the midnight sun. These circumstances favor the planting of a single perennial crop which by itself is capable of fully utilizing the short growing period. Fodder cropping thus gains in competition with cash cropping toward the pole, until it finally drives it out completely.

2. The *maritime fodder cropping farms,* which are most numerous in the northwest German and Dutch coastal areas, in Great Britain, and on the Cherbourg Peninsula. Here it is not the amount but the favorable distribution of precipitation, often in the form of fog and drizzle, and the high humidity that favors fodder cropping. For various reasons, the maritime climate is better suited for pasturing than for grassland cropping. This is particularly the case where the winters are mild enough to allow almost year-around grazing, while on the other hand the colder winters of the German marsh belt cause a feed shortage. Then types of livestock farming that make possible a seasonal adjustment of herd size to a drop in feed supplies (livestock raising and summer feeding farms) become desirable.

3. The *montane fodder cropping zone,* where a short growing season and damp climate make for a distinct agricultural system. Thus it is not surprising that fodder cropping assumes extreme forms in precisely the high mountain chains. The largest single montane fodder cropping area of Europe is the Alpine area, including the foothills. But fodder cropping also appears at specific altitudes in almost all of the European mountain lands. The mountain climate favors pasture mowing over grazing. This has its advantages for management because the cuttings help to overcome the deficit in feed during the long winter and thus facilitate the seasonal feed balance. Compared to the feed balance in the maritime fodder cropping zone, that in the mountains is thus less of a problem, so that here dairying can dominate while in the marine areas a greater share of young stock and feeder cattle must be kept on hand.

The system of combined fodder cropping and livestock raising is, apart from the raising of certain perennial crops, the only one in European agriculture that permits monoculture. Today one of the most important and extreme forms of specialization taking place on fodder cropping farms of up to 30 ha is the abandoning of the remaining crop areas and a shifting to pure grassland farming.

2. Annual-Cropping Systems

Annual-cropping systems are usually more intensive than grassland systems in the use of labor and fertilizers. Every year human hands must work the soil, seed or plant the crop, and then harvest it, whereas in grassland systems the last function is at least partly dealt with by the grazing animal itself. More nutrients are also exported from the farm with cropping and the materials must be replenished.

a) Primitive Rotation Farming

For primitive rotation farming, all this is true only marginally, for here labor and nutrients are saved through periodic alternation of cropland with the natural vegetation. Where this occurs in grass formations, the origin of crop farming is one of *steppe shifting cultivation,* which also existed in large parts of Europe at the beginning of a development of 2,000 years. Referring to this cropping system Tacitus wrote in his *Germania,* "Arva per annos mutant et super est ager," which according to Aereboe is to be understood as "The grain fields change every year, and there is enough land left over for cropping,".

So long as settlement on the steppes remains highly scattered and the land suitable for cultivation consists of predominantly natural grassland, the cultivator finds it sufficient to pursue only a very elementary cropping scheme. First he breaks up a small piece of grassland with a digging stick, a hoe, and later a plow, and then he plants crops, usually millet or sorghum, for two to four years. By that time, yields have fallen so low because of insufficient soil cultivation and a lack of any fertilization that the land is left to revert to its natural grass cover for as many years as are needed to restore fertility. Meanwhile another piece of land will be taken and worked in the same way. Later on, the farmer may again grow grain on the plots used previously, but only at irregular intervals of many years.

The same principle is followed, circumstances permitting, in the native forest areas. In the humid savannas and in the tropical rainforest belt, all the work of farm organization and management is one of battle against dampness and moisture, as much as 3,000 mm of rain a year.

Additionally, with the clearing of the native forest soils, erosion, weed growth, disappearance of humus, and breakdown of soil structure lead to a rapid decrease in harvests within a few years, and thus make necessary an ever more intensive working of the soil. Up to now there has hardly been a better way of countering this destructive sequence than to let the land that has been used for two to four years return to the original forest cover and to take instead a new piece of the forest for cultivation by clearing with fire. This is the *forest-burning system of shifting cultivation,* an extreme adaption to extreme natural conditions. Today over 200 million people spread over 30 million km² make a living with this type of economy.

Older forms of primitive rotation farming in Central Europe were represented by types such as the *Hauberg* and *bog-burning* systems. In all of these systems the objective was to save on labor and fertilizer by leaving to nature the task of rejuvenating the degraded soil, this to be done by using large amounts of land and cultivating any one patch only periodically.

b) Ley Farming

Ley farming, which consists of rotating annual food crops with fodder crops of several years' duration, also follows to a certain extent the same principle. The long period of fodder cropping reduces the need for land preparation, seeding, cultivating, and with the grazing, harvesting as well. At the same time, the weeds associated with the previous food crops are being eliminated, while during the cropping years, the weeds from the grasses are lost. Fodder crops also produce root humus, from which food crops subsequently benefit. The building up of soil fertility with fodder crops and their elimination through food crops are thus ingeniously alternated.

Ley farming is practiced under a variety of climatic conditions, where
— the growing season is short (northern Europe);
— the precipitation is high (mountain areas, East African Highlands);
— the precipitation, though not very high, is well distributed and accompanied by high humidity (North Atlantic coastal area of Europe and North America); and
— the precipitation, though not very high, is supplemented by irrigation.

Northern Europe clearly demonstrates that the useful economic life of leys becomes longer, and the forms of ley farming thus more extensive, with the shortening of the growing season on approach to the Arctic Circle. A life of only two years is typical for leys in Jutland and Scania, but the span is already three years in central Sweden. With advances into more northerly latitudes periods lengthen to four, six, and more years.

2. Annual-Cropping Systems

In *Great Britain*, not only is a north-south but an east-west differentiation recognizable in ley farming. The north-south differentiation is caused by the shortening of the growing season in the direction of Scotland, whereas the east-west differentiation results from the gradual rise of the terrain from the east coast to the west, up to the mountain areas of Cornwall, Wales, and Cumberland.

Especially extreme forms of ley farming may appear in the *Alpine area* because here high precipitation and a short growing season converge. Ley farming based on irrigation is represented on the *Po Plain*. The warm climate, combined with a rich supply of water flowing from the Alps, has allowed the development of an extraordinarily productive ley system, the many cuttings making possible a large investment in livestock. This agricultural zone is the most important dairy area of Italy.

c) Grain Farming

Primitive rotation, ley, and grain farming are essentially extensive cropping systems, and yet they possess a considerable intensity span:
- primitive rotation farming with its cultivation-cycle time and the choice of field crops;
- ley farming with its ratio between cropping and grazing years and its land use apportionment for the cropping years; and
- grain farming with its worldwide scope primarily in the frequency of its harvest.

There are dry steppes which farmers, in following the principles of dry farming for water economy, can use only with a fallow-fallow-barley rotation, thus extracting a harvest only every third year. With about 300 mm precipitation, as in the Columbia Basin of the U.S., a harvest can be taken every second year. In cases with 300 to 400 mm of rainfall, as in the extensive dry areas of Kansas, Cape Province, Iran, or Australia, fallow needs to be inserted in the rotation only every third year, so that two thirds of the cropland is available for seeding and harvesting. During the

Overview 9: Examples of Crop Rotations in West German Farming

South German corn areas	Rhine-Main area	Lüneburg Heath	Fehmarn Island
1. *Grass seed*	1. *Late corn*	1. W. wheat-catch crop	1. *W. rape*
2. *Grass seed*	2. S. barley, oats	2. S. barley-catch crop	2. *W. rape*
3. W. wheat	3. W. wheat	3. S. barley	3. W. wheat
4. S. barley	4. *Early corn*	4. W. rye-catch crop	4. W. wheat
5. *Grain corn*	5. W. wheat	5. S. barley-catch crop	5. Oats
6. S. barley-U.S.	6. S. barley	6. S. barley	6. W. barley

Middle Ages the three-course (or "three-field") fallowing system was customary in almost all of Europe and in large parts of Asia, though the causes were not ecological, but economic. Today a field in Central Europe can produce a grain harvest every year, and even higher yields are possible with more intensive inputs.

Crop rotations like those in Overview 9 have more recently met with high favor in West Germany because they allow full mechanization for higher economic efficiency, requiring just 2 to 3 workers per 100 ha FL and offering a correspondingly high labor productivity.

In warm countries, in which plant growth continues throughout the year, still another grain harvest can be extracted in the same year from the same field. Thus one finds in southern Japan the rotation:

    1st year, winter: w. rape
                summer: rice
    2nd year, winter: w. grain
                summer: rice

Two grain crops can be planted in the same year also in northern Chile, Egypt, and other warm countries. As a rule, however, irrigation is then a prerequisite for bridging the dry season and permitting year-around plant growth. Flooded rice fields in Taiwan and Guyana can even support three harvests in the same year.

If grain farming is classified according to the degree of cultivation, i.e. by the harvest-share of the cropland, then the intensity span ranges from 33.3% with a rotation of fallow-fallow-grain to 300% with a triple cropping in the same year.

d) Hoe Crop Farming

Grain cropping ensures higher labor productivity, hoe cropping higher land productivity. Under similar ecological conditions, therefore, densely settled countries must emphasize hoe cropping more than do sparsely settled ones. In the same country, smaller farms will practice hoe cropping more intensively than do the larger ones.

Hoe crop farming also exhibits intensity stages, which can be distinguished on the bases of cropping intensity and the types of hoe crops favored. Overview 10 shows a few examples.

If quick-growing crops are cultivated in areas with mild winters and sufficient water supplies, harvesting frequency rises considerably. In southern Italy a harvest of three to four types of vegetables in the same year is easily possible. In locations on Vesuvius where there is protection from north and east winds, even five to eight vegetables can succed each other annually, as for example:

2. Annual-Cropping-Systems

Nov. to Jan.: cauliflower
Feb. to March: brussel sprouts or kale
April to May: carrots
June to August: paprika
Sept. to Oct.: green peas

Hoe crops are a thoroughly heterogeneous group botanically. The designation, hoe cultivation, indicates a particular operational technique that has had to evolve, as have all annual- and perennial-cropping systems, by stages. In Southeastern Asia labor is extremely cheap and land and capital are extremely expensive. Rice is started in seed beds, transplanted individually to the fields, hoed, weeded, harvested with a sickle, and threshed with a flail or sledge. Here it is undoubtedly a hoe crop, though

Overview 10: Examples of Crop Rotations in Hoe Crop Farming with Varying Land Use Intensity [1]

Siberian coniferous forest zone shorter ←———	*Magdeburg Börde* ——— growing season ———	*Nile Delta* ———→ longer
1. Hoe crops or bare fallow 2. S. grain 3. Bare fallow 4. S. grain 5. S. grain	1. Potatoes and vegetables 2. S. beets 3. S. barley 4. W. wheat 5. S. beets 6. W. wheat	1. Cotton (Feb.–Oct.) 2. Vegetables, beans, clover (Nov.–May)-fallow 3. a) Wheat, barley (Oct.–May) b) Corn (July–Oct.)-catch crop (Nov.–Dec.)
70%	*Land Use Intensity* 100%	133 %
Uttar Pradesh/India shorter ←———	*Khuzestan/Iran* ——— growing season ——— ——— or increasing water supplies ———	*Southern Japan* ———→ longer ———→
1. Sugarcane 2. a) Corn b) Wheat/garbanzos 3. a) Cotton b) Fodder crops	1. a) S. beets (Sept.–May) b) Sudan grass (May–Sept.) 2. a) Feed rape (Oct.–Feb.) b) Alfalfa (Feb. seeding) 3.–4. Alfalfa 5. Wheat (Dec.–May) – catch crop	Spring: vegetable Summer: vegetable Winter: w. grain
167%	*Land Use Intensity* 180%	300%

[1] 1., 2., etc. = years; a), b), etc. = successive crops in the same year.

botanically it is classed with the grains (Gramineae). In the U.S. land and capital are comparatively cheap. Labor, though, is very expensive. As a result, rice in the Sacramento Valley of California is clearly raised as a grain: sowing, fertilization, and spraying of crops with insecticides and herbicides is done with aircraft and harvesting is accomplished with gigantic combines.

There can be just as little doubt that grain corn in the developing countries of the humid tropics has validity as a hoe crop, as there is about its character in the highly industrialized countries where it bears all the marks of grain cultivation. Even the sugar beet has for us lost the character of a hoe crop where its completely mechanized cultivation has found favor.

Many more examples could be cited that illustrate the fact that many farming enterprises are changing their operational characteristics in the course of national economic development. Thus what kind of systematic classification of farms is arrived at will depend on what one has most in mind, production orientation or operational technique.

3. Perennial-Cropping Systems

Perennial crops have a life span of several to many years. This has considerable consequences for farm management in that
- annual seeding or planting need not be done;
- there is a greater resistance to weeds;
- a deeper root system makes it easier to cope with drought;
- in an unprofitable early growth period costs accumulate which cannot be met by any productivity;
- the forms of bush or tree growth bring with them impediments to harvesting; etc.

The great intensity span within these agricultural systems is more expediently treated in the following three-type classification that one encounters:
1. gathering: only harvesting;
2. bush and tree crop farms: cultivating and harvesting; and
3. plantations: cultivating, harvesting, and processing.

a) Gathering

Gathering was the original form of human activity, no matter what the goal in the search for food, fruits, fish, or game. It is still found in original form among the Bushmen in southern Africa, even though the gathering of fruits of perennial plants is only partly practiced (melons, etc.). Gathering has persisted the longest with bush and tree crops. It is a purely acquisitive economy, in which wild-growing plants are harvested, often in combination with hunting and fishing, by the native population. Crop farming, livestock raising, or both, may sometimes supplement the activity. Besides providing self-subsistence in food, gathering frequently offers a monetary income. Examples are (Ruthenburg 1967, p. 123):

- gathering of wild palm nuts in parts of West Africa;
- gathering of wild honey in Tanzania;
- gathering of coffee beans on wild bushes in Ethiopia;
- gathering of gum arabic in the Sudan and in the Amazon Basin; etc.

All these gathering activites presuppose a rich supply of manpower, large unpropertied areas, and a lack of capital, thus the lowest stage of economic development. The almost complete displacement of the gathering of wild rubber in Amazonia by the Indonesian rubber plantations shows that the first activity is no longer able to compete with the latter as soon as wages significantly increase, capital goods become cheaper, and quality demands of the world market increase. Such demands have already promoted wide-scale substitutions of cultivated plant forms for wild ones, controlled growth for wild growth, homogeneity of the plantations for the heterogeneity of the virgin forest, and highly developed processing technologies for primitive forms.

b) Bush and Tree Crop Farms

The large number of bush and tree crops can be classified according to various criteria, such as

- whether they are foodstuffs or luxury items, and with that, in turn,
- whether they are of use for subsistence or export;
- whether they can be raised more easily on peasant farms or large farms, i.e.
- whether they are of a labor-intensive or capital-intensive character;
- what climatic and soil conditions favor them;
- whether they supply products that are sensitive to or capable of transport, i.e. how geographically restricted they are by transport capabilities; and
- whether their products require only a simple technical preparation such as cleaning, sorting, or the separation of the fruit from the skin or hull, or whether an additional industrial processing of a higher degree with the aid of physical or chemical measures is needed.

This last difference separates, in general, bush and tree crop farms from plantations. Some example crops are characterized in Table 6. Plantations

Table 6: Crops on Bush and Tree Crop Farms and on Plantations Compared in Management (Approximate Values)

Crops on Bush and Tree Crop Farms				Management Criteria	Plantation Crops			
Oil palms	Coffee (*C. robusta*)	Dates	Abaca		Sugarcane	Sisal	Pineapple	Tea
Humid Tropics	Humid Tropics	Subtropics, Deserts	Tropical Rainforest	Favored climatic zones	Humid Savannas	Trop. Highlands	Humid Tropics	Trop. Highlands
50	100	60–90	10–15	Useful economic life in years	1–9	5–9	4–6	over 50
350	740	1,000	850–1,200	Labor expenditures, MH/ha/yr.	630	630	600	3,200–5,600
1,600	800	–	–	Initial investments, DM/ha	2,000–2,300	2,500–3,000	–	4,000–12,000
M, (S)	S, (M)	S, (M)	S, (M)	Harvest processing in small facilities (S) or mills (M)	M	M	M, (S)	M, (S)
850	1,449	1,600	830	Land productivity, gross, DM/ha	3,000	1,200	14,200	1,300
2.06	1.95	1.60	0.83	Labor productivity, gross, DM/MH	4.75	1.90	23.67	0.51
Sources: same as for Table 9, p. 153.								

are always large-scale operations, whereas bush and tree crop farms can operate on any scale.

In Europe grapes, apples, plums, hops, cherries, apricots, olives, and almonds are normally raised on peasant farms. Where further processing is necessary, the good transport facilities of our area permit the use of cooperative installations. Thus the primitive winery in the farm cellar has been giving way to the large installations of the wine growers' cooperatives. In the tropics crops are overwhelmingly raised by peasants, e.g. oil and coconut palms as well as cacao. The highly varying quality of much of the exports, however, is detrimental if marketing boards cannot remedy the situation. Rubber and coffee are raised profitably on both peasant farms and large farms.

c) Plantations

Plantations should be disignated as only those large-scale commercial raisers of bush and tree crops that have processing facilities for their harvested products (tea and sisal factories, sugar and oil mills, coffee processing factilities, etc.). In countries with a tropical rainy climate, especially, bush and tree crop products form a substantial share of the total exports. Thus

    in Indonesia, 40% is contributed just by rubber;
    in Ghana,   60% by cacao;
    in Tanzania, 28% by sisal and coffee; and
    in Ethiopia, 56% by coffee.

Also to be cited is the paramount importance of the plantation crops of coffee for Brazil, rubber for Malaysia, tea for India and Sri Lanka, as well as sugarcane for Java and Cuba. Sisal and sugarcane are on the boundary between field crops and bush crops.

Plantation operations tend toward monoculture or at least certainly toward specialization. Specialization is encouraged not only by the mechanization of planting necessitated by the large size of the operation, but by the mechanization of the processing of the harvested crops. Because of cost digressions in mass production, the processing industry tends toward large-scale enterprises, which in order to save on transport costs force the surrounding agricultural zones into specialized production. This is nothing less than a concentration of agricultural outlets, which has as its consequence the strengthening of the differentiating force of transport site. This in turn leads to a regional concentration that is documented in the tea landscapes, sugarcane landscapes, sisal landscapes, and the like that surround the processing installations.

The concentration of a series of perennial crops in the humid tropics ought to be pursued even more in the future than it it now, for only through it or irrigation farming can the soil fertility problem which is so fateful for this area be conquered.

In all biological problems of agriculture the great teaching mistress of nature ought to be repeatedly consulted. The natural vegetation of the humid tropics consists of forest formations. There can be no doubt that the soil fertility of the humid tropics can be better utilized and preserved with bush and tree crops than it can with short-lived field crops. With bush and tree crops dependence is put on the vegetation cover best suited to this climate. Oil and coconut palms, rubber and cacao, and tea and coffee bushes are better able to preserve soil fertility than cassava (manioc), sweet potatoes, corn, sorghum, or millet. In Nigeria soils that have supported oil palms for seventy years have lost nothing of their natural fertility. Rubber and cacao cultivation also does not seem to have hurt soil fertility.

For its self-sufficiency, therefore, and with the decline of the forest-burning system of shifting cultivation, the rural population of the humid tropics ought to depend more on bananas. This crop qualifies as a basic food and ensures the preservation of soil fertility far better than do the short-lived field crops. Most of the other bush and tree crops, however, apart from the oil and coconut palms, pose a grave disadvantage for the agricultural states that are little integrated into the world economy in that they provide no basic food for the resident population. Consequently a concentrated cultivation of perennial crops so as to preserve soil fertility is still denied most tropical countries. Indeed what is required is a worldwide division of labor so that inhabitants of the humid tropics can widely exchange their bush and tree crop products for basic foodstuffs, especially cereals, from the mid-latitude climatic zone. Such an extreme division of labor, though, is still unrealizable with the currently completely insufficient access to transport, since high transport costs elevate farm gate prices for purchased goods far above world market prices and depress farm gate prices for saleable products far below world price levels.

For the more distant future, though, areal specialization on such an extensive scale appears to be a promising way of allowing farmers in each climatic zone, in the interests of cost reduction, to produce what is most appropriate to it, and of helping cultivators in the humid tropics to overcome their soil fertility problems.

VI. The Agricultural Geography of the Humid Tropics

From a world view, agricultural zones can be said to be determined mainly by climate and stage of economic development. However, in an overview of farming regions in the tropics only climate can be used for orientation, since in these lower latitudes there are, so far, only developing countries whose economies do not greatly differ.

In the following, population density will have to be considered as an additional factor in the casual relationship, though the primary concern is the connection between farming regions and climate. The interaction between natural and cultivated vegetation merits particular interest.

First, in order to illustrate the climatically-determined differences between the farming regions of the tropics, it is necessary to identify types of climate and types of farming. Here we can refer directly back to chapters II and V and summarize briefly those portions that are most vital to this section.

The tropical area spans the equator and extends to both the tropics of Cancer and Capricorn, thereby embracing the most arid as well as the most humid climatic belts on our globe. Fig. 27 gives a rough indication of how the seasonal climates of the tropics and subtropics are distributed over the earth. The representations of the climates on this well-known and highly significant agricultural-geographic map by Troll and Paffen deviate somewhat from those described in Chapter II, but identification is not difficult.

For our consideration here, however, a more detailed classification of tropical climates is needed, one that is oriented toward vegetation formations. With increasing proximity to the equator and with rising humidity, seven climatic zones can be distinguished. They are represented by seven vegetation formations (see Fig. 28):

1. the *dry-hot deserts,* with only sparse and sporadic rainfall, which lie mostly beyond the dry boundary of human habitation;
2. the similarly *arid semideserts,* partly inside and partly outside the dry boundary of livestock raising, which can only be used occasionally by nomads, hunters, and gatherers;
3. the *semiarid shrub savannas* (shrub, salt steppes), with two to four humid months, which extend up to the agronomic dry boundary.

    Apart from only minor irrigation farming and some dry farming, the shrub savannas are used almost exclusively for extensive grassland farming, in the New World by ranchers and in the Old World still frequently by nomads;

128 VI. The Agricultural Geography of the Humid Tropics

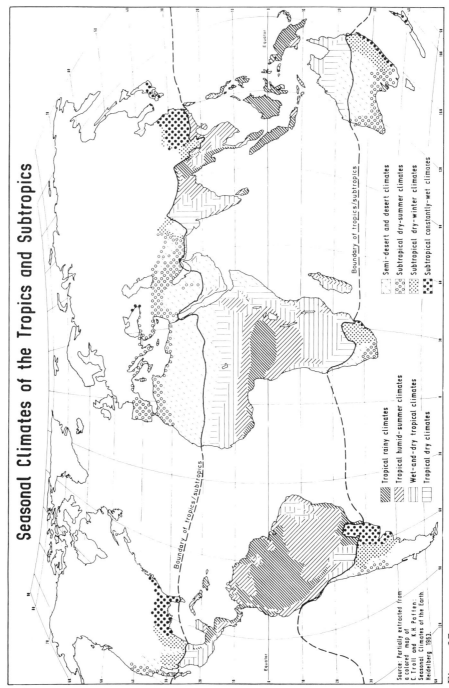

Figure 27

VI. The Agricultural Geography of the Humid Tropics

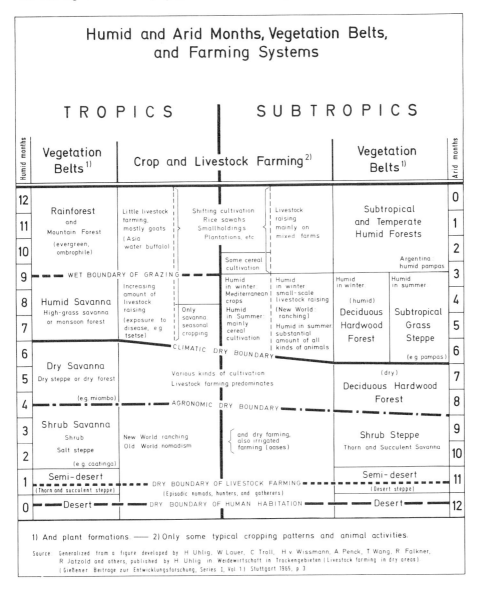

Figure 28

4. the *dry savannas*, with four to six humid months, located between the agronomic and climatic dry boundaries, thus capable of supporting rainfed cropping. The grassland is dry steppe and the woodland is dry forest which is green in the rainy season (e.g. miombo). A short rainy season is followed by a long dry season;

5. the *humid savannas,* with seven to nine humid months, located between the climatic dry boundary and the wet boundary of extensive grassland farming. The rainy season is now longer, the dry season shorter, the humidity higher. Annual precipitation ranges from approximately 800 to 1,500 mm. The grassland is high-grass savanna with gallery forests and the woodland is monsoon forest. This is the first climatic zone in which even the small rivers flow the year around, whereas in zones 1 to 4 they dry up soon after the rainy season;
6. the *tropical rainforest,* directly on the equator and with the wettest climates. There are two rainy seasons with a total of at least 1,500 mm precipitation. The mean temperature is 25° to 28°C and varies little throughout the year, and the humidity rarely falls below 90% (6.00 a.m.). This leads to a thoroughly humid, constantly wet type of climate giving rise to the evergreen ombrophiles, i.e. the tropical rainforest; and finally
7. those *tropical highlands* (more than 1,000 m above sea level) which though having additional and very diverse vegetation formations, can still be classified partially or even wholly with the above, especially the dry and humid savannas.

As in the organization of vegetation formations, we shall classify farming types in the tropics, as far as possible, by following a progression from the dry to the rainy climates. From what has been said in Chapter V, the most appropriate classification is the following, with only four types:

1. *grassland farming,* which in the tropics has so far been of a thoroughly extensive character since there is no cultivation of any kind. The population simply appropriates whatever nature will grow without human intervention and does this by using undemanding grazing animals. Stabling is completely unknown, as is the conservation of feed generally.
2. *rainfed farming,* which signifies the cultivation of short-lived crops based on rainfall, i.e. without irrigation.

   The problem of soil fertility often makes it necessary to alternate the cultivation of useful crops with bare, grass, bush, or even forest fallow. In extreme cases (e.g. the forest-burning system of shifting cultivation) only 15 to 25% of the available land can support a harvest at any one time.
3. *irrigation farming,* in which short-lived crops are given an increased water supply, either throughout the year or at least in the dry season, so that crop production can be expanded in those periods. It is then often possible to obtain two or even three harvests in the same year from the same field.
4. *farming of bush and tree crops,* which involves growth cycles of from several to a great many years, often several decades — as with the

VI. The Agricultural Geography of the Humid Tropics

cultivation of coconuts, oil palms, rubber, citrus fruit, coffee, cacao, and tea. Permanent ground cover and in certain cases necessary technical processing of the harvested crop are important characteristics of this type of farming. Most farms practice intensive cultivation. There is a widespread tendency towards monoculture and the plantation scale of farming.

Fig. 28 illustrates in outline what types of crop and livestock farming are native to the climatically-defined natural vegetation belts of the tropics. For agricultural-geographic orientation, it is especially important to observe that in this figure:

- the *dry boundary of human habitation* runs through the desert;
- the *dry boundary of livestock farming* passes through the semi-desert;
- the *agronomic dry boundary* (boundary of rainfed farming) separates the shrub savanna and dry savanna;
- the *climatic dry boundary* passes between the dry and humid savannas; and
- the *wet boundary of grassland farming* coincides with the contact zone between the humid savanna and the rainforest.

An attempt has been made in Fig. 29 to schematize the principal distributional characteristics of the four systems of tropical farming as they relate to the several climatic zones just described.

Since this Chapter VI has as its content the agricultural geography of the humid tropics, only three climatic zones,

    a) rainforest climate,
    b) humid savanna climate, and
    c) tropical highland climate; and

Farming Systems in the Climatic Zones of the Tropics
(Diagram)

Climax vegetation	Humid months	Grassland farming	Rainfed farming	Irrigated farming	Bush and tree crops
1. Deserts	—				
2. Semi-deserts	1	•			
3. Shrub savannas	1-4	•••	•	•	
4. Dry savannas	4-6	•••	••	••	•
5. Wet savannas	7-9	•	•••	•••	••
6. Rain forests	10-12		••	••	•••

• = infrequent   •• = moderately frequent   ••• = widespread

Figure 29

only three farming regions, those of
- a) rainfed farming,
- b) irrigation farming, and
- c) bush or tree crop farming,

will be treated. Therefore only the area extending poleward to the dry boundary will be considered. The outer, dry tropics will be dealt with in Chapter VII together with the dry areas of the mid-latitude climates, since they have a strong agricultural-geographic affinity for the latter group.

1. Regions of Rainfed Farming

a) Rainfed Farming in the Tropical Rainforest Belt

There are not many short-lived crops that prefer the hot and permanently wet climate of the tropical rainforests. Root crops such as manioc (cassava), sweet potatoes, and yams are grown, but these are typical crops for the farmer's household. As far as marketable grain crops are concerned, rice, with appropriate water supplies, and corn, because of its wide ecological distribution, are possibilities. Sugarcane grows well here. The marketing capabilities of rainfed farming are thus not very great, once the still widespread system of shifting cultivation practiced by subsistence farmers has been overcome. Most importantly, the state of soil fertility here becomes extremely precarious as soon as the soil-regenerating function of forest fallowing is no longer operative.

The system of shifting cultivation (Table 7) is the most important cropping system of the humid tropics and is still predominantly practiced by not quite 5% of the world population on over 22% of the land surface of the earth (Fig. 30). The system is based on cropland being periodically given over to fallow under the original vegetation. Annual rainfall in the humid tropics is at least 1,200 mm, and the natural vegetation is forest. Thus the primitive rotation system is clearly one of forest-and-field alternation, and since the forest can be cleared only by firing, it is called a forest-burning system.

The thousands of years of dominance of the forest-burning system of shifting cultivation in the humid tropics and the tenacious adherence of the tropical inhabitants to this system can be traced ultimately to the fact that the clay-humus aggregates, which are so important in temperate climates, disintegrate in the soil of the humid tropics. This is caused by rapid humus decomposition, once the natural vegetation has been removed. The clay by itself, unlike that found in the middle latitudes, possesses only a very limited power of sorption. Thus whereas in Central Europe the retention, storage, and subsequent release of plant nutrients are mainly functions of the clay-humus aggregates, in the rainy tropics they must be transferred, for the most part, to live or dead plant matter.

1. Regions of Rainfed Farming 133

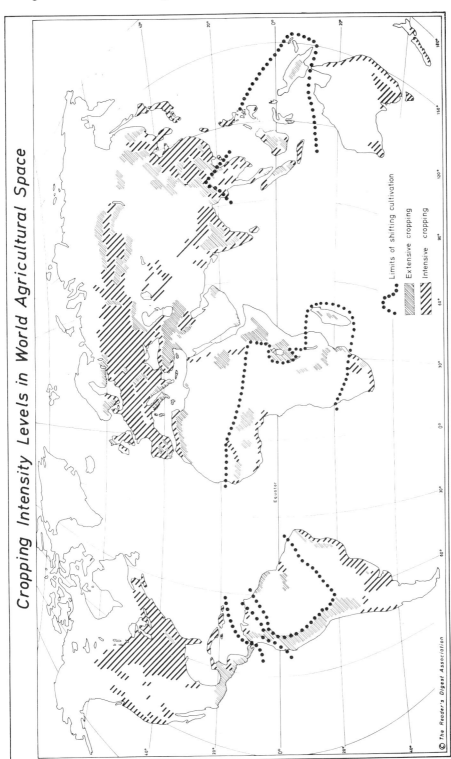

Figure 30

A large amount of raw material for humus is built up under forest fallow in the forest-burning system, and a large supply of plant nutrients is stored. Clearing the forest by burning mobilizes some of these nutrients in the form of ash, so that for the first year the crops introduced to the friable, loose, weedless, and nutrient-rich soil with hoes will produce relatively high and reliable yields. In the second and third years of cultivation there is still a supply of nutrients because of mineralization of the underground organic matter of the forest fallow. However, yields have already noticeably declined while resistance to cultivation, disintegration of structure, erosion, and weed growth have all increased. The input-output ratio becomes progressively worse, and after three (or at the most four) years of cropping it is no longer satisfactory. The cleared area is then left to the forest with the aim of restoring the soil fertility, while another, rested patch of forest is burned down and cultivated instead (see Fig. 31).

Biologically, the forest-burning system of shifting cultivation must be regarded as an ideal solution for farming in the humid tropics. It is economically reasonable and possible only in sparsely populated agricultural states. Its great disadvantage is its very low capacity for food production, since only about 20% of the available land yields a harvest

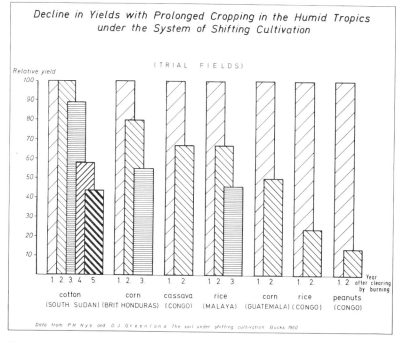

Figure 31

1. Regions of Rainfed Farming

in any one year. Even on the young and fertile volcanic soils of the island of Java, this system of shifting cultivation is able to feed only 40 to 50 people per km², and on poorer soils far less. Today many developing countries have passed the critical limit and are faced with the still largely unsolved problem of replacing the forest-burning system with more productive methods.

Table 7: Examples of the Forest – Burning System (Shifting Cultivation) in the Humid Tropics

Country or Region	Rainfall mm/yr.	Crops	Original Vegetation	Number of crop years	Number of fallow years	Cropped area in % of land available
Zambia	ca. 1125	...	miombo dry forest	2	up to 25	7 plus
Sarawak	ca. 3750	hill rice	rainforest	1	12 plus	at most 8
Liberia	2000–4500	rice, cassava	rainforest	1–2	8–15	11–12
Sumatra	ca. 2250	rice, tubers	rainforest	2	10–16	11–17
Assam	ca. 2500	rice, millet, corn	rainforest	2	10–12	14–17
Sierra Leone	2250–3250	rice, cassava	rainforest	1.5	8	16
Central Zaire	1750 plus	rice, cassava, corn	rainforest	2–3	10–15	17
Guatemala	3075	corn	rainforest	1	4	20
Philippines	2500 plus	rice, corn, tubers	rainforest	2–4	8–10	20–28
Nigeria, Umuahia	ca. 2250	corn, yams, cassava	bush (*Acioa barteri*)	1.5	4–7	18–27
Abeokuta	ca. 1250	...	bush thicket	2	6–7	22–25
West Africa	1500–2000	corn, cassava	semihumid forest	2–4	6–12	25
Gambia	ca. 1250	...	bush thicket	6–12	6–12	50

Source: Nye, P.H. and D.J. Greeland: The soil under shifting cultivation (Commonwealth Bureau of Soils. Technical Communication. No. 51). Bucks 1960, p. 128.

Since this change causes serious difficulties, an attempt should first be made to make the system itself more efficient in food production. A first step can be a shift from a more or less haphazard to a more organized forest-burning system of shifting cultivation (Fig. 32). The next step would be the introduction of artificial instead of natural afforestation, which with proper management could possibly allow the great technological leap from hoe to plow cultivation. Plow cultivation considerably increases not only soil but labor productivity, since each worker can now cultivate a much larger area of land.

As population continues to increase, however, even this farming system is no longer sufficiently productive and abandonment of the forest-burning system becomes inevitable. Forest fallow must now be replaced by a fallow under the kind of short-lived plants that can restore soil fertility more quickly, thus resulting in a higher proportion of land use. At first, legumes needing little fertilization are selected. Later, when an appreciable use of chemical fertilizers has become possible, deep-rooting grasses

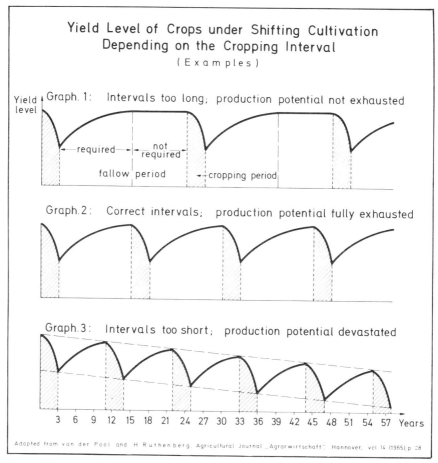

Figure 32

appropriate to the region can be planted. Admittedly, the surface growth cannot be used for cattle at first for several reasons — insufficient transport facilities in the tropics, very low purchasing power of the population, and also often widespread nagana disease. This means that the total cost of cultivating and planting the fallow areas must be regarded as a cost of conserving soil fertility. Nevertheless, the proportion of the available land that yields a harvest each year increases by a third to two-thirds in the course of this development. Only at a higher stage of development can all available land yield a harvest every year, for then quite different farming systems can be introduced. At that level, the increased earnings of the now predominantly urban population and the improved communications of the country will permit a change to ley farming in suitable locations. Other locations will be able, in the course of development of a worldwide division of labor, to shift to a strong emphasis on bush and tree crops. This can conserve soil fertility in the humid tropics far better than tillage; and as export trad becomes sufficiently developed, the products can be traded for basic foodstuffs from other climatic belts.

The succession of farming types described here is directed toward an increasing intensification of cultivation, first by means of increased labor inputs and later through more and more capital investment. As a development policy this evolution can be directed and accelerated by three different sets of measures:

— by an *infrastructure policy* aimed at developing the communications of the country, lowering transport costs, and reducing the differences between market prices and prices at the farm gate so that farmers in the humid tropics (who are producing mostly for very distant markets) can increase the exchange value of their products for commercially-manufactured inputs and wage labor;
— by an *industrialization policy* which raises general income levels and thus the level of agricultural prices. Industrialization also reduces the prices of commercial goods needed as agricultural inputs; and finally,
— by an *export policy* which leads to worldwide trading, so that the cultivation of bush and tree crops in the humid tropics can be expanded far more than at present. Soil and climatic conditions in the tropics are most favorable to these crops, which are much better able to maintain soil fertility there than are regularly cultivated crops.

b) Rainfed Farming in the Humid Savannas

The humid or high-grass savannas (also called savanna forest zones) locate in the tropical climatic areas where rainfed farming has its greatest role. Regular cropping is restricted by grassland in the drier zones and in the more humid ones by bush and tree crops.

Primitive rotation farming also forms the basis of the field system in the humid savanna. Depending on the natural vegetation these systems involve an alternation of forest with crops or grass with crops. Incentives for and aims in overcoming these practices are the same as in the rainforest zone. The central problem is the preservation of soil fertility without requiring long years of natural vegetative cover. Now it is no longer possible to use the traditional methods of forest or grass fallowing in the battle to combat humus losses, to limit erosion damage, to moderate at least the erosion of nutrient content, to master weed growth, and to prevent disasters caused by plant diseases and pests. All this is most successfully achieved if the soil is kept under a plant cover the year around, but it is precisely this practice that is prevented by the dry season.

As the population grows, an ever larger share of the usable land must be cultivated and the areas available for grass or forest fallow reduced. Thus in the humid savanna ley farming is best suited to take over the functions of fallowing under natural vegetation. Examples are shown in Fig. 33.

The percentage of land in grass that is necessary to conserve the soil is considerable and generally far exceeds the amount needed for feed. Ley farming is technically possible but at the moment is an expensive rotation

Ley Farming in African Wet Savannas

Kenya Highlands (Kericho District)[1]

LARGE FARMS		SMALL FARMS	
(A)	(B)	(C)	(D)
1.-3. Cultivated grass 4. Potatoes - **beans** 5.-7. Pyrethrum 8. Corn	1.-3. Cultivated grass 4. Potatoes - **beans** 5.-6. Wheat 7. Barley	1.-4. Natural grass 5. English potatoes - vegetables 6.-9. Corn	1.-5. Natural grass 6.-8. Corn 9. Sorghum, millet 10. Sweet potatoes - **vegetables**

Grassland % of arable land

38	43	44	50

Southeast Africa

RAINFED CROPPING		IRRIGATED CROPPING
(E)	(F)	(G)
Malawi[2]	North Transvaal	Mid-Transvaal (Rustenburg)[3]
1.-3. Cultivated grass 4. Tobacco 5. Cotton 6. Peanuts 7. Cotton 8-9. Corn	1.-3. Cultivated grass 4. Sorghum, millet 5. Sorghum, millet 6. Bechuana beans	1.-3. Alfalfa 4. Corn silage 5. Summer: Grass silage; Winter: Green oats 6. Summer: Bean hay; Winter: Wheat 7. Summer: Tobacco; Winter: Green oats 8. Summer: Tobacco; Winter: Wheat 9. Summer: Corn silage; Winter: Wheat

Cultivated grass or alfalfa % of arable land

33	50	33

1) H. Niederstucke, Bodennutzungsformen in tropischen Höhenlagen (Types of land use in tropical highlands). „Landwirt im Ausland", Vol. 4 (1970), p. 76. — 2) 6 ha useful agricultural land, 1300 mm rain, 4 workers per farm. — 3) 18 ha useful agricultural land, 600 mm rain, dairy cattle; 55 ar principal feedstuffs cultivation/cattle unit; gross return = 2,540 DM/ha, net return 1,000 DM/ha useful agricultural land.

Figure 33

1. Regions of Rainfed Farming 139

system. As long as the market for milk and meat in developing countries remains poor and the communications system incomplete, part of the fodder growth cannot be used. Therefore the total costs of the grass crop must be written off as the charge for promoting soil fertility. But this situation will change. Time is undoubtedly working in favor of the ley farming system of soil conservation in the humid savannas.

Overview 11: Crop Rotations for Rainfed Farming in the Humid Savannas

A West Africa	B Malawi	C India
1.–15. Forest fallow 16. Hill rice 17. Beans, yuca	1. Peanuts 2. Cotton 3. Rice	1. Sugarcane 2. Vegetables 3. Rice

Corn, rice, kidney beans, and peanuts are the short-lived field crops that dominate in the humid savannas. Manioc and yams also find acceptable ecological conditions here, though they are no longer optimal. Overview 11 shows examples of crop rotations.

Mixed cropping is very important on small farms in Africa and India. Examples of crop combinations given by Konnecke (1967) are:

    West Africa: yams and guinea corn; millet, sorghum, and guinea corn; corn and oil pumpkin; hill (dry) rice and cotton; corn and peanuts.

    India: cotton and Italian millet; bush peas and Italian millet; cotton and coriander (spice plant).

Table 8: Comparison of the Productivity of Some Important Tropical Field Crops

| Criterion | Unit | Grains | | | Tubers and Roots | | |
		Millet, Sorghum	Corn	Rice	Yams	Sweet Potatoes	Manioc (Cassava)
		A. Appraisal Data					
Growing season[1]	months	2–6	2–5	2–6	7–12	3–6	7–24
Labor input[2]	man-days (MD/ha)[a]	30	34	110 (75-150)	490 (375-600)	190 (175-200)	210 (200-225)
Yield[3]	q/ha	6.6 (1.9-12.6)	10.8 (3.8-15.0)	19.2 (7.0-31.2)	71.0 (20-130)	71.0 (20-130)	63.0 (20-140)
Energy content[1]	calories/100 kg	345	360	359	90	97	109
Protein content[4]	% edible substance	10.0	10.0	8.0	2.4	1.3	1.2
Producer's price[5] (Uganda 1968-9)	DM/q[b]	25.30	10.50	26.00	7.90	7.90	7.50

[a] 1 MD = 5.5–7.0 hours.
[b] 1 EASh (East African shilling) = 0.56 DM.

Table 8 (Continuation)

B. Productivity with One Harvest Per Year							
Caloric yield	kcal/ha	2270	3930	6880	6400	6900	6870
	kcal/MD	75.5	116.0	62.5	13.1	36.3	32.7
Protein yield	kg/ha	66	108	154	170	92	76
	g/MD	2200	3170	1400	347	484	363
Gross returns	DM/ha	167	113	500	560	560	473
	DM/MD	5.57	3.34	4.55	1.14	2.95	2.25
C. Productivity with Year-Around Production[c] (Labor Productivity as under B)							
Caloric yield	kcal/ha	4540	7860	13760	6400	13800	6870
Protein yield	kg/ha	132	216	308	170	184	76
Gross returns	DM/ha	334	226	1000	560	1120	473

[c] Two harvests per year are assumed for the grain crops, which is not possible for yams and manioc. With wet-rice cultivation, even three harvests can be obtained from the same field during the year in certain cases, and in exceptional situations, four.

Sources:
[1] Johnston, B.F.: The staple food economics of western tropical Africa. Stanford, 1958.
[2] Phillips, T.A.: An agricultural notebook (with special reference to Nigeria). London 1956.
[3] Average yields for Africa according to FAO, Production Yearbook, Vol. 23 (1969). Rome, 1970, pp. 35 and 60 ff.
[4] Jones, W.O.: Manioc in Africa. Stanford, 1959.
[5] Brandt, H.: Die Organisation bäuerlicher Betriebe unter dem Einfluß der Entwicklung einer Industriestadt: Der Fall Jinja/Uganda. Diss. Berlin, 1971, Table IV/15–16.
[1,4] Nicholls, L.: Tropical nutrition and dietetics. London, 1961.

Table 8 gives the returns per hectare and per worker in calories, protein, and income for six important food crops. A comparison of these figures for the various crops also shows their competitive positions, as well as several kinds of yield equilibrium that lead to the same productivity. Depending on whether the farm operator is faced with insufficient food production capacity, must make up a protein deficiency, or is operating on a more commercial basis, the competitive relationship among the six crops will swing in favor of one crop at one time, and in favor of another in a different period.

c) Rainfed Farming in the Tropical Highlands

Rainfed farming in the tropical highlands has many forms. Moisture frequently, but not always, corresponds to that of the humid savannas. Temperatures are lower. Slope, exposure, and the like play a special role as locational factors. Thus in some cases crop combinations and field systems exhibit the greatest diversity in some of the most confined areas. Illustrative of this variety is the wide range of the following farming types:

1. shifting hoe cultivation with burning, which still takes up 80% of the land surface of Borneo;

2. ley farming, which e.g. plays a large role in the Kenya Highlands (see Fig. 33, p. 138);
3. regular cropping of the common group of tropical cultivated plants: corn, rice, tobacco, cassava, sugarcane, bush beans, pole beans, cotton, etc.;
4. regular cropping with crops specific to the tropical highlands. Thus Ethiopia with its highlands has important cultivated plants such as teff (*Eragrostis abyssinica*), which furnishes a popular flat loaf (Indjera) and whose straw is rich in nutrients and makes a welcome feed in the dry season; or the oil plant, nug, which is excellent for combatting weeds and whose oil does not turn rancid even under the most primitive storage conditions.
5. cultivating plants of the mid-latitude climates, such as wheat, barley, and vegetables like peas and potatoes, in tropical highland climates with cooler months. Wheat is cultivated in Kenya up to 2,000 m above sea level and in Ethiopia, though more distant from the equator, still up to 1,700 m. All African wheat countries also produce barley, but not vice versa. Where the rainy season is shorter and the climate drier, as in the countries between the Sahara and the Mediterranean, barley with its special weapons of short growing season and drought resistance forces wheat from the fields. However here it is no longer a matter of tropical highland climates.

2. Regions of Irrigation Farming

An essential function of irrigation farming in the inner tropics is the maintenance of soil fertility. The layer of surface water assumes to some extent the soil protection function of the forest. The water also prevents organic matter from decomposing too rapidly and supplies the soil with nutrients. Thus irrigated, or "wet", rice is the only crop that has been grown in the tropics for centuries without rotation or fertilization and without destroying the soil. If the rainforest is the vegetation climax of the *natural* landscape of the humid tropics, then irrigated rice can be considered the ultimate vegetation of its *cultural* landscape (von Uexküll 1969). This function of irrigation water, plus the fact that the inner tropics have a rich water supply and temperatures are sufficient for plant growth the year around, explains why land is irrigated in the tropical rainy climates as well as the tropical dry climates. In South America 0.7% CL is irrigated; in Africa, 10.7%, and in Asia, 21.8%. Fig. 34 shows that humid tropical Southeastern Asia, because of its high population density, is in fact numbered among the most important irrigated areas of the world. Irrigated rice is emphasized, as can be seen in its intensity gradations shown in Fig. 35, p. 145: land use intensity rises from 50% CL to 300% CL.

142 VI. The Agricultural Geography of the Humid Tropics

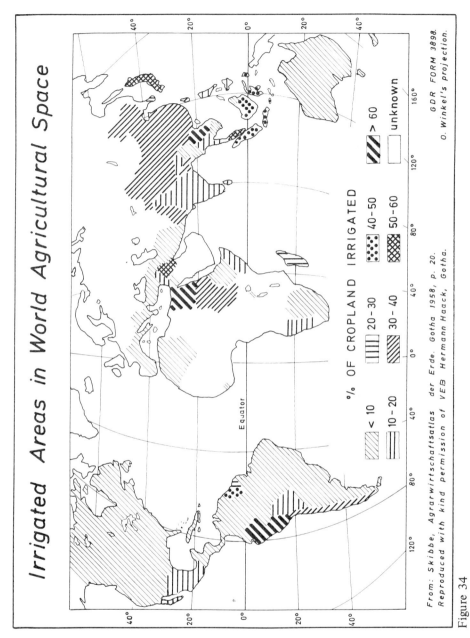

Figure 34

2. Regions of Irrigation Farming

a) Farm Management Functions of Crop Irrigation

Irrigation farming has the following farm management functions (Ruthenberg 1967, pp. 153 f.):

1. *increasing gross yields* per hectare by
   - increasing hectare-yields of the individual crop,
   - cultivating several crops per year,
   - cultivating of heavier-yielding crops, and
   - applying additional labor and production inputs;
2. making possible *permanent and constant land use* without interruption by forest, grass, or bare fallow;
3. *reducing harvest fluctuations* and thus providing a more dependable food supply;
4. *providing greater flexibility* in production orientation and intensity than is true for rainfed farming or bush and tree crop farming;
5. *increasing food production capacity*, i.e. reducing the minimum size of farm necessary for a family. Using hoes, a family can hardly cultivate more than two to three hectares. In the rainfed farming system this means a poor existence, but with irrigation it means an ample one.
6. *expanding cultivation* into dry areas not otherwise capable of cropping.

These advantages, though, must be bought at the cost of considerable technological and material expense. Gross investment in water supply, canalization, land leveling, and the like amounts to 4,000 to 10,000 DM/ha. Irrigated wheat growing in China needs 58% of its total labor for routine maintenance of the water supply (Ruthenberg 1967).

b) Irrigation Methods in Geographic Comparison

Overview 12 contains some examples of irrigation farming. Above all, it shows that this system is by no means limited to the tropical rainy climates, nor indeed to the tropics as a whole, but rather extends far beyond them into the subtropics (Turkey, Egypt, southern Japan). It is especially in the subtropics where summers are dry that irrigation can increase land productivity immensely. The same applies to the wet-and-dry tropics, where irrigation allows year-around crop production. Long-lived perennial crops such as perennial fodder crops and sugarcane can then be successfully cultivated, or if short-lived crops (rice, vegetables, sweet potatoes) are grown, several successive harvests can be obtained in the same year. Thus there are paddy fields in the tropics producing three and even four harvests annually (see Fig. 35, p. 145). In Overview 12, the example of southern China shows three harvests, and the example from Taiwan even four harvests for the same year. This increase in the frequency of harvesting through irrigation is of great value to overpopulated

Overview 12: Examples of Crop Rotations in Irrigation Farming

A. One-year Rotations		
Egypt (200) a) Summer: rice b) Winter: grains, legumes, fodder crops, vegetables	*Southern China (300)* a) Rice b) Rice c) Wheat	*Taiwan (400)* a) Sweet Potatoes b) Rice c) Vegetables d) Rice
B. Two-year Rotations		

Turkey (100)
1. Rice
2. Wheat
	India (150)
	1. Sugarcane
	2. a) Cotton
	   b) Fodder crops
		Southern Japan (200)
		1. a) Rape
		   b) Rice
		2. a) Renge grass
		   b) Rice

C. Three-year Rotations		

Uttar Pradesh (100)
1. Sugarcane
2. Ratoon
3. Pearl millet
	India (100)
	1. Vegetables
	2. Rice
	3. Sugarcane
		Taiwan (167)
		1. Sugarcane
		2. a) Sweet potatoes
		   b) Peanuts
		3. a) Peanuts
		   b) Rice

Notes: (...) = land use intensity = annual harvest area in % CL. – 1., 2., etc. = year, a), b), etc. = successive crops in the same year.

Sources: Ruthenberg, H.: Organisationsformen der Bodennutzung und Viehhaltung in den Tropen und Subtropen. In: Handbuch der Landwirtschaft und Ernährung in den Entwicklungsländern, edited by P.V. Blanckenburg and H.D. Cremer, Vol. 1, Stuttgart, 1967, pp. 159 and 167. – Tsuzuki, T.: Die Fruchtfolgen des japanischen Ackerbaues. "Berichte über Landwirtschaft," Hamburg and Berlin, Vol. 41 (1963), p. 837. – Wang, Y., F. Nagel, und H. Ruthenberg: Bodennutzung und technische Fortschritte auf Taiwan. (Zeitschrift für ausländische Landwirtschaft, Sonderheft 7) Frankfurt/Main, 1969, p. 18. – Franke, G., et al.: Nutzpflanzen der Tropen und Subtropen, Vol. II, Leipzig, 1967, pp. 189 ff.

agricultural states and to small family farms, and indeed to areas and farms under all conditions of production where the price-cost ratio makes increased land productivity the first economic goal.

Cotton is the most important cash crop to be irrigated. In 1970, for example, it represented 65.4, 63.3, and 44.6% of the total value of exports from Chad, Sudan, and Egypt, respectively.

c) Rice Growing as a Representative of Irrigated Cropping on Family Farms

Wet rice is the irrigated field crop par excellence. It is thoroughly compatible with monoculture, as Fig. 35 shows.

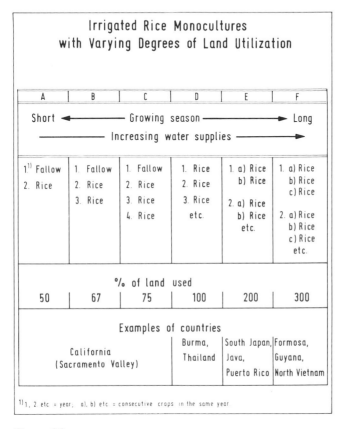

Figure 35

The longer the growing season and the greater the water supply, the more intensive land use can be. In Guyana the Bluebelle variety of rice produces three or four harvests annually with a yield of 30 to 40 q/ha per cutting, thus an annual yield of about 120 q/ha. This is 3.6 times the hectare-yield of grain in West Germany.

On the other hand, wet rice is also grown in rotation (see Fig. 36), e.g.
– when water supplies are not sufficient for the entire year, so that wet rice and unirrigated crops must be alternated; or
– when the cooler season is not warm enough for rice but still mild enough for other crops; or
– when diseases, pests, and weeds preclude rice monoculture.

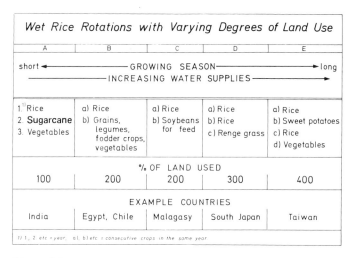

Figure 36

The great flexibility of irrigation farming is also reflected in the very different production methods associated with the same crop. Here, too, wet rice offers a good conceptual model. Particularly drastic differences in the minimum cost combination can be expected when one compares two countries which are clearly different from each other in not only degree of industrialization, but population density. Thailand and the United States furnish such an example. In Thailand 321 inhabitants live on 100 ha AL, in the U.S. only 43. The industrialization factor, the value of industrialization as a multiple of the value of agricultural production, comes to only 0.5 for Thailand, but rises to 9.2 for the U.S. Thailand belongs to the overpopulated agricultural countries, whereas the U.S. is counted among the sparsely settled but strongly industrialized countries.

If we now compare the production methods of rice farming for these two countries, we can see that even the same farm enterprise is capable of drastic changes in the combination of production factors. The price ratios among the production factors for the two nations are sharply contrasting: in Thailand labor is cheap, while land and capital are expensive, whereas in the U.S. labor is expensive, while land and capital are available in abundance. Thus in order to achieve the minimum cost combination in agriculture, Thailand must strive for high land and capital productivity by mustering considerable manpower. In the U.S., contrarily, primary dependence is on high labor productivity, which is attained through expenditures of large land areas per worker and a high capital investment per worker.

Overview 13 shows most interesting differences in the production methods of rice growing for the two countries. A family farm in Thailand averages only 0.75 to 4 ha AL, whereas those specializing in rice in the U.S. range from 75 to 250 ha AL. In Thailand, with its rich labor supply but limited land area and high prices for capital goods, rice is cultivated as a hoe crop; in contrast, in the U.S., with its high wage levels but cheap and abundant supplies of land and capital goods, the crop is tilled as a grain. The result is that a hectare of rice in Thailand demands 600 to 1,200 MH/ha while in the U.S. 20 to 30 MH/ha suffice.

Overview 13: Comparison of Production Methods in Rice Growing in Thailand and the U.S.

Characteristics	Thailand	California and Arkansas
Climate	wet-and-dry tropics	dry-summer or warm-summer subtropics
Production-factor price ratios	cheap labor, expensive land and capital	expensive labor, cheap land and capital
Size of family farm	0.75 to 4.0 ha AL	75 to 250 ha AL
Farming methods correspond to	hoe cropping	grain cropping
Labor requirements	600 to 1,200 MH/ha	20 to 30 MH/ha
Land preparation	with digging stick and hoe; 0.05 ha per worker per day	tractor plows; 6.0 to 7.5 ha per worker per day
Planting	starting plants in seedbeds; transplanting seedlings	airplane seeding
Fertilization and plant protection	little applied	spraying by airplane contractors
Irrigation	with bucket wheels and winches driven by water buffalo	with large pumping plants driven by diesel engines
Harvesting	panicles cut by knives or sickles; drying in the shock or on racks; threshing sleds drawn by water buffalo	large self-propelled combines, 10 km/h, up to 6 m cutting swath; largest combine needs only one operator and harvests 300 to 400 ha of rice annually
Yields, 1975	17.7 q/ha	51.1 q/ha
Producer's price for rice in U.S. cents/kg (1974)	9.9	23.0

One full-time worker in Thailand, using digging stick and hoe, can cultivate only 0.05 ha a day. In contrast, one full-time U.S. operator with a tractor plows 6.0 to 7.5 ha per day. The plant stock in Thailand is started by planting seedlings in seed beds and subsequently transplanting them to the field. In California, on the other hand, contractors are hired to carry out airplane seeding. Fertilizers and insecticides are little applied in Thailand, while they are used in quantity in the U.S. and contractors are hired here too to spray from aircraft. Rice fields in Thailand are irrigated by means of bucket wheels and winches driven by water buffalo. In the U.S., however, gigantic pumping plants powered by diesel engines are installed for irrigation.

The harvesting methods of the two countries also are substantially different from each other. In Thailand the panicles are cut by a knife or sickle, then dried in the shock or on racks, and finally threshed with wooden sleds drawn by water buffalo; in the U.S. self-propelled combines are used. These machines have a speed of 10 km/h, a cutting swath of up to 6 m, and a harvesting capacity of as much as 100 q/h. The large combine harvests 300 to 400 ha annually.

There are also many other farming enterprises in which production factors can be recombined so as to adjust to changing factor costs in the course of national economic development, thus ensuring the most economical production. A good example is furnished by another significant irrigated crop, sugarcane, which as a perennialy cultivated crop lies on the boundary between field crops and perennial crops. Sugarcane can be raised either as an annual or perennial. Highest hectare yields are attained when the cane is harvested only once and after that a new crop is planted. But this practice is also burdened with high labor costs since planting commands an especially large share of the labor expenditures in sugarcane growing. However, sugarcane can also be allowed to resprout from the rootstocks six and even more times, so that additional cuttings (ratoons) can be obtained for several years. Hectare yields decline from harvest to harvest, but so do labor expenditures, since only a certain portion of the total cane area must be replanted each year.

What production method best achieves the minimum cost combination and thus merits preference depends again on the prices of the production factors. On the extraordinarily densely populated island of Java labor is extremely cheap, but land is dear. Hence it becomes a matter of obtaining the highest land productivity, even if this can be achieved only at the expense of labor productivity. Thus when sugarcane is harvested only once, harvests of approximately 170 q of sugar per hectare are attained. Some yields even reach 250 q/ha.

In Cuba or Ethiopia, in contrast, land is cheaper but labor is already more expensive. Thus it is advantageous to increase sugar production per worker, even if sugar production per hectare must be allowed to fall. In

Ethiopia, with an hourly wage of 22 Dpf., cane is cut four times, the first time 20 to 22 months after the planting of the cuttings and thereafter the three additional harvests at intervals of 18 months. Sugar yields per worker, then, are higher than in Java; sugar yields per hectare, however, come to barely 120 q compared to 170 q in Java.

3. Regions with Predominantly Bush and Tree Crops

Finally, the third major category of farming systems in the humid tropics, those emphasizing bush and tree crops, must be described. They occur largely in the inner humid tropics, thus in the equatorial highlands, in the subhumid savannas, and especially in the humid and evergreen tropical rainforest. Peasant farmers in developing countries do not concentrate exclusively on bush and tree crops, but also practice rainfed or irrigation farming or both, at least to the extent of supplying their own food (see Fig. 37).

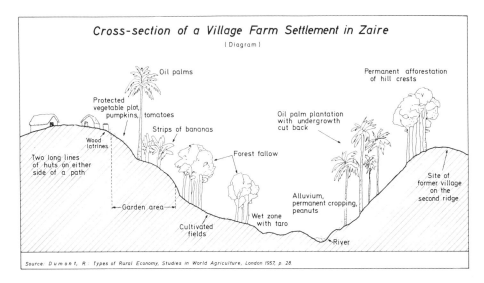

Figure 37

a) Farm Management Characteristics and Geographic Distribution

In countries with tropical rainy climates, considerable proportions of the total exports are products of bush and tree crop farming. Examples are Indonesia, with 40% contributed by rubber; Ghana, 60% by cacao; and Ethiopia, 56% by coffee.

Also worthy of mention is the critical importance of coffee for Brazil, rubber for Malaysia, tea for India and Sri Lanka, sisal for Kenya, and sugarcane for Java and Cuba (though sisal and sugarcane are on the borderline between field crops and perennial crops).

Sisal, sugarcane, and tea tend to be grown on a large-scale (plantation) level, so that the expensive factory plant can be fully employed by a small collecting area. For sisal and sugarcane an additional problem is the necessity of transporting a crop with so much bulky waste. This makes it imperative to grow the crop close to the mill. The fiber yield of sisal is only 3 to 4%, but the processed product is well worth transporting all over the world. Almost the same applies to sugar.

On the other hand, oil palms, coconut palms, and cacao are truly peasant crops. The widely varying quality of the large amounts that are being exported is a disadvantage, though, if marketing boards cannot provide remedies. Finally, rubber and coffee can be raised on the peasant farm as well as on the large unit. Coffee is predominantly a plantation crop in Angola, Kenya, Tanzania, and Zaire, whereas in Uganda and Ethiopia it has been developed on the peasant farm. Because of the drying technique for coffee, it is possible to produce it economically from even the smallest area of land and just a few bushes or trees.

The particular farm management characteristics of bush and tree crops are largely attributable to their character as perennial crops. The *useful economic lives* of these crops are, according to Masefield (1948), as follows:

Pineapple	6 to 13 yrs.	Oil palms	50 yrs.
Bananas	20 yrs.	Tea	over 50 yrs.
Pepper	25 to 30 yrs.	Coconut palms	80 yrs.
Coffee (*C. arabica*)	30 yrs.	Date palms	75 to 100 yrs.
Rubber	35 yrs.	Coffee (*C. robusta*)	over 100 yrs.
Citrus trees	40 yrs.		

The consequences of these lengthy land use cycles are that:

a) high initial investments must be made to cover expenses during the early growing period when the young trees yield little or no income, and until costs are recovered by the gradually increasing production. Costs for the oil palms are recouped about four years after planting, for rubber after about six years, and for cacao only after about eight years (Ruthenberg 1967);

b) relatively dependable yields, however, can be obtained once bearing begins, since the deep penetration by the root systems of the now well established bush and tree crops makes them more resistant to annual and seasonal variations in precipitation;

c) the permanent ground cover, which resembles the natural forest vegetation of the humid tropics, maintains soil fertility or at least conserves a good amount of it;

3. Regions with Predominantly Bush and Tree Crops 151

d) harvesting problems afflict many bush and tree crops because the work requires high labor peaks and thus leads to overall high labor expenditures. These costs can be absorbed only when there are correspondingly high gross returns or low wage levels. Most bush and tree crops must therefore be regarded as distinctly intensive crops; and

e) the minimum farm size sufficient for a peasant family of a given size is, as in irrigation farming, much less for the raising of bush and tree crops than for rainfed farming and certainly extensive grassland farming.

Bush and tree crops tolerate monoculture better than most field crops because:

— they require no annual crop rotation;
— the need to reduce the costs of the processing plant forces concentration on a single crop, thus, economies of scale;
— soil fertility in the humid tropics can be better preserved by the year-around ground cover of bush and tree crops than it is by most types of cropping;
— the humid tropics are extremely poor in livestock, and thus questions of feed balance and production of manure play no role in the sense of a diversified farming operation;
— labor peaks can be accomodated by the seasonal help of the commonly underemployed workers of the developing countries; and
— most bush and tree crops are receptive to large-scale operations, and the ever greater capital investment constantly stimulates specialization.

And yet monoculture by no means dominates on the plantation,

first, because often the varied soils, microclimates, and topographic conditions of the farm area require an adaption that can be areally suitable only with the planting of different crops;

second, because the risk of monoculture is great, for the crop (plant diseases and pests) as well as the market; and

third, because many bush and tree crops, after they are pulled, need an interim crop before they can be reestablished on the same piece of ground. Thus in West Africa cacao is followed by 10 to 20 years of bush fallow and a few years of rainfed cropping before it is replanted. In southern Brazil two growth cycles of coffee are interrupted by pasturing. In Nyanza Province, Kenya, the land use sequence of ca. 10 years of bananas — millet — ca. 10 years of bananas is common. In Central America one perennial crop follows the other, bananas often being replaced by cacao or oil palms (Ruthenberg 1967, p. 176).

Above all, however, the peasant farmers are as a rule more or even highly diversified in their farm organization, especially since they have their own supply needs in mind and most bush and tree crops are cash crops which thus often can be given only a small share of the farm area. The competition of the subsistence and commercial crops for the meager land area assumes, according to Ruthenberg (1967, pp. 177 f.), primarily two forms:

1. Perennial and subsistence crops are *raised on separate plots.* Expansion of the sales crops then restricts the subsistence area, which the farmer attempts to counteract by reducing the share of the farm in fallow. This in turn initiates a degradation process in the soil. Accordingly, areal share and areal yields for subsistence crops fall to the same extent, until sooner or later the benefit of cash receipts from bush and tree crops is no longer an adequate compensation. The farmers then find themselves in a conflict situation. The determination of the optimum areal limits of subsistence and cash crops thus becomes a cardinal question of farm organization.

2. Perennial and subsistence crops are *raised together on the same plot.* Now the competitive situation of the two enterprises is a matter of relative shares in the same mixed cropping operation. In the mixed cropping of coffee and cooking bananas the share of bananas must be all the greater the larger is the farm family that must be fed from a given area, and vice versa. Mixed cropping avoids the degradation of crop farming, but is an impediment to the application of chemical fertilizers, insecticides, and the like.

b) The Principal Bush and Tree Crops and their Locations

It can be seen from the large number of farm enterprises being considered here that the farm management characteristics stated as being associated with bush and tree crops apply only in general principle, and hold less rigidly in individual cases. A selection of six bush and tree crops in Table 9 illustrates the great differences between crops, differences which also signify the flexibility of farm management (for further examples see Table 6, p. 124).

The amount of initial investment in a particular crop is governed by the law of minimum cost combination. This makes imperative appropriate changes in the production procedure as economic development proceeds. However, the ratios between the initial investments for the six crops are probably essentially type-specific. The longer the expected productive period, the more diligent and conscientious the operator can be in his planting.

The ratio in labor expenditures between coconut palm and banana cultivation is almost one to four, while that for other crops lies in between.

3. Regions with Predominantly Bush and Tree Crops

Table 9: Examples of Tropical Bush and Tree Crops (Approximate Values)

Tropical Rainforest			Farm Management Characteristics (approx. values)	Tropical Highlands		
coconut palms	rubber	cacao		export bananas	coffee (*C. arabica*)	passion fruit
80	35	over 20	useful economic life in years	5–20	30	5
2,800–8,000	2,400	3,700	initial investment, DM/ha	ca. 800	1,350–3,000	600–900
560	960	300–2,000	labor expenditures, MH/ha/yr.	2,200	1,600–2,600	750
600	1,300	1,300	gross returns, DM/ha/yr.	2,160	1,500	1,500
S, (M)	M	S	harvested crop processed in small facilities (S) or mills (M)	maximum port distance: 150 km	S, (M)	M, (S)
F	F, P	F	specially suitable for peasant farms (F) or plantations (P)	P, (F)	F, P	F, (P)

Sources: Franke, G., et al.: Nutzpflanzen der Tropen und Subtropen, Vol. I and II. Leipzig, 1967.– Ruhr-Stickstoff AG: Schriftenreihe über tropische und subtropische Kulturpflanzen, Bochum, 1953–57. – Ruthenberg, H.: Organisationsformen der Bodennutzung und Viehhaltung in den Tropen und Subtropen. In: Handbuch der Landwirtschaft und Ernährung in den Entwicklungsländern, ed. by P. von Blanckenburg and H.-D. Cremer, Vol. 1, Stuttgart, 1967.

This range in variation is basically attributable to the different methods of harvesting and different annual proportions of new plantings. It shows that bush and tree crop cultivation can be adapted to the most varying farming conditions through selection of appropriate forms.

However, the particular suitability of the individual enterprises for the peasant farm on the one hand and the large-scale farm (plantation) on the other cannot simply be deduced from the amount of labor required. This is because of the very different technical processing requirements of the harvested crops. As long as tea processing by hand was sufficient for standards of product quality, tea was grown as a garden crop on small peasant farms, especially in Japan and China because of its high labor requirements. Since then the world market has demanded and been able to pay for a better and more uniform quality tea, which can be processed only in the strict cleanliness of large tea factories. Thus tea has now become a definite plantation crop in India, Pakistan, Sri Lanka, and Java in spite of the amount of labor it requires.

The relatively small labor requirements of sisal and its need for expensive processing plants operating at full capacity and from a small supply area (transport costs) clearly call for the large-scale farm. Oil palms in West Africa are a typical peasant crop, though they tie up relatively little labor. The small initial investment is highly esteemed and the problem of processing is being solved by a shift from the extraction of oil at home to its production in cooperative oil mills. The reverse situation applies to the West African peasant crop, cacao, where allowances have to be made for the high initial investment in labor that is needed to bring a small plot into production.

c) Peasant Farms or Plantations?

Today farm types with bush and tree crops are undergoing considerable change. Basically three questions stand out in this dynamism:
– Home processing or factory processing?
– Peasant farm or plantation?
– Monoculture or diversified production?
All are closely related to one another.

Even in developing countries capital inputs are gradually becoming cheaper and thus more profitable. Since wages are climbing at the same time, the rates of substitution of capital for labor are becoming more favorable. The situation for many bush and tree crops is like that of coffee, with harvesting contributing a labor peak that surpasses all others and yet on the whole defies mechanization. In the cropping of hevea 5/6 of the total labor expenditure is devoted to tapping, and for oil palms the share for harvesting is still 2/5. If capital is to be substituted for labor in such crops, it must be done during the processing and not the harvesting stage. Further, many products of bush and tree crops such as sisal fiber and raw sugar can generally not be processed in small facilities. The third thing that speaks for the technologically mature large-scale plant is that many bush and tree crops furnish export products that are encountering ever greater quality demands on the world market, a result of the rising living standards of the industrial states; the technologically imperfect processing by the small operator can no longer satisfy these demands (coffee, tea, cacao, pineapple). Thus development is forcing small-scale processing to give way to the factory, though not to the same extent for all bush and tree crops. The relatively strong pressure for further industrial processing of the harvested product is therefore increasingly influencing the decision as to whether the individual bush or tree crop shall be raised on a peasant farm or a plantation.

3. Regions with Predominantly Bush and Tree Crops

aa) Typical Peasant Crops

Typical peasant crops are those that can yield marketable products with simple methods. Examples are olive trees or coconut palms, the latter if copra rather than oils is to be extracted. Other examples are the perennial crops requiring processing methods that, though more differentiated, can also be accomplished on a cooperative level. The palm oil producers of West Africa are now successfully struggling to extract, via peasant cropping, oil batches that are large and of uniformly high quality. This is being made possible by a gradual replacement of home extraction facilities with new and large, and in part cooperativ oil mills. The advanced development that has taken place for the last two decades in Ghana, the premier cacao producer of the world, is primarily attributable to the successful activity of the Cocoa Marketing Board, which also is of an extensive cooperative character.

bb) Typical Plantation Crops

Typical plantations are those that require efficient transport facilities for the shipping of harvested products to the port (export bananas) or plant (sugarcane, sisal, pineapple), and (or) whose further processing cannot be done well by cooperative enterprises. For sisal and sugarcane both criteria apply. The harvest has a great amount of bulky waste and thus is sensitive to transport (fiber content in sisal is only 3 to 4%; sugar yield in cane is 7.5 to 13.4%). The profitability of cultivation of the two crops therefore increases rapidly with approach to the mills. Pure agave landscapes form around the sisal mills just as cane landscapes do next to the sugar mills, all for the sake of obtaining production from the smallest possible radius. Both crops do not fit in well with peasant farming operations, where farms are small and production is diversified to ensure self-sufficiency. Thus the tendency is strongly toward large-scale plantings.

Large sugarcane plantations have their own mills, whose operating seasons the planters try to lengthen by shifting planting times. Even a small sugar refinery with a daily capacity of approximately 800 tons of cane and a yield of 750 q/ha of cane needs 2,000 ha of the crop with monoculture and 4,000 to 5,000 ha with a rotation (Gnielinski, p. 285). The supply area of a mill ought not to exceed a radius of 20 km, for to do so would defeat the purpose of establishing large and continuous sugarcane zones in the vicinity of the mills: concentration of cultivation so as to save on transport costs.

How very strong the compulsion can be, not only for an existing large plantation to have a processing plant but for the inevitable plant to have a large plantation, can be well seen in pineapple cultivation (Schendel 1971, pp. 43 f.). The Hawaiian Islands produce today about 80% of the world's canned pineapple. One reason for this almost unchallenged position is the climate. As a tropical crop pineapple thrives under the uniformly warm temperatures and high humidity of the islands, which lie just above 20° N.L. and have a mean annual temperature of 23.9° C and a mean annual relative humidity of 69%. The economic life of a single planting is six to thirteen years and yields average 400 to 800 q/ha.

Not satisfied by nature, though, are the great demands of pineapple for water, since the potential evaporation comes to 2,030 mm a year, whereas only 650 mm of precipitation falls during the same period. This substantial water deficit is made up by irrigating with large field sprinklers and covering about a third of the fields with 60 cm-wide black plastic sheets, which by restricting evaporation loss conserve soil moisture and store additional heat in the soil. This production technique favors the large plantation over the peasant farm for several reasons:

1. The sizable additional water demands of pineapple cultivation can be satisfied most rationally by the large field sprinklers. The sprinklers are mounted on mobile stands and eject water in a widely radiating pattern of about 50 m in radius. The water is under 6 to 7 atmospheres of pressure, and approximately 50 m^3 is consumed per ha.
2. Preparation of the field for planting is done in efficient fashion with large machines, which build at intervals of approximately 1.80 m, dikes that are up to 10 to 15 cm high and about 1 m wide. At the same time the machinery is also unrolling the strips of plastic sheeting.
3. Transport and preservation of the crop require large investments and can be achieved much more cheaply in fully utilized large facilities than on small farms.
4. In the sale of the product on the world market, the large plantation has a better overview of that market and a stronger market position than does the peasant farm.

Thus the plantation has also gained general acceptance in the pineapple industry of Hawaii. The largest share of the pineapple fields belongs to the Dole company, no less than 13,000 ha.

Sugarcane and bananas, as well as small amounts of coffee and rice, are also raised on the Hawaiian Islands. The tropical bush and tree crops therefore easily predominate.

cc) Crops Suitable for both Peasant Farms and Plantations

It is questionable whether tea cultivation should still be classed in this category or whether it should already be assigned to the unqualified plantation crops. Growing demands of the world market for quality are forcing hand processing to increasingly give way to the more rational procedures of the factory. One can probably say that tea for developing

3. Regions with Predominantly Bush and Tree Crops 157

countries (including producing countries) can be produced on peasant farms, while tea for industrial countries is raised on plantations.

Table 10 shows the size structure of tea farms and plantations in the two most important tea-producing countries of the world. In India peasant farms of up to 40.5 ha comprise 80.4% of the operating units but include only 4.2% of the tea area. In contrast, 74.2% of the tea area is worked by plantations of more that 202 ha. In Sri Lanka the peasant farms stand out far more.

Table 10: The Tea Farms and Plantations of India and Sri Lanka, by Farm Size Classes

Farm Size Classes in Acres[1]	Tea Farms and Plantations			Tea Land
	Number	% of Total	Acres	% of Total Tea Land
	A. India, 1953			
up to 100	5,283	80.4	32,835	4.2
100–300	365	5.6	69,677	8.8
300–500	255	3.9	100,842	12.8
over 500	666	10.1	584,161	74.2
Total	6,569	100.0	787,515	100.0
	B. Sri Lanka, 1964			
up to 10	99,310	97.0	89,836	15.0
10–100	2,274	2.1	62,450	11.0
100–500	525	0.6	145,292	25.0
over 500	330	0.3	289,797	49.0
Total	102,439	100.00	587,375	100.0

[1] 1 acre = 4,047 m^2.
Sources: FAO, Tea-trends and prospects. Commodity Bull., Series 30, Rome, 1960. Kullak-Ublick, H.: Die Teewirtschaft in Ceylon., "Zeitschrift für ausländische Landwirtschaft," Frankfurt/Main, No. 1, 1965.

Here the operating units in the size class of up to 40 ha include 99.1% of all farms and plantations and 26.0% of the tea land. Undisputed peasant farmers, those with up to 4 ha, account for 97% of all tea raisers and 15% of the cropped area. Though the plantation prevails here areally, it is clear that tea raising is conspicuous in all farm size classes.

The same situation applies to coffee raising, as shown by Table 11. It is evident that with increasing size of coffee farms, fertilizer expenditures per hectare of coffee climb rapidly and coffee yield in kg/ha noticeably

VI. The Agricultural Geography of the Humid Tropics

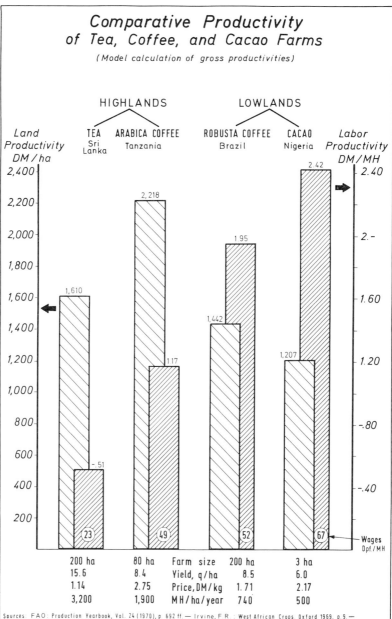

Figure 38

increases. The differences in gross returns among the farm classes are certainly not compensated by additional expenditures of farm inputs that increase yield, so that a rent differential in favor of the large operation will remain, thus making the plantation competitively superior. Fig. 38 offers comparative figures for tea, coffee, and cacao farms.

Table 11: Coffee Raising in Sao Paulo State, by Farm Size Classes, 1960.

Size of Coffee Farms in ha[1]	% of All Coffee Trees	Coffee Crops	Food Crops	Fertilizer Expenditures, kg/ha of coffee	Coffee Yield in kg/ha	% of Total Coffee Production
		in % AL				
under 8	13.6	2.2–8.8	4.3–4.4	13	349	10.0
8– 32	36.4	14.1	2.1	36	441	35.1
32– 64	18.6	13.4	2.0	46	434	17.8
64–128	14.2	20.4	2.0	101	480	15.5
128–512	14.8	20.5–21.6	2.1	102	506	18.4
over 512	2.4	23.5	1.3	152	506	3.2
Mean	(100.0%)	–	–	61	446	(100.0%)

[1] Average of 834 trees per hectare.
Source: Franke, G., et al.: Nutzpflanzen der Tropen und Subtropen. Vol. I, revised edition, Leipzig, 1975, p. 49.

VII. The Agricultural Geography of the Dry Areas

The "dry areas" are outlined by the semiarid and arid climatic zones. Fig. 39 makes their locations clear. It shows that dry areas are to be found in the tropics as well as the outer tropics. The uneven distribution of precipitation over the year in semiarid areas can be agriculturally advantageous, for the more a given small amount of rain is concentrated in the growing season of short-lived cultivated plants, the higher their yields will be. Sixty millimeters of rain in a long dry season is of little use, but it is highly valuable for crop production if it falls in the short rainy season.

Nevertheless, the fact that precipitation in semiarid areas is usually distributed unevenly over the year is one of the most serious problems of agriculture. It brings with it the burden of risk and requires passive adaption if the handicap cannot be actively overcome by irrigation.

The rainfall curve for Grootfontein (Fig. 40) has, over a period of 63 years — so far as they are known, exceeded the 800 mm limit eight times and has fallen below the 300 mm-limit six times. With a long-term annual precipitation mean of 534 mm, dryland farming is still possible here, but undoubtedly it would be more practicable and profitable, even with only 450 mm precipitation, if that amount could be relied upon with certainly every year.

The strong year-to-year fluctuation of precipitation in the contact area between dry and shrub savanna results in the termination of dryland cropping in regions that are wetter than they would be were these fluctuations absent. The farmer certainly can make allowances for a harvest failure every fifth year, also probably every fourth year, but certainly not every other year. The less precipitation varies from year to year in a region, the drier are the zones into which dryland crop farming can venture and the smaller then is the annual precipitation mean on the agronomic dry boundary. The more precipitation varies from year to year, the more cultivation is forced back into the more humid regions by extensive grassland farming. Extensive grassland farming is of course less sensitive to precipitation oscillations because during dry years the worst can be averted by taking such measures as migrating, selling off livestock, or purchasing feed. In cropping, though, the total expenditure must be written of as a loss when drought keeps the crop from sprouting; corn, millet, and wheat from heading; and the like. The greater precipitation varies from year to year, the more extensive grassland farming competes with rainfed farming and forces it back to the wetter locations. Meanwhile, the agronomic boundary comes increasingly closer to the climatic dry boundary.

VII. The Agricultural Geography of the Dry Areas 161

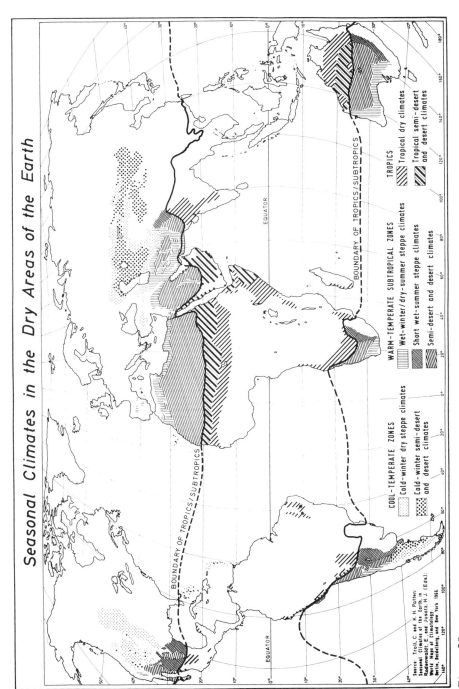

Figure 39

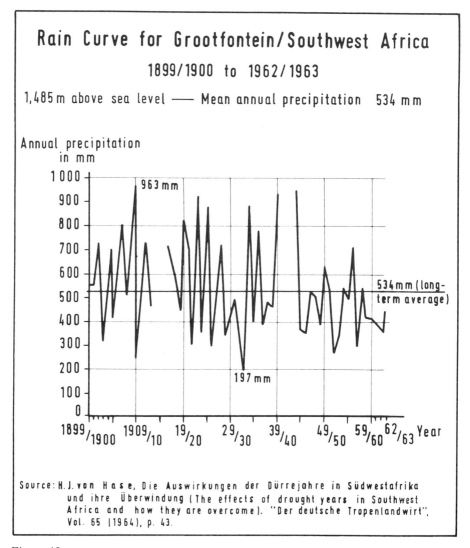

Figure 40

All farming types on the agronomic dry boundaries are subject to a threefold inflexible obligation:
1. The meager water supply must be utilized to the highest possible degree by employing farming types, farming enterprises, cultivated plants, and useful animals that will produce the most under these conditions. The greatest farming success is achieved when the scarcest factor, water, is brought to the highest levels of productivity.

VII. The Agricultural Geography of the Dry Areas

2. Precipitation fluctuations from year to year must be countered, which in most cases means a flexible defensiveness but in a situation where the instruments available to farm management are limited.
3. Adaptations to a very short rainy period must be made, if necessary with appropriate forms of cattle herding, selection of short-lived cultivated plants, and the like.

All measures taken by the farmer must submit to this threefold constraint: selection of plant species and strains with the shortest growing season and greatest resistance to drought and aridity; cultivation, fertilization, and plant protection with the highest possible consideration for the sparse water supplies; extreme concentration of production in the rainy season; exhaustion of all the possibilities for spreading risk, including the building of physical and monetary reserves for drought years.

Several different paths are open to the farmer who must deal with the water deficiency in dry areas through the selection of farming types. The *most obvious possibility* is the increasing of the water available to plants through *irrigation* and the distributing of it more evenly over the year in relation to plant needs. Yet as desirable as it is to make use of this operational technique in the dry areas, it is precisely this practice that faces the greatest obstacles. The same water deficiency that makes irrigation seem so attractive also notably impedes it. All medium and small rivers disappear during the long dry season, lakes and ponds are few, and the groundwater table is very low. Part of the irrigation water for the dry San Joaquin Valley in California is brought in over more than 800 km from the northern Sacramento Valley and is correspondingly expensive.

But where large streams are flowing out of high mountains the year around, irrigation can foster an intensive agriculture in dry areas. The Nile delta is an example par excellence (see Overviews 10 and 12, pp. 121 and 144). In the semiarid Peshawar Basin of Pakistan, which has only 350 mm annual precipitation, the Indus and its tributaries permit year-around crop production with rotations such as those noted below. Because it has already been treated in Chapter VI-2, irrigation farming will not be taken up again in this chapter, though it is of great significance for the dry areas.

A *second means* of adjusting to aridity that is available to the farmer is the emphasizing of *bush and tree crops* resistant to drought. With their deeply penetrating root systems, they are in a better position than short-lived field crops to tap the groundwater and the moisture that is left from the rainy season and has settled in the lower layers of the soil. Thus these plants can survive many dry months without serious damage. Among the drought-resistant bush and tree crops that are regionally important in this function are the olive tree in the Mediterranean area; the date palm in places like North Africa, the Arabian Peninsula, and Irak; and sisal in the tropical highlands near the equator.

Crop Rotations in Irrigation Farming in the Peshawar Basin of Pakistan

Year	Summer Crop	Winter Crop	Year	Summer Crop	Winter Crop
		Extensive Rotations			
	A.			B.	
1.–3.	Sugarcane	(for the sugar mill)	1.–3.	Sugarcane	(for the sugar mill and gur)
4.	Grain corn	Wheat	4.	Soybeans	Wheat
			5.	Grain corn	Sugar beets
	1,508	MH/ha/yr.		1,675	
	645	Financial Input in DM/ha/yr.		810	
		Intensive Rotations			
	C.			D.	
1.–2.	Sugarcane	(for making gur)	1.	Grain corn	Sugar beets
			2.	Jute	Wheat
3.	Grain corn	Tobacco	3.	Grain corn	Tobacco
4.	Jute (or cotton)	Wheat	4.	Jute (or rice)	Wheat
5.	Grain corn	Sugar beets			
	2,390	MH/ha/yr.		2,668	
	940	Financial Input in DM/ha/yr.		1,115	

A *third means* of adapting to severe aridity is the selection of *field crops* that can combine low water demands with such a *short growing season* that the production process can be completed in the brief rainy period. Here plants like barley and wheat in the subtropics and millet, sorghum, and peanuts in the outer tropics stand in the foremost rank, and following a little farther back, sesame, various legumes, cotton, tobacco, and corn, among others.

A *fourth possibility* is offered by the *dry farming system*, in which bare fallow is introduced every fourth, third, or even second year. In the most extreme cases, only one wheat year follows two fallow years, so that each field produces a harvest only every third year. Here the farmer takes advantage of the fact that bare ground loses less moisture through evaporation than a surface with plant cover, so that part of the rain falling during the fallow period is stored in the soil for the wheat crop that follows the next year.

These last two possibilities can be grouped under the heading of *dryland crop farming*, which extends to the agronomic dry boundary and here comes into competition with the fifth possibility, *extensive grassland farming*, which also is guided by, among other things, the principle of giving priority to perennial plant types. Since natural pastures have, in part, perennial grasses with significant depth of root penetration, they can begin their assimilation activity immediately after the start of the rainy season and can continue the process even beyond the end of the wet period.

1. Regions of Extensive Grassland Farming

Grassland farming is found in all tropical climatic zones except the rainforest belt. The main centers, though, are the shrub and dry savannas, thus the outer tropics. These regions are mostly semiarid, with about 200 to 400 mm precipitation, though in most areas it is between 250

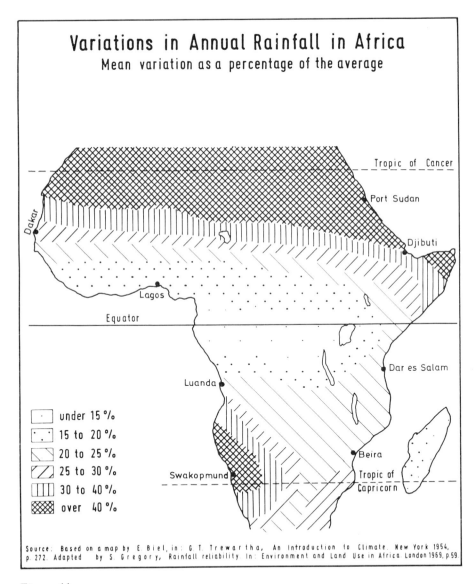

Figure 41

and 350 mm. The greater the variations in precipitation from year to year, the more the agronomic dry boundary shifts into the wetter zones. Thus in the Grootfontein area (Fig. 40, p. 162), despite a long-term mean annual rainfall of 534 mm, grassland farming is still predominant because it is better able to withstand the drought years than is dryland cropping. The poleward increase in variation of annual rainfall (Fig. 41) especially fits the outer tropics for extensive grassland farming, since the unreliability associated with large rainfall fluctuations is added to the burden of an already meager moisture supply.

Schickele (1931) distinguished three main categories of extensive grassland farming: nomadic herding, representative cattle farming, and commercial sedentary grassland farming.

a) Zones of Nomadic Herding

In the nomadic economy man and animals periodically migrate. Production is predominantly for domestic consumption. Thus milk, as a basic foodstuff, is the most important product. All kinds of domestic animals are milked: camels, horses, donkeys, cows, goats, and sheep.

The *desert nomads* must resort to especially long journeys. They therefore need animals that can range widely and survive long periods of drought. They keep mainly camels (simultaneously for milk, as beasts of burden, and for riding), but few smaller animals and no cattle. Camels need to be watered only every two to three days in the warm season and only every two to three weeks in the cool season, and as a result can be taken up to 80 km away from their drinking place (Ruthenberg 1967). Cattle, in contrast, must go to their drinking place twice daily during the warm season in the Sahara, and every two to three days in the cool season. Cattle therefore can only graze within a radius of four kilometers from their watering place (Ruthenberg 1967). Desert nomads produce little for market, but they often contribute significantly to transport with their camel caravans.

Higher marketable production is achieved by the *steppe nomads,* who more often keep sheep and goats rather than camels. Horses also often play a larger role, as in many Arab countries. Because fodder vegetation grows better in the steppe than in the desert, some cattle can also be raised. Nomads often barter and trade with neighboring peoples, and in this way procure grain, tea, tobacco, alcoholic beverages, and other items in exchange for their own products. The Mongols, Kirghiz, and the Berbers of the Algerian plateau are steppe nomads, as were the hordes of Genghis Khan.

The *seminomads* are native to areas where fodder vegetation is somewhat more plentiful. Their nomadic journeys are therefore much shorter.

This means they are in a better position to keep cattle, which now predominate over sheep, camels, and horses. The North African and Near Eastern mountain nomads belong to this category. They are able to practice some crop farming at the foot of the mountains bordering the steppe or on the banks of rivers. However, the seminomads also include purely pastoral peoples such as the Fula in central Sudan, the Masai in Kenya and Tanzania, and to some extent the Hereros in Southwest Africa.

b) Zones of Sedentary Grassland Farming

As his second category of extensive grassland farming, Schickele designates savanna grassland farming with representative cattle herds. It is practiced on a sedentary basis. Cattle predominate but have little economic importance as they are kept more for reasons of ritual, sociability, religion, and prestige.

The savanna peoples usually obtain their main food supplies from cultivated crops, and not from their cattle. Representative cattle herding is common among such groups as the many Bantu tribes in East Africa and the Sudanese blacks of the Lake Chad area.

Finally, the third and most important category of extensive grassland farming to be named is *commercial steppe grassland farming* (or *ranching*). This is also sedentary farming, and products for the market are of a kind that is easy to transport, such as wool, hides, skins, and feeder stock. Here again Schickele distinguishes three intensity stages:

Wild steppe grassland farming is the most labor- and capital-extensive form, since protection of the herds is left to the stallions and bulls that lead them, keep them together, and defend them against predators. Wild steppe grassland farming is only possible with cattle and horses and today no longer plays an important role.

Open-range steppe grassland farming imposes greater labor demands, since the herds must be kept under constant supervision of a herdsman. This mostly involves wool sheep, which can walk for long periods and so cover great distances to reach their drinking places. Thus only a few wells are necessary. The strong herd instinct makes it easier for the herder to keep the animals together.

Finally, there is *fenced-range or enclosure farming,* in which much of the labor of the herdsman is replaced by capital investment. The pasture area is divided up by wire fences, and this requires the installation of numerous drinking places. Fenced-range or enclosure farming is therefore more labor *ex*tensive and more capital *in*tensive than open-range steppe grassland farming. The result is a relatively high market productivity for sheep and cattle. This system is often combined with the open-range system on the same farm. In this case, the larger animals are kept in enclosures near the farmstead, while the smaller ones are herded on the more distant and unfenced pastures.

The precipitation situation determines the type of pasture, and every pasture type has a particular carrying capacity (see Fig. 42) as well as a specific suitability for this or that animal type and use orientation. There is then, for every pasture type, a specific production orientation in the raising of ruminants that gains competitive superiority over all others.

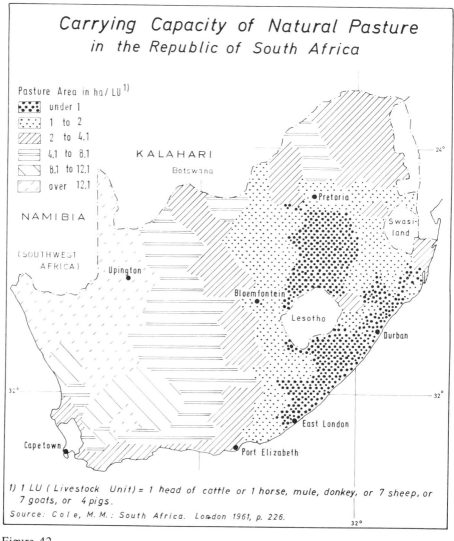

Figure 42

As to where the dry boundaries of the individual lines of production locate, that depends not only on the amount of precipitation, but on its regime, the wind velocity, the soil conditions, the relief, and other factors. For central Southwest Africa, the following generalizations can be made:

With 100 to 200 mm precipitation, vegetation is so sparse that only a type of sheep that demands little and can range over long distances can use it. Sheep thus have the advantage over cattle here, since they are able to graze more selectively with their pointed muzzles and to pick out the most nourishing parts of the plants. Production efforts concentrate exclusively on obtaining *wool or skins.*

With *250 to 300 mm precipitation,* cattle raising becomes possible, but only for producing feeder stock, Calf production is still impossible because the sparse growth of fodder vegetation is insufficient for cows to be capable of developing a fetus or raising a calf at the udder. We thus have a *feeder stock farm that depends on the division of labor,* one that buys calves from zones with better fodder growth.

With *350 to 450 mm precipitation,* the typical operating unit becomes the *self-supplying cattle fattening* (or *finishing*) *farm.* Pasture is now so improved that both calves and feeders can be produced. To avoid the unfavorable transport situation associated with the areal division of labor and to reduce risk by diversifying operations more, both systems are combined on the same farm. The operator keeps a breeding herd, raises all the calves, and sells them as feeders.

Finally, *beginning with about 500 mm of precipitation,* cultivation and with that, feed production, becomes possible. This is very important for fodder cropping, both in seasonal balance and in improvement of production volume and nutrient concentration. Now high quality can be stressed: *quality feeder stock and quality milk production.*

That the production orientation of extensive grassland farming is not only conditioned by physical factors, but also strongly influenced by market access, has already been described in connection with Fig. 16, p. 73.

c) Seasonal Feed Balance as the Central Problem

In extensive grassland farming not only is every kind of cropping generally precluded, but the farm may have to limit itself to one kind of animal and be dependent on a single product. This means that monoculture, with all its dangers, is widespread. The scope of production is narrow, its depth small, and the *market risk* correspondingly high.

In addition, there is a high *production risk* related to the annual rainfall variations, a problem especially affecting areas where it is impossible to produce feed or to purchase feed concentrates. Here the farm operation is strikingly dependent on nature in its sole dependence on whatever plant growth is produced by the rain. The full severity of this problem can only be appreciated when it is remembered that in the dry and shrub savannas, even in years of normal rainfall, a very short rainy season is followed by a long dry season. To carry the herds through this dry season without great losses is the number one concern of every grassland operator. This is also the key to any improvement in productivity.

Seasons with feed shortages can be countered by basically two different kinds of farm management measures, active and passive. The methods of *actively overcoming seasonal feed deficiencies* are those generally applied in the industrial countries. They consist of either storing reserves during the fodder-growing seasons in forms like hay and silage or purchasing feed for these times of short supply. All these measures require high expenditures and therefore presume a favorable price-cost ratio.

If this condition is not fulfilled, i.e. if the purchasing power of animal products for wage labor and commercial feed is so small that neither the storing nor purchasing of feed can be contemplated, then there only remains the *passive adaptation to the seasons when feed is scarce*. This consists of accepting the limits set by nature and simply attempting to avoid undue losses at the times of feed shortage by a careful selection of the livestock raising system. Passive adaptation involves essentially five groups of measures:

1. Species of animals are selected which are capable of surviving periods of feed scarcity. Thus the desert nomads keep many camels and sheep, which can cope with long periods of hunger and thirst. In the highlands of Ethiopia there are many goats, since they can support themselves on bush foliage during times of drought.

2. Undemanding animal breeds are chosen. Fat-tailed sheep or zebu cattle can withstand long periods of hunger because they accumulate fat in times of good feed supply, thus stockpiling nutrients that the farmer for economic reasons cannot supply.

3. Lines of production are selected which are least dependent on seasonal feed balance. Thus with cattle, raising of feeder stock is favored over dairy production, and with sheep, raising for wool and skins rather than meat is preferred.

4. Breeding is coordinated with the seasons when feed is plentiful. Thus it is planned that the young should be born just before the beginning of the rainy season, so that the dams do not go hungry during the suckling period. The final fattening of the animals is also done during the rainy season.

5. Animal numbers are adjusted to the seasonal variations in fodder growth. Herds are expanded, if possible, for the rainy season, and

1. Regions of Extensive Grassland Farming

reduced for the dry season. All births and purchases are therefore planned for just before the beginning of the rainy season, and all slaughtering and sales take place at the beginning of the dry season.

In Fig. 43 nine stages of extensive grassland farming are presented in relation to increasing national economic development. As long as labor and capital are scarce and land is abundant, the feed balance must be achieved by means of land reserves.

Stage I is represented by *nomadic grazing*. This system is found in extremely arid areas where communications are poor. For example, this is the type of economy of the steppe Bedouins, the Kurds, or the Berbers. In Central Asia nomadic herding is still widespread, such as among the Mongols and the Kirghiz. In East Africa the Masai, the Fula, and the Somalis could be termed at least semi-nomadic. Nomadism is necessary to achieve a feed balance, which is sought through extensive migrations. This form of economy therefore develops most favorably where the annual vegetation rhythm varies from region to region because of mountain chains, hills, and valleys. The migrations of the mountain nomads are more vertical, those of the nomads on the plains more horizontal.

When the population later turns, in *Stage II,* to a sedentary way of life, permanent ownership of the land develops. However, as long as population density is small, a land reserve still remains in the hands of the tribe or the state. When a feed shortage occurs in dry seasons, the still unoccupied areas of land stand available as reserve pasture.

This situation changes with *Stage III.* Now all the land is in private hands. The farms are now fenced in, though the pastureland is not yet subdivided. In the diagram of Stage III it is assumed that there is only one drinking place within the area belonging to the farm. Vegetation zones form around this center. *Near the drinking place* there is only bare soil, since the herds kill off all plant growth by the concentration of their excrement and the severity of their grazing and trampling. In the *second zone* there is only a small amount of vegetation, mostly inferior weeds. Even the *third zone* is overgrazed, though the vegetation is more plentiful. Only in the *fourth zone* does one find an almost natural vegetation with many valuable grasses. However, this zone is as a rule the outermost ring of those areas grazed by the animals within reach of the watering place. The *fifth zone* is not used during the good fodder growth of the rainy season since the walking distance to the drinking spot is too great. Here natural vegetation predominates. This fifth ring, therefore, stands available as a feed reserve for the dry season when feed is scarce.

There is also, though, the tendency for land in the course of economic development to become scarcer and costlier, so that its productive use must be intensified. This has to be done primarily with increased capital inputs, which meanwhile have become cheaper. Additional wells are sunk

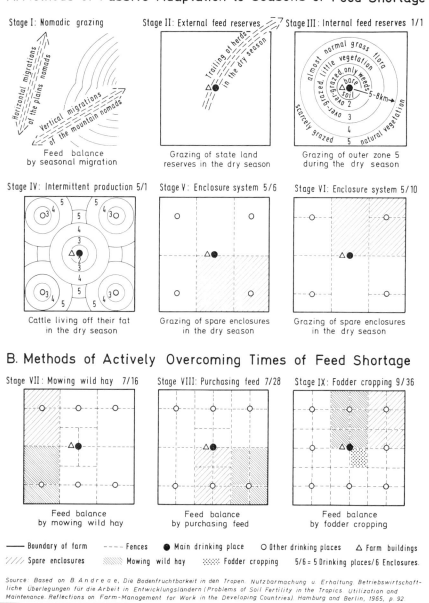

Figure 43

during *Stage IV*. Increasing the number of drinking places steps up animal production in three ways:
a) The distances animals must travel to drink are reduced. The energy thus saved can be channeled into market performance.
b) The inner zones without any vegetation now completely disappear, since each watering place is less used.
c) The little-used outer zone disappears too, since all areas are grazed, even during the rainy season.

The vegetation cover of the farm thus becomes more evenly distributed and more productive in the rainy season, so that stocking can be increased. However, the feed situation in the dry season has become worse because now feed reserves both outside of the farm and on it are exhausted. The nutrient reservoir for times of feed shortage, which was previously available in the form of alternative pasture and which in developed countries is stored in barns, stacks, and silos, must now be stored in the animals themselves as fat deposits.

This stage IV is thus the one in which the animals suffer from hunger the most during the dry season. But this must be accepted because of the prevailing price-cost ratio. Land is still cheap compared to labor and capital, so everything depends on managing production in order to save as much labor and capital as possible, even if this means using large areas of land. This is achieved by intermittent production. Not only is no meat produced during the dry season but weight losses have to be accepted, which have to be made good once the rainy season begins. Intermittent production of this kind means that cattle do not reach slaughter weight until they are four or five years old. Yet this form of grassland management is economically sound since it is in only this way that slaughter cattle can be produced with a minimum of labor and capital, and thus at minimum cost.

Only when land becomes even more expensive and capital still cheaper will it pay to make larger capital investments that ensure better land use. This heralds *Stages V and VI*, when fencing the range becomes possible. Fenced tracts allow controlled grazing and thus a more complete use of the land. Different animal species as well as animals of different ages can be assigned to pastures of different quality. Reserve or spare enclosures are set up for the dry season, rotated annually, and left ungrazed during the rainy season. By bridging the times of feed shortage in this way, the animals at least are not allowed to suffer a weight loss during the dry season so that the age of oxen ready for slaughter is brought down to about three years. A three-year old ox consumes far fewer maintenance rations than a five-year old animal of the same weight class, and thus the proportion of feed for production increases. The higher productivity of land can also be explained by the fact that soil fertility suffers less than in Stage IV, so that the nutrient yield per hectare increases.

Stage VII differs from Stage VI in two ways: by degree, in that the number of drinking places and enclosures is increased in the course of intensification; and in principle, in that a limited amount of wild hay is produced. Thus besides a further increase in capital investment, there is an increase in the labor input. The hay is stacked in small enclosures, *kraals*. This hay is richer in nutrients than standing hay because it has been cut at a time when the nutrients have not yet moved from the stalks and leaves into the rootstocks or seeds. Production of wild hay represents the first active step in the economic development process of overcoming the seasonal feed shortage. This can begin only when the purchasing power of animal products for wage labor and technical inputs has attained a high level, for the cost in wage labor, machinery, and traction power is great.

At *Stage VIII* the number of enclosures has risen to 28, so the grazing technique is even further improved. The usefulness of the spare enclosures in bridging the seasons of feed shortage is now reinforced not only by the production of wild hay, but by the purchase of feed. With a higher degree of industrialization, land prices tend to rise so that it is even more important for grassland farmers to increase land productivity. Indeed, in this situation, the exchange value of meat animals as against feed concentrates has become so favorable that it is now possible to supplement the pasture forage with concentrates. Corn from distant rainfed farming zones is used for this purpose.

The transport of grain corn has become possible because at this stage of development the transport infrastructure of the country has improved. The feeding of concentrates has also become profitable for another reason. The mass income of the population grows with increasing industrialization, and with the rise in consumption levels, better-quality meat animals command higher prices. Since such high-quality animals cannot be raised on natural pastureland, final fattening with concentrates is undertaken in the dry season.

The principal difference between *Stage IX* and Stage VIII is that now a third and additional method of actively overcoming times of feed shortage has been introduced: irrigated fodder cropping. This method of fodder cropping presupposes a higher stage of national economic development: a high purchasing power of animal products not only for wage labor, but for farm inputs coming from the industrial sphere, such as machinery and chemical fertilizers.

2. Regions of Dryland Crop Farming

The origin of crop farming along its dry boundaries has been in the form of savanna or steppe shifting cultivation, which also characterized large parts of Europe at the beginning of more than two thousand years of cropping development.

2. Regions of Dryland Crop Farming

As long as settlement on the dry savannas and steppes remains highly scattered and the land suitable for cultivation consists of natural grassland, simple farming patterns suffice. A small piece of grassland is turned up, first with a digging stick and hoe and later with a plow, after which a domesticated plant, preferably millet or sorghum, is planted for two to four years. By that time yields have dropped because the soil has not been thoroughly worked and no fertilizers have been applied. The farmer then lets the land revert to the original grass cover over a long period of years in order to restore fertility. Meanwhile, he takes over another piece of land and crops it in the same manner. Later on, he may complete the cycle by returning to cultivate the previously farmed land, though at irregular and long-term intervals.

The further development of the steppe or savanna shifting cultivation system into more productive forms of agriculture

– is made necessary by population increase,
– is simplified by technical progress in soil cultivation, fertilization, and the like, and
– is aimed primarily at increasing the proportion of land used productively.

If three years of cultivation follow fifteen years of grass fallow, the proportion of land used is only one-sixth. However, if four years of cultivation alternate with twelve years of grass fallow, then the share of land used increases to one-fourth. Finally, if four years of cultivation alternate with eight years of grass fallow, then a proportion of one-third is attained. A prerequisite for such a development, though, is that the restorative effect of grass fallowing on soil fertility be supplemented by instituting other measures having similar beneficial effects, and that is a difficult assignment in the dry savannas.

The more the grass fallow is reduced, the more the moisture deficiency becomes a limiting factor and the more it determines the farm organization. Eventually the farmer is faced with the alternatives of

– either resting content with crops that need very little water but have low yields, in order to ensure an annual harvest,
– or choosing, within the framework of the dry farming system, crops that need more water but also yield more produce, but at the same time accepting the need for fallow areas to store water.

Crops that require little water and are suitable for the first course include millet, sorghum, peanuts, cowpeas, garbanzos, and sesame. This explains the basic importance of millet and sorghum growing for all the dry regions of Africa and also the predominance of peanut cultivation in the states bordering the Sahara, such as Niger, Chad, Upper Volta, Mali, and Senegal.

The second course, via the dry farming system, generally imposes higher technological requirements and thus is easier to take in the more industrialized states. The major distribution areas are consequently the drylands of the U.S. and Canada, Argentina, Australia, Republic of South Africa, and the U.S.S.R. This system is also found north of the Sahara and in Iran and Turkey, as well as in many other countries. The major crops are mostly wheat, barley, millet, and sorghum, and also to some extent, cotton.

a) Wheat-Fallow Farming

The function of bare fallow in the dry farming system is not to combat weeds and to help space work (Germany in the Middle Ages), or to raise hay or silage, or to complement a crop rotation (central Sweden, medium heavy marsh soils), but rather to increase the yield of grain crops. Yields are increased primarily by saving water and secondarily by

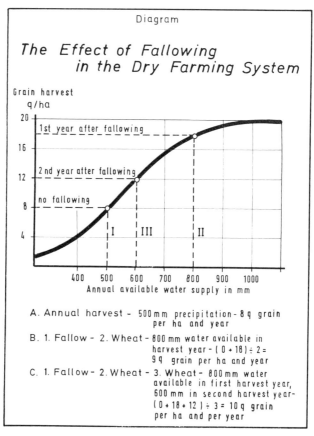

Figure 44

2. Regions of Dryland Crop Farming

releasing plant nutrients. Indeed, the whole organizational principle of the dry farming system is based on the fact that bare soil evaporates less water than a plant cover, so that some of the rain falling in the fallow year can be stored in the soil and used to increase the yields of the following grain crop.

If the production function proceeds as in Fig. 44, above, then the rotations

A.	B.	C.
1. Wheat	1. *Fallow*	1. *Fallow*
2. Wheat	2. Wheat	2. Wheat
3. Wheat		3. Wheat

produce average annual grain yields of 8, 9, and 10 q/ha respectively. Thus rotation C deserves precedence since the slightly larger expenditure than that of rotation B in the form of a 17% larger area sown to wheat does not cancel out its additional output.

Zones of Intensity of the Dry Farming System
(Diagram)

Rainfall mm/year	Long Rainy season	Short Rainy season	Fallow % arable land
ca. 300 mm	(A) 1. Bare fallow 2. Wheat	(B) 1. Bare fallow 2. Barley (four-row)	50 %
ca. 350 mm	(C) 1. Bare fallow 2. Wheat 3. Millet	(D) 1. Bare fallow 2. Peanuts 3. Millet, sorghum	33 %
ca. 400 mm	(E) 1. Bare fallow 2. Wheat 3. Wheat		33 %
450–500 mm	(F) 1. Bare fallow 2. Wheat 3. corn, blue lupine 4. Wheat	(G) 1. Bare fallow 2. Peanuts 3. Sorghum, millet 4. Millet	25 %
Transition to wet savanna	(H) 1. Sunflowers, peanuts 2. Corn	(J) 1. Dried peas 2. Wheat	—

Figure 45

Actually, type C rotations are commonly found in the vicinity of the agronomic dry boundaries on all continents. However, the production function can proceed differently than pictured in Fig. 44, especially when precipitation amount and distribution are different. The larger proportions in fallow of as much as 67% CL or smaller proportions of as little as 25% CL can achieve competitive superiority. Some of the rotations in this category are shown in Fig. 45, above.

Three things in this diagram bear particular attention:

1. The crops are practically all grains, since all leafy crops except peanuts demand considerable water.
2. If rainfall increases from 300 to 500 mm/yr., then the proportion in fallow drops from 50% to 25% CL, excluding extreme cases.
3. If the rainy season becomes shorter with the same amount of annual precipitation, then the farmer must fall back on crops that are especially short-lived, though less valuable. Corn and wheat are then displaced by millet, sorghum, barley, and peanuts.

Farms of this kind can hardly be expanded through intensification, for precipitation conditions establish the necessary fallow quotient, allow scarcely anything else but grain to be raised, and set early limits to all yield-increasing measures (cultivation, fertilization, etc.). Production and income potentials can be significantly increased only with extensive methods, and this means especially the enlargement of the farm.

Thus grain-fallow farming is in principle, though certainly not in degree, similar to extensive grassland farming in the tendency to use large areas, provided that a growing population does not put obstacles in the way. Wherever an agricultural society is established with the dry farming system, there prosperity can only be achieved with a large-area economy, i.e. with a low population density. However, the economic capacity of this cropping system for food production is so small that population growth must sooner or later lead to impoverishment, though by then employment opportunities may be provided by a flourishing industry.

Table 12 describes wheat-fallow farms in three natural regions of the Unites States. A comparison of the three farm groups can only be done with caution since the locations are so generalized. Nevertheless, one can recognize that as precipitation increases (from left to right), the following takes place:

– the proportion in fallow drops from 46.5% to 33.3% CL;
– cropping becomes more diversified;
– the pasture requirement drops from ca. 4.5 to ca. 2.1 ha/cow;
– labor expenditures, because of intensification, increase from 6.5 to 9.8 MH/ha AL; and
– the cultivated land for a family farm, because of increase in productivity with increasing annual precipitation, is able to be reduced from 474 to ca. 225 ha.

2. Regions of Dryland Crop Farming

Table 12: Wheat Farms in the Dry Areas of the U.S., 1968.

Characteristic	Unit	Columbia Basin (Ore., Wash.)[1]	Northern Plains (Mont., N.D.)[1]	Southern Plains (Colo., Neb., Kan.)[1]
Precipitation	mm/yr.	250–350	300–520	300–750
Farm size	ha AL	635	378	348
CL/farm[2]	ha	474	266	ca. 225
Share of cropland:				
Fallow	% CL	46.8	40.2	ca. 33.3
W. wheat/s. wheat	% CL	42.4/0.4	–/36.8	40.8/–
S. barley and other grains	% CL	6.4	11.3	...
Millet/flax	% CL	–/–	–/ 2.0	10.6/–
Fodder crops	% CL	1.9	8.4	13.4
Normal crop rotation:	–	1. *Fallow* 2. *W. wheat* (plus some irrigated alfalfa)	1. *Fallow* 2. *S. wheat* 3. *Fallow* 4. *S. wheat,* flax 5. *S. barley,* feed crops	1. *Fallow* 2. *W. wheat,* feed sorghum 3. *W. wheat,* grain sorghum
Yields/harvested ha:				
Wheat	q/ha	21.3	16.5	14.8
Barley	q/ha	21.0	...	...
Sorghum/flax	q/ha	.../...	.../5.1	23.0/...
Livestock raising	cattle/100 ha AL	6.0	8.5	20.5
Pasture requirement	ha/head	ca. 4.5	ca. 4.1	ca. 2.1
Labor expenditure	MH/ha AL	6.5	7.5	9.8
Working capital	$(DM)/ha AL[3]	325(1,190)	245(900)	425(1,560)
Gross cash returns	$(DM)/ha AL[3]	54(198)	44(162)	59(216)
Wage and capital expenditures	$(DM)/AL[3]	22(80)	18(68)	27(99)
Gross income	$(DM)/AL[3]	32(118)	25(94)	32(117)
Ditto	$(DM)/farmer[3]	20,470 (75,500)	9,650 (35,600)	11,200 (41,300)

[1] Averages for at least 80 to 100 farms.
[2] Remainder of AL consists of permanent pasture, in part wasteland.
[3] Computed on a rate of exchange of 1 U.S. $ = 3.6883 DM (N.Y. quotation for 1969).

Source: Farm Costs and Returns. Commercial farms by type, size, and location. Agricultural Information Bulletin. U.S.D.A. No. 230. Washington, D.C., 1969, pp. 16–18 (final edition).

Overview 14: Changes in Wheat-Fallow Rotations with Increasing Precipitation

A. Tunisia north of the line Hammamet − Le Kef [1]			
Precipitation in mm/yr.			
below 350	350 to 400	400 to 500	over 500
A.	B.	C.	D.
1. *Fallow*	1. *Fallow*	1. *Fallow*	1. Feed crops
2. Wheat or barley	2. Wheat	2. Wheat	2. Wheat
	3.–4. Rotation pasture	3. Barley, oats, feed crops	3. Legumes
	5. Wheat	4. Legumes (peas, garbanzos, field beans)	4. Barley, oats
		in % CL	
50	20	Fallow 25	–
–	40	Feed Crops 12	25
–	–	Legumes 25	25
50	40	Grains 38	50
B. State of Kansas, U.S.A., in a west-to-east direction [2]			
Precipitation in mm/yr.			
300 to 450	500 to 600	600 to 650	650 to 750
E.	F.	G.	H.
1. *Fallow*	1. *Fallow*	1. *Fallow*	1. Corn
2. W. wheat	2. W. wheat	2. Alfalfa	2. Sorghum
	3. W. wheat, sorghum	3. Corn	3. W. wheat
		4. W. wheat	
		in % CL	
50	33	Fallow 25	–
50	50	W. wheat 25	33
–	17	Other crops 50	67

[1] Whyte, R.O.: Milk Production in Developing Countries. London, 1967, pp. 134f.
[2] Personal data.

Although the wheat-fallow farms in the state of Kansas are indicated as more intensive than those of the Columbia Basin, they are still, with 9.8 MH/ha AL, extremely extensive. Yet this labor expenditure means that one operator with full work spacing and at 2,400 MH/year can farm 245 ha, five times more than is possible today in the FRG on farms specializing exclusively in combine-harvested crops, and this under the best of conditions. Among major farming systems, the grain-fallow economy is the most labor-extensive cropping system in the world. The heavy contribution of technology to this prominence is especially clear in the U.S. (highest stage of mechanization).

Moreover, the annual amount of precipitation in Kansas increases rapidly from west to east, so that the rotations change drastically, as outlined in Overview 14 (above). The proportion of CL in fallow, which is 50% along the agronomic dry boundary in eastern Colorado, first drops to 33% and then to 25%, and finally, with still higher rainfall, disappears completely. In the same direction, wheat monoculture gradually becomes a crop complex, as first the still very drought-tolerant grain sorghum comes in, and then, with higher precipitation, corn and finally alfalfa join the group.

Very different precipitation amounts also fall in Tunisia (Overview 14), diminishing from the Mediterranean coast to the Sahara. The crop rotations are comparable with those of Kansas, when allowances are made for the two years of rotation pasture that take over part of the functions of fallowing in example B and for the different companion crops of wheat because of differing climatic conditions. With less than 350 mm precipitation, bare fallow becomes necessary every second year. Beginning with about 400 mm, part of the fallow is cultivated with legumes for food, green manure, or feed, the most important crop being field beans, especially near Tunis and Bizerte (Jentzsch 1965, p. 55). At over 500 mm, fodder cropping gains increased acceptance. There are even more extensive forms of rotation than that of example A in the more arid central and southern part of the country. There two to three years of fallowing are needed to improve the water budget before grain can be sown. Wheat is then increasingly replaced by the less-demanding (shorter growing season) barley, which must be planted at the beginning of winter and harvested in May. The differences in yields between northern and southern Tunisia, as well as between modern and traditional agriculture, can amount to 50% or more (Jentzsch 1965, pp. 59 f.). On the average, only 2 q/ha of grain can be harvested with a seed expenditure of 60 kg/ha in the central and southern areas. This might even be unprofitable in the situation of a pure subsistence economy and thus takes us beyond the agronomic dry boundary as understood economically.

b) Millet-Sorghum-Peanut Farming

Wheat and barley are cultivated plants of the higher latitudes. They only rarely advance beyond the tropics of Cancer and Capricorn. Only in India have the two crops, above all, wheat, conquered large areas south of Cancer. In the tropical highlands, as in parts of East Africa, wheat can advance to the equator. In general, however, wheat and barley are principally represented poleward of the two tropics. The wheat-fallow economy is thus the typical dry farming system on the agronomic dry boundaries of the subtropics and middle latitudes.

Millet, sorghum, and peanuts have a greater ecological range and are far more characteristic of the lower latitudes. Though China and the United

States are, with their subtropical zones, among the six most important peanut producers of the world, the other four leading areas, India (especially the southern tip), Nigeria, Senegal, and central Brazil, are definitely tropical in character. A clear concentration of cultivation for sorghum and common millet is recognizable in the subtropical areas of the Americas, Asia, and Africa, yet the crops also play a great role in the outer tropics, including Nigeria, Upper Volta, Niger, Mali, and Sudan. If the drought-favoring millet-sorghum-peanut economy is almost never found along the agronomic dry boundaries in the subtropics but rather along those in the tropics, then this is not because the subtropics form an ecological barrier to these cultivated plants; instead it is because

- the wheat-fallow economy, for reasons already given, is only possible poleward of the tropics of Cancer and Capricorn and thus offers no competition in the outer tropics;
- the developing countries of the tropics often favor millet and sorghum over wheat as the basic food grain, whereas in strongly industrialized countries such as the U.S., Canada, South Africa, and Australia, it is just the reverse;
- millet, sorghum, and peanuts have generally been better suited to the subsistence farms that dominate in the tropics, and wheat and barley have been more adaptable to the cash crop farms that are already widely distributed in the subtropics (though sorghum and peanuts now form important centers in the U.S. too);
- the cost ratios between labor and machine capital and the farm size structure in the outer tropics have favored the millet-sorghum-peanut economy, which has been technically more difficult to mechanize, while in the subtropics these influences have favored the grain-fallow economy, which has been easier to mechanize.

Table 13 compares a small Senegalese peanut-millet-sorghum farm with a large Kenyan cattle ranch, both with the same amount of precipitation, 600 mm/year. It shows the typically drastic economic differences between ranching and dryland cropping:

1. Ranching calls for a higher proportion of its gross returns to be devoted to capital expenditures, thus it is more capital intensive.
2. Dry crop farming requires a labor intensity that is many times higher.
3. Ranching is clearly even more superior to dryland cropping in labor productivity (farm income in DM/MY) than dryland cropping is to ranching in land productivity (farm income in DM/ha AL).

The peanut-millet-sorghum farm in Senegal stands out with its high productivity, for despite the widely distributed hoe method of cultivation, the average worker toils long days (7 to 9 hours) during the cropping season. His approximately 460 MH/ha normally yields about 0.5 ha of peanuts and 0.5 ha of millet, sorghum, and barley (Ruthen-

2. Regions of Dryland Crop Farming

Table 13: Comparative Competitiveness of a Peanut-Millet-Sorghum Farm and Cattle Ranch (Calculations for a typical unit)

Indices	A. Cattle Ranch KENYA 600 mm Precipitation	B. Peanut-Millet-Sorghum Farm SENEGAL 600 mm
Production factors:		
Farm size, ha AL	32,000 [1]	7.9
MY/100 ha AL	0.16	38
CLLU/100 ha AL	22	ca. 28 [2]
Cropping:	No cropping of any kind	7.9 ha;
Crops in % CL, q/ha, and DM/q		Peanuts 38% CL, 7.5 q/ha, 32.0 DM/q; millet & sorghum 38% CL 5.0 q/ha, 36.0 DM/q
Fallow, % CL		24%
Livestock raising:		
Stocking, head	8,040	3
% in cows and heifers	30	100
% in young stock and feeders	69 [3]	
Price of oxen, liveweight, DM/q	85.0	ca. 70.0
% calving	80	60 to 70
Gain in liveweight, kg/ha	19	8 to 12
Farm profits:		
Gross returns, DM/farm	510,000	1,430
% in capital expenditures	25.5 [4]	10.7
Gross returns, DM/ha AL	16	181
DM/MY	10,000	476
Farm income, DM/ha AL	12	162
DM/MY	7,700	426

[1] 3.1 mill. DM fixed capital, of which 22% is in land, 17% in watering places and fences, and 48% in cattle.

[2] Pasture includes common pasture.

[3] Boran zebus, sales at 3.5–4 yrs., with a weight of 350 kg.

[4] Capital expenditures: labor expenditures = 1 : 1, farm capital bearing interest of 8%.

Source: Ruthenberg, H.: Beitrag im Handbuch der Landwirtschaft und Ernährung in den Entwicklungsländern, Vol. 1, Stuttgart, 1967, pp. 146 and 201.

berg 1967, p. 148). The various millet and sorghum types serve as food crops and peanuts provide the cash crop. Peanut raising in monoculture affects the structure and nutrient content of the soil unfavorably and can lead to a rapid spread of leaf-spot and leaf-mosaic diseases. Therefore it is recommended that peanuts be raised in a rotation (Franke 1975, Vol. I). Of the especially favorable rotation partners for peanuts, namely sorghum, corn, cotton, and various green manure crops, it is often only sorghum that can be used along the agronomic dry boundaries; in other cases, only fallowing is possible. Bally (Franke 1975, Vol. I) found in his work at the Bambey Experiment Station in Senegal that when peanuts followed sorghum instead of another peanut crop, yields increased by 57%, and that this figure rose to 67% when peanuts followed a fallow period, and increased to 104% when the crop came after a green manure crop. Thus the following peanut-millet-sorghum rotations have become common on the agronomic dry boundaries (Overview 15):

Overview 15: Peanut-Millet-Sorghum Rotations on the Agronomic Dry Boundaries

A Central African Republic	B Senegal	C India	D Nigeria	E Senegal
1. *Sesame* 2. Sorghum/*peanuts*-mixed planting 3. Sorghum 4.–20. Fallow	1. *Peanuts* 2. Millet 3. *Peanuts* 4. Millet 5. Millet 6.–10. Fallow	1. *Peanuts* 2. Fallow with green manuring 3. Millet/*cowpeas*-mixed planting	1. *Peanuts* 2. Millet 3. Millet	1. *Peanuts* 2. Millet
		in % CL		
Peanuts and sesame	8 20	34	33	50
Millet and sorghum	7 30	33	67	50
Fallow	85 50	33	–	–

——— increasing precipitation ——→
——— increasing scarcity of land ——→
——— increasing commercialization ——→

At the stage of pure hand labor, peanut raising is burdened with an extremely unfavorable work distribution since 70% of an already high labor expenditure of up to 130 MD/ha is allotted to the harvest. Thus the sooner rural underemployment permits reduction of the high labor peak, the easier it becomes to put a high share of the cultivated land into the crop.

c) Other Forms of Dryland Cropping

If the agronomic dry boundary is viewed in the strictest sense, then the forms of dryland cropping can be considered almost exhausted with the treatment of the grain-fallow and millet-sorghum-peanut systems. However, if the boundary is seen as a more or less broad — and this is advisable, then the picture becomes much more varied. Now crops behind the "boundary," those that are not quite so drought-resistant and short-lived as wheat, barley, millet, sorghum, and peanuts, can also be added to the farm program.

Increasing precipitation does not bring a change in the basic lead-crop elements of dryland crop farming, wheat and barley on the subtropical dry boundaries and millet, sorghum, and peanuts on the tropical dry boundaries. The cropping system, though, can be expanded with some companion crops. These provide more supplies for domestic use, encourage rotations, and help to spread risk and to space work. Sesame advances to the subtropical and tropical dry boundaries with a precipitation minimum of only 400 mm/yr. and a growing season of just three to four months. The plant has great value as a preparatory crop in both its minor soil demands and its improvement of conditions for the lead or "hub" crop. Its seeds are also rich in oil content (51%) and quality. Yet despite all this, the crop has attained a world hectarage that is only 30% of that planted to peanuts. On large farms, cultivation of the crop has been hindered by the difficulties of mechanizing the harvest and processing the product. On small farms, the problem is one of sesame cropping furnishing only about half as much of the land productivity of peanut raising. Sesame cropping often takes place in mixed planting. Characteristic partners are sorghums, millet, cotton, peas, corn, and dry rice (Franke 1975, Vol. I).

All forms of legumes are valuable near the agronomic dry boundary. They make a grain-dominant rotation more flexible, provide protein for human nourishment, and through their symbiosis with certain bacteria (bacterium radicicola) promote the fixation of nitrogen, which is so expensive in the developing countries. Garbanzos are distributed along the subtropical dry boundary and bush peas are found along the tropical dry boundary. Both have a short growing season and both are highly resistant to drought. Both are also good preparatory crops and furnish human nourishment, with garbanzos also providing a cash crop and bush peas supplying feed.

Roselle hemp and kenaf require a somewhat higher precipitation of 450 or 500 mm/yr. They are especially sensitive to temporary drought. Their fibers are suitable for the manufacture of a yarn that is similar to jute and can be used to make sacks and other woven articles. The Africans usually use only the bast and twist it into coarse, thick ropes for tethering domestic animals, securing grass mats to the roofs of huts, and the like. Both of these fiber plants thus serve almost exclusively home needs.

The sweet potato is cultivated when rainfall reaches about 500 mm. It is the most drought-resistant of the roots and tubers, and can get along if necessary with a short rainy period of only three to four months. Although its labor productivity is the smallest of the field crops that have been named so far, its land productivity is by far the highest. This makes it a natural subsistence crop for small family farms that suffer from food shortages but have a rich labor potential, so that the required 175 to 200 MD/ha can be satisfied. On somewhat larger farms, sweet potato cultivation diminishes, though it is still carried on in small areas to supplement the food supply and for as long as moisture permits.

Beginning with the 500 mm precipitation boundary, the range of potential cultivated plants greatly expands. With more than 500 mm moisture and a sufficiently long rainy season, two completely new cash crops appear, cotton and tobacco. They can be raised as a companion, or support crop or as a hub crop determining the orientation of farm operations. Both are also amenable in their cropping techniques to widely differing combinations of labor and capital, so that their production can be adapted to extremely constrasting farm sizes and stages of economic development. To this widespread economic distribution can be added a sizable ecological dispersion, which extends practically from the equator up into the subtropics. Overview 16 lists some cotton rotations (unirrigated).

Overview 16: Cotton Rotations in Tropical Rainfed Farming

A Gezira/Sudan	B Central African Republic — increasing precipitation →		C Uganda
1.–2. Fallow 3. *Cotton* 4. Fallow 5. Sorghum 6. Egyptian bean 7. Fallow 8. Cotton	1.–16. Bare and bush fallow 17. *Cotton* 18. Peanuts with corn and manioc 19.–20. Manioc		1.–2. Fallow 3. *Cotton* 4. Sorghum-millet- *cotton* 5. Manioc and sweet potatoes
	in % CL		
Fallow	50	80	40
Cotton	25	5	40
Grain	13	2	20
Roots and tubers	—	12	20

Small proportions of the cropped area are devoted to cotton even in the very dry areas, and indeed at one time the crop was largely centered in the semiarid tropics. It is raised for domestic needs or for cash on the distant markets, an option favored by the transport capability of the cotton fiber. In Mali and Niger the northern boundary of cotton raising runs closer to the Sahara than does the northern boundary of peanut cultivation.

3. Regions of Combined Extensive Grassland Farming and Dryland Crop Farming

a) Risk as a Hindrance to Diversified Production

Extensive grassland farming and dryland crop farming are in fact found only rarely in pure form on the agronomic dry boundary, and as a rule are combined. It is immediately evident that dryland cropping does not give ground in the direction of the agronomic dry boundary in abrupt fits and starts, but rather in a gradual and step-by-step form. The very last positions are also the very places where the last bit of cropland is stubbornly defended. This is because the dry zones are usually quite isolated from markets and the farmer therefore is highly reluctant to do without cropping for all his family needs. Also, the longer the dry season becomes on approach to the semideserts the more necessary it is to produce feed to carry livestock through times of distress. Subsistence cropping is also still worthwhile, despite its much smaller yields than those of commercial cropping, because the farmer saves on marketing costs. Nor do small returns prevent subsistence farming for food and feed from competing with systems depending on purchased supplementary supplies, since the latter are encumbered by high purchasing costs.

Similarly, looking in the reverse direction, i.e. from the natural grassland of the shrub savannas to the cropped areas in the dry savannas, one sees no abrupt withdrawal of grazing in favor of dryland crop farming. Rather, it is first a cautious penetration of small fields into the rangeland, on the order of probably 1% AL, and then expanding with increasing precipitation to 3, 8, 15, 30, and probably also 40 to 50% AL. Pure crop farms in the vicinity of the agronomic dry boundary are highly unusual.

b) The Example of Southwest Africa

What is at first a very timid establishment of some dryland crop farming is necessary because cropping near the dry boundaries is subject to high risk and thus can only acquire the status of a small subsidiary enterprise.

Grain corn yields in Southwest Africa can fluctuate between 0 and 45 q/ha. Table 14 shows the yield oscillations for some sample years, based on national averages. The influence of the two drought periods of 1943 to 1945 and 1957 to 1962 is clearly recognizable, for the years 1946 and 1960 show large fallow areas, low hectare-yields, and smaller proportions in ripened grain.

Table 14: Grain Corn Cultivation in Southwest Africa, 1921–1970

Year	Cultivated Land in 000s of ha			Corn in % of Cultivated Area	Corn Harvests	
	Cultivated Land	Fallow	Corn		% ripened	q/ha
1921	...	...	7.9	73.3	...	3.84
1946	...	...	13.9	81.1	62.3	1.08
1950	31.9	11.9	17.9	91.6	95.5	6.40
1955	39.8	11.8	24.4	88.5	90.1	5.94
1960	66.5	34.5	24.7	77.1	72.3	2.43
1970	...	...	32.0	39.0	...	3.80

Source: Bähr, J.: Kulturgeographische Wandlungen in der Farmzone Südwestafrikas (Bonner Geographische Abhandlungen, No. 40), Bonn, 1968, pp. 52 ff.

Today, in the great South African "Maize Triangle" of the Transvaal and Orange Free State, corn grain on the farms of whites does not become profitable until yields of 11 to 12 q/ha are reached. Thus, with the yield levels and fluctuations as shown in Table 14, a farmer whose main enterprise was corn cropping could in no way survive. Corn cultivation is of course justified to a minor extent in ranching operations, where the grain furnishes produce for the workers and provides feed. If the crop fails to mature, the cobs can be fed to cattle or hogs and, if necessary, preserved as silage for times of feed shortage. The Uitkom experimental farm at Grootfontein is trying to introduce suitable forms of such a complementary system.

There are grassland farming systems in the extreme northeastern part of the farming zone in Southwest Africa that use more than 200 ha of cropland. In 1960, farm operators generally devoting some land to cropping cultivated an average of 41.9 ha in the Gobabis District, 53.3 ha in the Otjiwarongo District, 56.6 ha in the Grootfontein District, and 61.3 ha in the Tsumeb District, the figures rising with improvement in precipitation. Yet even in the last two named districts in which 40% of the total cultivated land was located, the proportion of farmland in cropland came to only 0.8 or 0.9% (Bähr 1970, p. 53). Even in densely

3. Regions of Combined Extensive Grassland and Dryland Crop Farming

settled Ovamboland (South-West Africa), only 10 to 20% of the land is cultivated, while the remaining area is used for cattle grazing, the typical activity of the Bantu tribes. Many goats are also raised. Millet and to a degree sorghum, the principal food crops, almost completey dominate the cropland, with the rest in peanuts (protein supply) and a little corn (consumed fresh). The millet consists largely of pennisetum. The supply of protein is furnished primarily by cow's milk and secondarily by peanuts.

The main food supply comes from cropping. Millet-sorghum monoculture is widespread. Grass fallow is introduced at irregular intervals and for varying periods.

As precipitation in Ovamboland increases from 450 mm in the southwest to 600 mm in the northeast, the character of the landscape changes from one of bush to tree savanna. Cultivation, not even represented in the southwest, increases in a northeasterly direction. Distinctly cash crops, however, are also not to be found in the northeast. Instead, principal cash sources are the proceeds from the sale of wickerwork and especially the wages for contract labor for Europeans beyond the homelands.

c) Modifications by Economic and Ecological Variants

Though both are in the same areas, the northern and northeastern parts of Southwest Africa, native farmers emphasize millet and sorghum and European farmers concentrate on corn. The reasons for this are primarily the differences in farm size and pastureland-cropland ratios.

On European farms, and especially those in the Otavi Valley, cropland is devoted almost exclusively to corn, though it is less drought-resistant than millet or sorghum. This is because

- an especially efficient marketing organization exists for corn;
- corn is needed for food for the farmworkers; and
- the feed value of corn stover is higher than that of millet and sorghum straw, so that
- this aspect weighs especially heavily in the years when the grain harvest fails.

In Ovamboland, contrarily, millet and sorghum are heavily favored in areas with the same amount of rainfall because

- the soils here are lighter and more saline;
- millet and sorghum require less cultivation (hoe cultivation) and fertilization;
- the Ovambos prefer millet and sorghum as their basic foodstuffs, in which the brewing of beer plays a role;

- millet and sorghum can do better by themselves than corn, so that cropping can concentrate near the huts (village settlements); and
- cropping is more prominent on the Ovambo farms than on the European farms, and thus Ovambo operators cannot bear as high a risk with crop failure in grain farming as can European operators.

As to which of the various forms of dryland crop farming on the agronomic dry boundary combines with livestock raising into a single farming system, we can, ignoring special conditions, make the following generalizations about agricultural zones and operational motivations:

In the temperate or mid-latitude climates it is the w. wheat-fallow economy, or in higher latitudes (Dakotas, Canada), the s. wheat-s. barley-fallow economy, since millet, sorghum, and peanuts do not grow here.

In the subtropics with winter rain it is also the w. wheat-fallow economy, since the growing season falls in the cool winter, which is favorable for wheat but fails to satisfy the heat requirements of millet, sorghum, and peanuts.

In the dry savannas, which basically receive summer rain, peanuts, millet, and sorghum are areally suitable because they are very much heat-loving plants.

Small native farmers prefer peanuts, millet, and sorghum, even if the location is less favorable for them, for the simple production techniques, food habits (also beer), and the possibility of dealing with the high labor peaks for peanuts all speak for these cultivated plants.

European farmers, in contrast, tend more toward the grain-fallow economy, which can be more easily mechanized and thus less burdened with sharp labor peaks. That production techniques are more difficult does not matter, and wheat is a better cash crop.

With greater remoteness from markets, grain corn becomes the first field crop to penetrate the extensive grassland farming system. In market-oriented operations, increasing transport costs cause the agronomic dry boundary to shift into the more humid and prolific areas, so that corn becomes more competitive. This applies to all farm sizes, since corn is not only a good subsistence crop, but also a good commercial crop. This also holds for both the dry savannas and the wet-summer subtropics.

4. The Macrospatial Structure of a Dry Area: The Example of Australia

We shall conclude the chapter on the agricultural geography of the dry areas by describing the agrospatial structure of a large dry area, with Australia as the example.

4. The Macrospatial Structure of a Dry Area: The Example of Australia

a) Physicospatial Structure

For the illustration of the physicospatial structure we can turn to the world precipitation map (Fig. 5, p. 35) and the map sketch of the seasonal climates in the dry areas of the earth (Fig. 39, p. 161).

The Tropic of Capricorn passes through Australia, and around this axis are grouped the climatic zones. On both sides of the tropic can be found a large landmass of *deserts and semideserts* with only 130 to 250 mm annual precipitation. Even less than 130 mm fall in the center of this block, north of Adelaide. A large part of this section, especially the western part, permits no agriculture of any kind.

Bordering this desert block on all sides but the northwest is a *semiarid ring*. A steppe climate with wet winters and dry summers and about 250 to 500 mm annual precipitation locates on the subtropical southern side, while a steppe climate with a short summer rainy period and the same amount of rainfall marks the southeastern side. Shrub and dry savannas with similar precipitation amounts characterize the northern arc.

Only the outer ring on the coasts of the country is strongly differentiated. On the southwestern tip and also between Adelaide and Melbourne we find a *Mediterranean climate,* with approximately 510 to 760 mm annual precipitation, which is concentrated in the winter. The southeast coastal strip, in contrast, is distinguished by a *subtropical wet-summer climate* and, directly on the Pacific, even by a *constantly humid subtropical climate.* In the latter case, precipitation increases from 760 to 1,270 mm, as in the vicinity of Sydney or Brisbane. Humid savannas with the same precipitation span are found on the tropical north coast, and regionally they are more extensive. Annual precipitation at Darwin reaches 1,780 mm, and at Townsville it goes even beyond that.

The agricultural effects of a decline from at least 760 mm precipitation on the northern and eastern coasts to less than 130 mm in the interior of this continent are heightened by a related spatial trend, an increase in variation of annual precipitation (mean deviation) of from 15 to 25% to over 40% of the average.

b) Agricultural Zones

What agricultural zones have now been formed in these physical areas, and why? To begin with, there are, according to Fig. 46, extensive parts of the interior of the country that lie beyond the area of human habitation and permit no agriculture whatsoever.

1. Proceeding outward to the boundary of livestock grazing, the first activity to be encountered in wide areas is *extensive sheep grazing.* It claims a semicircular belt that is interrupted only on the southern

coast, extending from northwestern Queensland and through New South Wales and the southeastern part of South Australia to the western part of Western Australia. This large sheep grazing belt can again be subdivided into three zones (Dahlke 1973, pp. 403ff.):

a) A zone of *pure sheep grazing* forms the innermost ring, has the greatest extent, and allows no rainfed farming of any kind. Semi-arid or even arid climates produce such a scanty vegetation of grasses and bushes, and walking distances to watering places are so great, that cattle raising usually remains excluded and sheep raising must assume its least demanding form, raising animals only for wool. This form also copes best with the great distances to market, for only the excellent transport and storage qualities of a product like wool can overcome the problem. Since a profitable farm operation requires from 5,000 to 10,00 sheep and the carrying capacity of

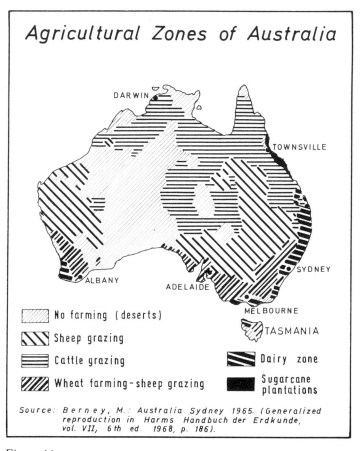

Figure 46

4. The Macrospatial Structure of a Dry Area: The Example of Australia

the pasture varies anywhere from 3 to 30 ha per sheep, depending on the density and type of vegetation, farm sizes are large. They range from 15,000 up to 300,000 ha and in some places as high as 400,000 ha. It is to be expected that the boundary of livestock raising will retreat from the desert to the same extent that labor productivity must increase. Today there are still regions used in which 40 ha of range is needed for one sheep. We estimate 30 ha to be marginal. With continued rapid development of Australia, the limit will soon become 20 ha per sheep, or in other words, the unpopulated center of Australia will expand.

b) Next to the inner ring is a more humid, *sheep grazing-wheat belt.* Wheat-fallow farming is established on smaller areas. Although the system is associated with heavy risk, it does provide many an advantage, not only in providing sheep with stubble and pasture fallow, but in spreading risk and spracing work. And since range vegetation is now more plentiful, sheep can be raised for both wool and meat. The combined effect of all these conditions is to make possible a typical farm flock of only 1,000 to 5,000 head.

c) The coastal termination of the sheep grazing belt is identified by a combination of *sheep raising with cattle raising.* The normal flock size comes up to only 500 to 2,000 sheep, for precipitation has increased from 500 to 1,250 mm per year. The double use of the sheep (wool and meat) is now so pronounced that lambs are fattened for the market.

2. Just as the desert area of interior Australia is bordered on the south by a belt of extensive sheep grazing, so is it bordered on the north by a zone of *extensive cattle grazing.* Although the cow is superior to the sheep in competition for use of the range in the greater part of the world, the competitive position of sheep raising in Australia is far stronger. This is due to, among other things, the fact that

— wool is more capable of transport than beef, which for a very isolated continent with sizable agricultural surpluses is of great significance. Of the total current production, some 35% of the beef, 33% of the mutton, and 84% of the wool is exported, mainly to Europe;

— the dry areas of the continental interior have so low a carrying capacity, so few watering places, and so much open range, that they place a premium on the characteristics in which the sheep is superior to the cow: ranging capability, herding instinct, and thirst tolerance; and

— there are dryland pastures in which no competition exists, because if they could not be used by sheep they would become wasteland.

The result is that extensive cattle raising can only become important in regions unsuitable for sheep raising. This is the case in the coastal strip of the tropical north. The high-grass savannas are not suitable for sheep raising because of their low nutrient value, and because the sheep are difficult to control here. The dingoes (wild dogs) endanger the sheep here in much the same way that jackals do in Africa.

A cardinal problem of the Australian ranch (called "station" in Australia) is one of supplying the market with an extremely wide-meshed transport net. Usually extensive distances must be negotiated in cattle drives before the animals can be loaded for shipment or slaughtered. After slaughter, export needs require that the meat be preserved and transport weight reduced as much as possible. Sixty percent of the total beef export is frozen and the rest canned.

3. The more intensive *wheat farming-sheep grazing system* forms the third category of drought-resisting farming types in Australian agricultural space. It outlines a small, elongated zone in the winter-rain area of the southern continent, lying approximately between the isohyets of 250 and 635 mm. A typical wheat farmer farms about 400 ha, depends on his own or an expanded family labor supply, sows about 12 q/ha of grain in the fall or early winter, and harvests it in the early summer. In the more humid areas of the wheat belt, the operator has broken away from the dry farming system of earlier periods and has inserted drought-resistant clover varieties between for sheep raising, protection against erosion, and improved soil fertility. The following are typical rotations (Dahlke 1973, pp. 198f.):

A.	B.	C.
1.–2. Clover	1.–3. Clover	1.–4. Clover
3. Wheat	4. Wheat	5. Wheat
4. Wheat	5. Wheat	6. Wheat.

The farmer generally cultivates wheat two years in a row and then follows it with two, three, or four years of clover pasture so that the clover takes up 50, 60, or 67% CL. Clover can be left longer on the lighter soils than on the heavier types. Rotation B has the best prospects for a wide distribution.

4. The other agricultural zones of Australia lie on the wetter east coast, which has at the least 760 mm annual precipitation and is the major populated area of the country. A *dairy zone* has developed in the southern, subtropical coastal strip to supply the large cities, and sugarcane is raised in the northern and tropical coastal belt. But discussion of these and other agricultural zones exceeds the limits of this chapter on the agricultural geography of the dry areas.

c) Water Reclamation Projects

As in all dry areas, the expansion of the irrigated area plays a special role. Also, as everywhere else, efforts are concentrated not so much in the regions where precipitation is the lowest and water could be used to the maximum, but rather in the areas where irrigation water can be obtained most easily and cheaply.

Of the numerous water reclamation projects in Australia, only the most important and also boldest one will be mentioned here: the "Snowy Mountains Scheme," currently in construction. It will help to stimulate development in what is already the economic heart of the continent, the Sydney-Adelaide-Melbourne triangle of the southeastern region. Water will be conducted from the 1,000 to 1,500 m-high Snowy Mountains into the Murray River and its tributaries, which drain an area of 100 million ha, or four times the size of the FRG. This will be accomplished with a complex of 17 dams, 160 km of aqueducts, seven hydroelectric power plants, and one pumping station. The rivers will then receive an additional volume of 235 million m^3. This will permit an expansion of the irrigated area in Australia, now up to about 1.69 million ha, by 0.25 million ha. About three quarters of the irrigated land in Australia will then be in the drainage area of the Murray River (Schendel 1971, pp. 8 f.). The heavy losses of water through evaporation from the reservoirs will be reduced by 30% by covering the surface with a film of acetyl alcohol. This in turn will increase the irrigation capacity of the project.

VIII. The Agricultural Geography of the Middle Latitudes

To the two large climatic realms of world agriculture already described, the humid tropics and the dry areas, we can now add a third, that of the middle latitudes. Fig. 30, p. 133, shows that intensive cropping concentrates in the mid-latitude climates, and Fig. 47 makes it clear that the regional differentiation of world nutrition is one that favors the higher latitudes.

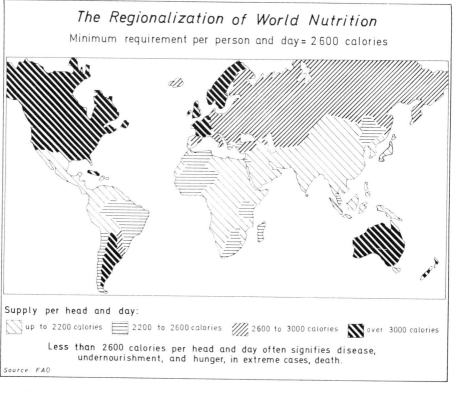

Figure 47

However, one should not conclude from these two facts that the mid-latitude climates favor agriculture more than do the humid tropics. Rather, it is the much higher stages of development of the national economies of Europe and North America that stimulate agricultural production and the food economy. The agricultural regions of the higher latitudes, which we shall discuss now, are at such a high development stage in spite of, rather than because of the mid-latitude climates. They are thus conditioned far more economically than ecologically.

No attempt will be made here to present a complete agricultural geography of the middle latitudes, so that the most important and instructive agricultural areas with mid-latitude climates can be treated more thoroughly. They are the area of the European Common Market of the Nine, North America, and the East Bloc.

1. The Agricultural Geography of Western Europe

The following attempt to delimit farming types and structural zones in European agriculture and to interpret their spatial picture lies in the boundary area between agricultural economics and agricultural geography.

Since there is no economic enterprise that is areally bound to such a high degree as agriculture, an attempt should first be made to give the place-specific characteristics of European agriculture. After that, a systematic representation of farms, which takes into account the objectives of scientific regional analysis, can be developed. Only then is the way made clear for considering in turn the spatial patterns of systems of crop rotation, land use, livestock raising, and farming. In concluding, we shall at least allude to some of the problems of competition among agricultural regions stemming from Europe's internal boundaries.

Although the Common Market of the Nine is the object of our regional investigation, and the statistical and cartographical information is restricted to it, examples will now and then be drawn from neighboring regions when this is factually possible and usefull for understanding relationships.

a) Place-Specific Characteristics

On a macrospatial scale, climate remains the most important place-specific factor for agricultural production. Five major climatic zones can be distinguished in Europe (see Fig. 48):

1. *Temperate summer, cold winter:* The largest part of Scandinavia belongs to this climatic belt. The mean annual temperature in southern Sweden is +6°C. It lowers toward the north, falling to -2°C in Lapland.
2. *Temperate summer and winter:* Here one can speak of a Central European transitional climate. Hoe cropping and winter-grain cropping play a significant role.

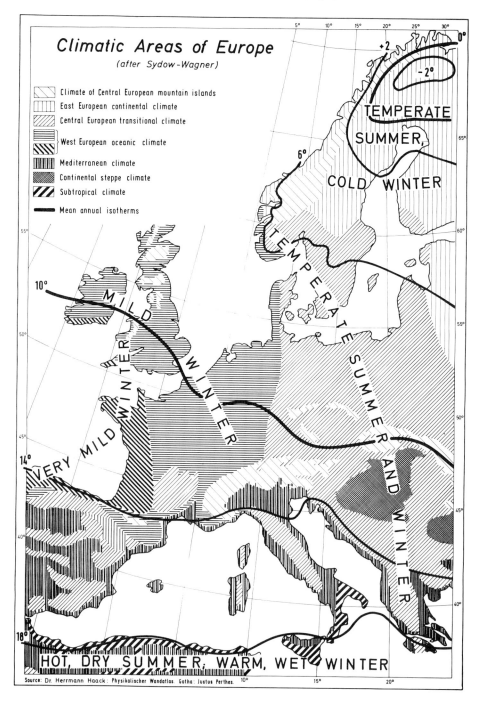

Figure 48

3. *Mild winter, cool summer:* This climate is typical for Great Britain and a large part of France. Fodder growth is vigorous and consequently so is the development of cattle raising.
4. *Very mild winters and warm summers* dominate the regions around the Bay of Biscay, i.e. northern Spain and southwestern France.
5. *Hot, dry summers and warm, wet winters* distinguish the Mediterranean area. These conditions become more emphatic the farther one proceeds to the south, until finally agriculture is possible only during the winter rainy season and the summer heat and drought are hostile to all cropping. The mean annual temperature ranges from 14° to 18°C.

The coldest climate is that of the high mountain chains of Europe, especially in Norway and the Alps. Fodder cropping clearly reigns in this *mountain climate*. Less cold is the *East European continental climate*. Then comes the *Central European transitional climate*, in which hoe cropping and winter-grain cropping play a significant role. It is the major zone of hoe crop-grain crop farming. The *West European oceanic climate* is distinguished by its very mild winters, which favor fodder cropping and permit almost year-around grazing. While mowing of meadows characterizes fodder cropping in the highlands, grazing of pasture prevails here. The *Mediterranean climate* is marked by mild winters and hot, dry summers, which provide strong support for bush and tree crops. Finally, the *subtropical climate* of southern Italy makes it possible to raise, among other things, fruits, olives, almonds, and cotton.

To delineate farming types it is important to know the duration of a temperature of at least 5°C, since most cultivated plants begin to assimilate nourishment at this level (Fig. 49). Phenological dates are also important, e.g. the beginning of the winter wheat harvest. Near its northern cropping boundary wheat cannot be harvested until August, whereas in Italy all of the crop must be brought in by the end of May.

Fig. 7, p. 49, shows the *northern cropping boundaries* of important cultivated plants. Potatoes are cultivated farthest north, even beyond the northern boundary of Sweden. Cropping of summer barley stops a little farther south, strains of which in their quadrilocular forms can mature in 60 to 65 days. Also important for cropping is the northern boundary of winter wheat. It runs a little north of the Stockholm-Oslo line and extends still farther north in the warm Gulf Stream climate of western Norway, a phenomenon also to be observed in the cropping boundaries of spring barley and potatoes. The northern cropping boundary of the sugar beet coincides approximately with the northern boundary of the Swedish East Göta Plain.

Then comes a broad belt running through Central Europe, in which all the important cultivated plants of the mid-latitude climates, with the exception of specialty crops and decidedly heat-loving crop types, can be raised.

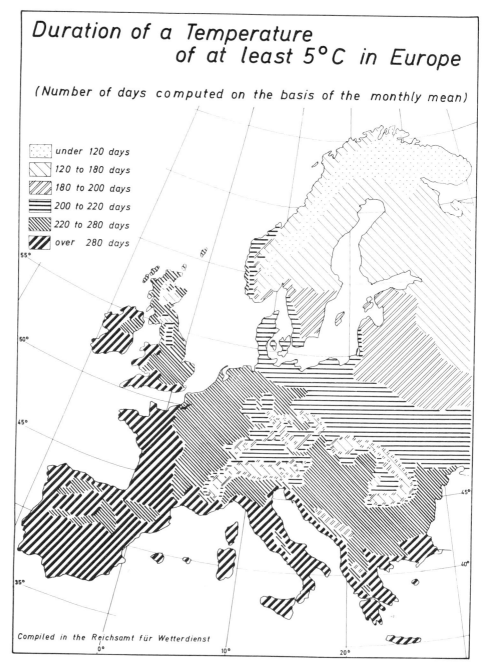

Figure 49

The cropping boundary of grapevines is to be sought approximately along a line running from the mouth of the Loire to the Mosel, after which it drops southward in Hesse and Bavaria. The cropping of grain corn has become possible much farther north than was the case earlier, when earlier-ripening hybrid strains still had not been developed. Already the Rhine-Main area must be considered capable today of producing grain corn, and noteworthy proportions of land devoted to the crop can be found on scattered farms almost up to the Mittelland Canal.

The cropping boundary of rice runs still farther south. We have, of course, only two significant rice cropping areas in Europe, one in the Rhone Delta and the other on the Po Plain in Lombardy. The boundary of the olive tree is important in the southern European countries. In Italy it practically coincides with the northern boundaries of Liguria, Tuscany, and Umbria. Farther south is to be found the cropping boundary of citrus fruits. The only areas in Italy favorable for citrus crops are Sicily, Apulia, Calabria, and a strip on the west coast extending up to the latitude of Rome.

Mean annual precipitation in Europe is highly variable. In the horizontal dimension, it will generally increase the closer we come to the Atlantic coastal area. In the vertical dimension, it normally increases with altitude. However, for the farmer the amount in itself is not decisive; seasonal distribution is also important. Uniform distribution over the entire year, which is especially favorable for fodder growth, is found in the maritime area. The continental climate, in contrast, is partially handicapped by pre-summer drought. Under such conditions fodder cropping cannot claim any sizable areas, and in grain cropping as much emphasis as possible must be put on types that ripen early and use the winter moisture efficiently, as is true of our winter grain types.

The Mediterranean climate is characterized by a dry summer. Only 86 mm of rain falls in northern Apulia in the June-to-August quarter, while on the coast of the Ionian Sea the amount is but 71 mm. Here annual cropping must either be restricted to the winter months or supported by irrigation. However, the farmer in the Mediterranean countries adjusts to the pronounced summer drought in still other ways, namely with bush and tree crops. With their deep root system, they can utilize the winter moisture and groundwater better than annual crops can, and consequently are able to surmount the many dry months without much damage.

Amount and seasonal distribution of precipitation still do not completely account for the formation of farming types. Both must be seen in relation to temperature conditions. With high temperatures, a relatively rich amount of precipitation is taken up by the atmosphere relatively quickly through intense evaporation; the cultivated plants thus benefit from only a small part of the moisture. Conversely, with lower temperatures and thus less evaporation, a smaller amount of precipitation can penetrate deeper into the soil and there become available for plants. Seen in this relationship, abundant precipitation in warm areas gains quite a different aspect. Dry periods occur there in spite of the precipitation. But if, as in the Mediterranean region, the summer months have relatively meager rainfall with high temperatures, the result is a pronounced drought for several months, a special climatic feature of this portion of the EC area.

Thus, when the efficiency of precipitation is considered, we get a completely different picture from the one presented when precipitation amount is considered in isolation. The first one shows, for example, that in the Atlantic coastal area, with its relatively low temperatures, high humidity, and much of its precipitation in the form of fog and drizzle, even relatively small amounts of precipitation are sufficient to give fodder cropping the upper hand. In the Allgäu, on the other hand, a much greater amount of precipitation has to fall to ensure the predominance of fodder cropping, especially since rain here falls in shorter periods and more torrential form, humidity is lower, and solar radiation is stronger. Finally, in the Mediterranean area 1,000 mm of precipitation still does not mean much, when one considers the hot climate, the very dry air, and the unfavorable seasonal distribution of this precipitation.

What, now, is the situation for the *soils of Europe?* If the mountain soils are left out of consideration, then we have the following picture (Fig. 50). Great Britain and the Baltic Sea coastal area have many loam soils. South of that is a broad zone represented by sandy soils, and it embraces Jutland, the Geest region of Schleswig-Holstein and Lower Saxony, the province of Brandenburg, and large parts of Poland. South of that is the Central European loess belt, which thrusts between the northern sandy soils and the southern sub-Alpine mountains. It begins in the Paris Basin, extends over Flanders and the Cologne-Aachen Bay, then narrows in the Westphalian Hellweg region. Continuing east, it widens again in the Hildesheim-Brunswick and Magedburg portions of the fertile plain of the *Börde* and then almost disappears, only to become prominent again on the Wroctaw (Breslau) Platform, after which it finally expands into Eastern Europe as a very broad zone in the Ukraine. South of this loess zone, a great variety of soil types prevail. A high proportion of the Mediterranean countries is in clay soils. Other place-specific factors of a physical nature that are important here include relief, exposure, and slope. All are especially important for highland agriculture. An effect of the physical location factors is reflected in the grassland-farmland ratio. Fig. 51 shows to what extent permanent grassland in Europe is distributed. The Atlantic coastal area is especially prominent in grassland, though of course still surpassed by the Alpine area. Permanent grassland in most parts of Europe is physically deter-

1. The Agricultural Geography of Western Europe

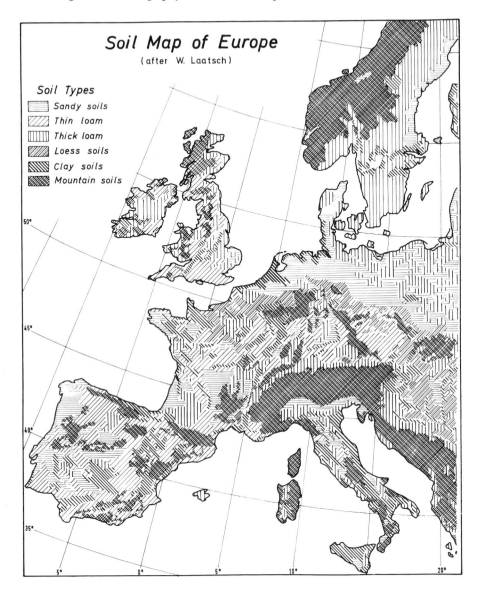

Figure 50

mined and little capable of conversion to another land use. Only in Great Britain and France can a part of the permanent grasslands be explained by economic reasons.

Generally one can say that for modern agriculture a smaller amount of land in permanent grassland is more favorable. In most parts of Europe a larger proportion is a strain on farm management. Roughage can be

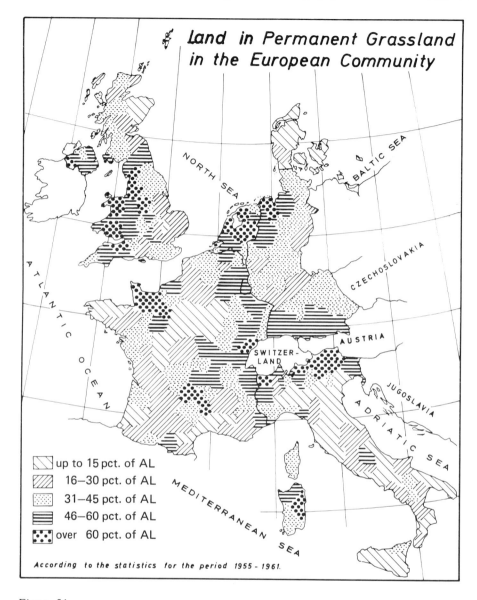

Figure 51

produced much more efficiently through cultivation, because then fodder cropping can better support commercial farming with its crops that improve the soil and require little fertilizer. The areas with little permanent grassland, such as Denmark, Sweden, and parts of Italy, can avail themselves of the great advantages of cultivating crops for roughages: better soil conditions preceding the crop; rich supplies of root humus, which allow the farmer to keep the manuring operation within bounds; higher productivity of fodder cropping; and so forth.

Next to be characterized are the *economic place-specific factors* for European agriculture. If Europe did not have such a strongly varied relief, one would be able to recognize intensity rings in European agriculture. They would lie concentrically around the large industrialized area bounded approximately by the four cities, Manchester, Brunswick, Karlsruhe, and Paris. Here lies the center of gravity for industrial production and the great population concentration that is the principal customer for European agriculture. It is next to this area that, so far as economic place-specific factors are concerned, highly intensive farming must develop. However, the farther the agricultural zones are from this population concentration, the more agriculture is burdened with production and transport costs and the more extensive the types of farming become. In this connection, the high density of the transport net marks another economic prerequisite for agricultural production in Europe.

Also important is the distribution of farm sizes. The smallest farms are in Norway and in a broad zone that extends through southwestern Germany to the Riviera. Areas with the largest farms are most extensive in Spain and Italy, followed by smaller regions in central France, and then eastern England. Between these two large zones lie areas with medium to large family farms (see Fig. 52).

The distribution of farm sizes cannot be explained by inheritance customs alone. Yet it can be observed that the areas with predominantly divided inheritance are northern France, the Riviera, Italy, and southwestern Germany, while those with a prevailing undivided inheritance are southern France, Bavaria, Westphalia, Lower Saxony, Schleswig-Holstein, Denmark, Scandinavia, and Great Britain.

Another product of inheritance customs is the average size of farm parcel. Areas with predominantly divided inheritance generally have only small parcels, while in areas with mostly undivided inheritance the individual farm parcels are larger. The importance of parcel size in the era of technology is well established, particularly in areas with divided inheritance, which because of the smaller parcels are at a great disadvantage for mechanization.

The size of the labor force in European agriculture is a consequence of all the factors of the agricultural structure that have just been noted, but primarily of farm size and also parcel size. The Federal Republic, Italy, and the Benelux countries especially stand out as labor-intensive areas. In the Federal Republic and Benelux countries, the high labor intensity may be attributed most immediately to large livestock numbers and then to extensive hoe cropping. In Italy, contrarily, it is more the large share of bush and tree crops that determine the large work force. Southern Italy also has rural overpopulation, which is rooted, on the one

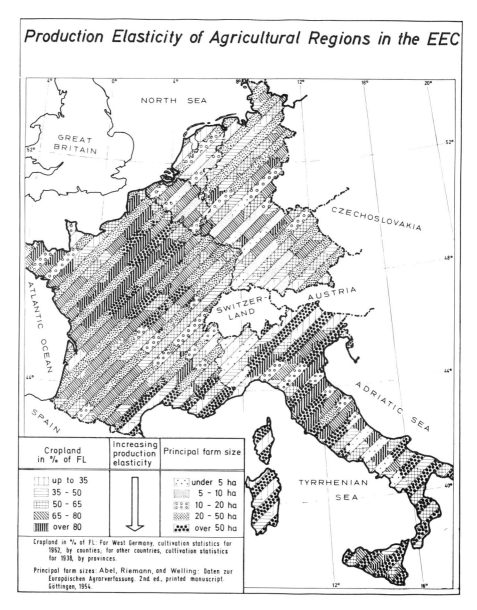

Figure 52

hand, in a latifundia system that is still not completely eliminated, and on the other, in a lack of industrial potential. But the size of the labor force per 100 ha FL cannot always be viewed as an expression of farming types. The reverse relationship is perhaps stronger, in that where a large rural labor force exists because of small family farms, there intensive farming types can be expected to achieve a breakthrough. As everywhere in agriculture, relations here are also reciprocal.

b) Methodology of Delimitation and Statistical Derivation of Agricultural Regions

The possibility of delimiting and mapping agricultural regions, and thereby facilitating spatial investigation by means of the map model, was provided with the introduction of newer methods in regional agricultural statistics in the late 1870s. In Germany, Engelbrecht (1930) was the first to recognize the possibility of depicting cartographically the regional distribution and concentration of individual crops. Thus he provided, as a major founder of agricultural geography, the stimulus for an approach whose foundation up to today is still not fully laid. The crop atlases of Engelbrecht were of course still one-dimensional. If the maps were to represent more fully the agricultural conditions of an area, they would have to be able to show complex phenomena, or systems. For that purpose, the delimitation of land use systems based on crop ratios offered especially good possibilites, since the necessary statistics were available for the small administrative units and could be systematically processed for larger areas with a reasonable expenditure. This led to the development of the cartographic representation of land use systems, in which Busch (1936) and Woermann (1943) were especially active.

European farming systems thus have been defined for a long time by land use systems. These were distinguished by designating the lead crop and first companion crop. At first these two crops were simply determined by their share of the land. Only a well-read work by Brinkmann (1930, pp. 559 ff.) sparked interest in a new and better delimitation method. Brinkmann determined the lead and companion crops by the product of the areal share in % AL and a crop-specific weighted index number. This weighted index number was then used to compure the farm enterprise weight of the individual crop-type groups. Busch (1936) adopted this method and produced a weighted-index system that was appropriate to Central Europe at the time. He then proceeded to delimit the land use systems as follows (Overview 17):

The most important crop types were first combined according to their relationships in the farm operation into a few major groups — fodder, grain, and hoe cropping. The crop data for these principal groups were computed in % AL. These data were then multiplied by the weighted index numbers, which take into account the varying farm management weights of the individual crop groups. The product of the crop share of AL times the weighted index number yielded the crop weight. Fodder cropping, grain cropping, and hoe cropping were now ranked by their

Overview 17: Determination of Land Use Systems with the Help of Weighted Index Numbers (Example)

Crop Area in ha	% AL x	Weighted Index Number =	Crop Weight
Fodder cropping (F) 20 ha permanent grassland	40	0.5	20
Hoe cropping (H) 6 ha potatoes 4 ha sugar beets	12 8	2.0 2.5	24 ⎱ 44 20 ⎰
Grain cropping (G) 10 ha w. rye 10 ha s. grain	40	1.0	40
Total 50 ha AL	100%		H–G

crop weight, and the land use system was classified with the designation of the lead and first companion crop (e.g. H–G = hoe and grain crop farming). In this manner, economic enterprises and areas with related farm management structures were combined into groups.

This method of delimiting land use systems has long been the dominant one. Various authors (Rolfes, Woermann, and Blohm, among others) have basically taken over the Busch method, though they have partially refined the discriminations by modifying the weights in the index numbers. The resulting more sharply differentiated graduations of the index number values have thus made possible a more exact delimitation of the land use systems. However, the refining of the method also increased the work necessary for determining the systems, so that macrospatial analysis using fine areal detail became increasingly difficult.

The statistical determination of land use systems became still more laborious with the application of a method devised by Hoffmann (1954, pp. 263 ff.) which involved multiplying the crop area data for the crop groups by weighted index numbers based not on expenditures, but on yields. These weights were obtained by averaging yields over several years, and were then made comparable by converting the values to grain units.

In point of fact, standardized weighted expenditure indices have lost their value in our age of technology, since every farm enterprise today can be encumbered with very different amounts and composites of expenditures depending on the stage of technological development. On the other hand, the use of weighted index numbers for yields requires so much work that they are hardly suitable for macrospatial investigations, since statistics for sufficiently small administrative units or regions must now be prepared for both land use and harvesting.

1. The Agricultural Geography of Western Europe

None of the methods so far described is adequate for the regional analysis that is planned here for the agricultural areas of the European Community of the Nine. The functional relations between land use and livestock raising have become so relaxed, that only a very incomplete picture of the livestock enterprise can be obtained from the land use system. Thus as a criterion of the farming system (the objective of all the regionalization methods), the land use system is in most cases no longer capable of being more than a partial and only indirect indicator. While the necessity of using weighted index numbers is undisputed, all of them up to now have more or less failed to provide a universally verifiable scale unit. In the following presentation, weighted index numbers based on the total hand labor expenditures for crop and livestock enterprises are offered as the most characteristic expenditures factor in the agriculture of the industrial states. Now livestock systems as well as land use systems are distinguished (Andreae 1964). Thus the procedures for delimiting land use and livestock raising systems have been selected so as to be fully compatible and consequently now capable of defining farming systems. Eight farming enterprises have been selected as the most important for inclusion in the computations:

Farming Enterprise	Symbol	Weighted Index Number
Fodder cropping	F	1
Grain cropping	G	1
Hoe cropping	H	5
Specialty crops	B	20
Dairy cows	D	3
Young cattle	Y	1
Small ruminants	R	0.2
Swine	S	0.3

As for the rest of the method, Overview 18 shows the stages in which the land use system, the livestock raising system, and finally, as their summation, the farming system, are determined. First, the crop weights of the four land use enterprises were determined, with the result indicating a pure hoe cropping system. Then the livestock weights for the four livestock enterprises were computed in the same manner, the result denoting a swine-sheep raising system. Finally, the farming system was arrived at, now taking into consideration all eight farming enterprises. Hoe cropping achieved the highest weight in this farm operation and consequently was listed as the leading farm enterprise. Swine raising attained the second highest weight and thus became the companion enterprise. The third highest weight was for sheep raising, which also was listed as a supplementary enterprise. The farming system was therefore characterized as a hoe crop-swine-sheep raising operation (H–S–R).

The coordination of the individual procedures is thus complete, since both land use and livestock enterprises are evaluated on the basis of a

Overview 18: The Method of Determining the Farming System (Example Farm)

Land Use Enterprise	Symbol	Cultivated Area in % AL	Weighted Index Number	Crop Weight	System Designation [3]
Fodder cropping	F	5.7	1	5.7	
Grain cropping	G	53.0	1	53.0	
Hoe cropping	H	40.0	5	200.0	
Specialty crops	B[1]	1.3	20	26.0	
Total crop weight				284.7	
Land Use System:	Pure hoe cropping system				H
Livestock Enterprise	Symbol	Head per 100 ha AL	Weighted Index Number	Livestock Weight	System Designation [3]
Dairy cows	D	9.1	3	27.3	
Young cattle	Y	4.0	1	4.0	
Sheep and goats	R[2]	296.0	0.2	59.2	
Swine	S	287.0	0.3	86.1	
Total livestock weight				176.6	
Livestock Raising System:	Swine-sheep raising system				S–R
Farming weight				461.3	
Farming System	Hoe crop-swine-sheep raising farm				H–S–R[3]

[1] B = bush and tree crops.
[2] R = small ruminants.
[3] The first letter indicates the leading farm enterprise, the second the companion enterprise, and the third the supplementary enterprise. No companion and supplementary enterprises are named if the lead enterprise attains at least two thirds of the total crop, livestock, or farming weight.

weighted index of 1 being equal to 45–50 MH per 100 ha AL. Also, no companion and supplementary farm enterprises were named if the lead enterprise attained at least two thirds of the total crop, livestock, of farming weight.

c) Crop Rotation Regions

The overview of farming types in the agricultural area of the European Community of the Nine will first be restricted to just cropping, and later will be gradually expanded to include more complex phenomena. The scope of cropping is shown by Fig. 53. The two most important European Community of the Nine will first be restricted to just cropping, and later will be gradually expanded to include more complex phenom-

1. The Agricultural Geography of Western Europe 211

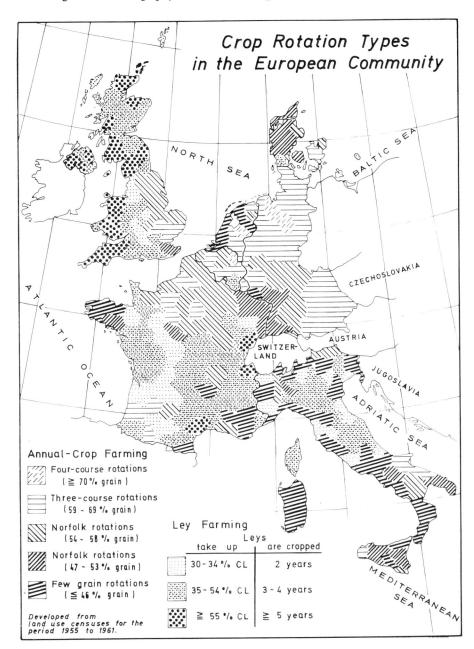

Figure 53

ena. The scope of cropping is shown by Fig. 53. The two most important European rotation categories are ley farming and annual-crop farming. The value chosen for separating the two is 25% CL in fodder cropping. Rotation systems that include this proportion or more in fodder cropping feature leys with a useful economic life of at least two years and crops that conserve and enrich the soil and save labor. All of these characteristics make up the essence of ley farming types and differentiate them from types of annual-crop farming.

The regions of ley farming are ranked by the number of years a ley is cropped after it is sown. Regions of annual-crop farming, thus rotations with crops harvested only once and of no more than a year's duration, are ordered by the ratio of grain crops to leafy crops. The class graduations signify a decrease in grain cropping and an increase in cultivation of the leafy crops, and with that, an increasing realization of the principles of crop rotations. Ley farming regions stand out in Fig. 53 where

— the growing season is short (northern Europe);
— the precipitation is high (mountain areas, especially the Alpine area and the mountains in western Great Britain);
— the precipitation, though not very high, is well distributed seasonally and accompanied by high humidity (Great Britain and the Atlantic coastal area of France); and
— the precipitation, though not very high, is supplemented by irrigation (Po Plain).

Northern Europe shows very clearly that as the growing season becomes shorter with approach to the Arctic Circle, the cropping period for leys lengthen and the regions of ley farming consequently assume an increasingly extensive character. A useful economic life of only two years for leys is typical of Jutland and Scania, while in central Sweden it is already three years. With advances into still more northerly latitudes periods lengthen to four, five, six, and more years.

In *Great Britain*, ley farming exhibits an east-west as well as north-south differentiation. The north-south contrast is caused by the shortening of the growing season in the direction of Scotland, whereas the east-west differentiation results from the gradual rise of the terrain from the east coast to the west, up to the mountain areas of Cornwall, Wales, and Cumberland.

Somewhat unexpected is the marked prominence of ley farming in *France*. The reason, besides the moist Atlantic climate and the highlands of the French Central Plateau, may be the relatively extensive type of agriculture that characterizes almost all French agricultural zones.

Especially extreme forms of ley farming may appear in the *Alpine area* because here high precipitation and a short growing season converge. Since the place-specific factors are numerous (altitude, slope, exposure, etc.) and vary sharply within small areas, a minutely compartmentalized and variegated picture of widely contrasting agricultural regions is typical.

Ley farming based on irrigation is represented on the *Po Plain*. The warm climate, combined with a rich supply of water flowing from the Alps, has allowed the development of an extraordinarily productive ley system. The many cuttings make possible a large investment in livestock, the majority of which are kept in the stall the year around. This agricultural region is the most important dairy area of Italy.

Two large regions stand as *symbols of annual-crop farming*. Fig. 53 clearly shows the influence of a changing grassland-farmland ratio *north of the Alps*. In general, it can be said that areas with a large proportion of land in permanent grassland are characterized by rotations poorer in leafy crops. Thus the great grassland girdle of the North Sea coastal area stands out like the pre-Alpine areas with a heavy emphasis on grain cropping.

Even more influential is the size of farm. The West German states (*Länder*) with relatively large farms, such as Schleswig-Holstein, Lower Saxony, Westphalia, and Bavaria, are characterized by four- and three-course rotations of annual crops. In contrast, in the federal states settled by small farmers, such as Hesse and Baden-Württemberg, Norfolk rotations or their related forms are still practiced to some extent.

Fig. 53 shows the *South European region* of annual-crop farming as having overall a large proportion in leafy crops. Usually at least the stage of Norfolk rotations, and often even that of practically no grain rotations, is attained. The reasons, among others, are:

— The correlation between grassland share and crop rotation, so obvious in Central Europe, is less clear in Southern Europe. This is because the mild climate provides a long, in part year-around grazing season, so that a larger grassland proportion and more livestock do not necessarily require that a large amount of land be given to grain for supplying feed and straw.

— Also to be kept in mind is that the hot, dry climate allows barn manure to have only a minor effect. Grain cropping with an emphasis on manuring is typical only of locations in mid-latitude climates.

— In addition, since there is only a minor operational need for grain hectarage, more land can be given over to the cultivation of leafy crops, here in the form of grain corn. Corn is an ideal leafy crop for warm climatic areas. Also, all of its cropping procedures can be mechanized for maximum effect, and thus it does not have nearly the labor-intensive character of the most important representatives of leafy crop cultivation in Central Europe, sugar beets and potatoes.

Thus countries and regions with climates suitable for grain corn, when faced with rising wage levels and increasing pressures for mechanization, do not depend as heavily on rotations emphasizing small grains as does Central Europe. The Klagenfurt Basin and Krappfeld areas of Carinthia are outstanding examples of this.

If one looks at the crop rotation regions of Europe as a whole and views them in strongly simplified outline, three large rotation regions can be seen to stand out:

– *A region with heavy emphasis on fodder cropping.* It is characterized by ley farming and it embraces, besides Scandinavia, the entire Atlantic coastal area and the mountain regions.
– *A region with heavy emphasis on grain cropping.* It is characterized by four- and three-course rotations and it embraces the entire block of Central and Eastern Europe.
– *A region with heavy emphasis on leafy crops.* It lies in the Mediterranean area and comprises the countries with a climate suitable for grain corn, such as Spain, southern Italy, and Jugoslavia.

d) Land Use Regions

Fig. 53, showing crop rotation regions, can only offer insights into the organization of crop farming. If one wishes to obtain an overview of the organization of the entire agricultural area, permanent grassland and bush and tree crops must also be considered. Land use systems must be delimited, as is done in Fig. 54.

The methodology for obtaining this more comprehensive view was illustrated above (Chapter VIII-1-b). First all land use enterprises were assigned to one of four major groups (fodder cropping, grain cropping, hoe cropping, specialty crops). Then the share of each of these major groups in % AL was determined, using mean data for administrative units. These figures were then multiplied by a weighted index number based on normal work procedures. The product of crop share and index number yielded the crop weight. The crop group with the highest crop weight represents the lead crop, or principal cropping emphasis, and the second highest represents the companion crop, or secondary cropping emphasis. Lead and companion crops give the land use system its name. If one of the four crop groups obtains two thirds or more of the total crop weight, the land use system is identified only by the lead crop.

Fig. 54 allows us to recognize in macrospatial form a threefold division of land use regions in Europe:

1. In *Northern Europe* we find *a region of fodder cropping farms.* It is supplemented by the greater part of Great Britain, which in the Atlantic coastal areas is especially favorable for fodder growth, and the

1. The Agricultural Geography of Western Europe

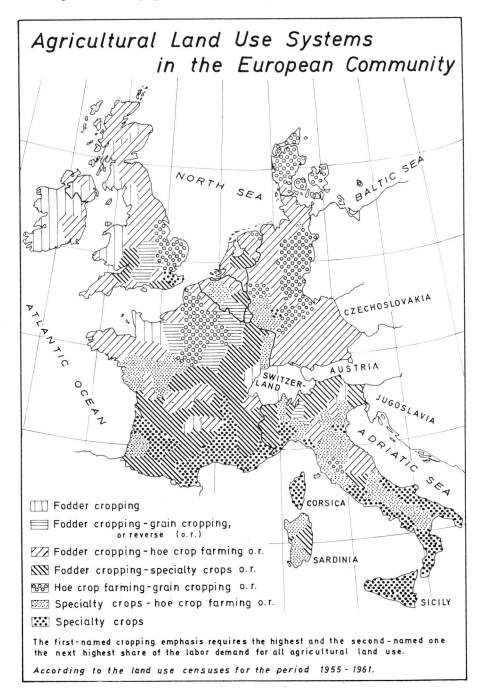

Figure 54

European highlands, especially the Alpine area, but also such places as the French Central Plateau and the higher locations in the Black Forest. Grain cropping occurs as a companion enterprise in fodder cropping with a shorter growing season or on larger farms, or in both situations (Central England). In contrast, with a longer growing season or smaller farms, or both situations, hoe cropping becomes the secondary operation (Norway, Black Forest). Finally, if the growing season is still longer and the terrain rugged, fodder cropping can appear in combination form with specialty crops (central France).

2. In *Central Europe* we have *a region of hoe cropping farms.* The greater parts of northeastern France, western Germany, and Poland stand as a symbol of hoe-grain crop farming. Much less common is the combination of the system with a secondary emphasis on fodder cropping (eastern Scotland, Brittany), and even more rarely, on bush and tree crops.

3. *Southern Europe,* finally, stands as *a symbol of specialty crop farms.* Wherever a warm, hot climate has favored bush and tree crops, there the specialty crop system with its many variants has arisen. Fodder cropping appears as a secondary enterprise in rugged mountain areas, while on the dry plains the role is taken over by grain cropping and in Italy by hoe cropping.

If we compare this picture of land use regions with the maps depicting the place-specific conditions for Europe, the following can be said:

— There is a clear relationship between land use systems and the *temperature and the length of the growing season.* Fodder cropping systems predominate where the growing season is short and the climate cold. The group of hoe cropping systems begin to appear as the growing season becomes longer and the climate more moderate. In a very warm and hot climate, it is the group of specialty cropping systems that dominates.

— Land use systems are also clearly related to the *amount and seasonal distribution of precipitation.* Fodder cropping systems are definitely prominent in a very moist climate, as are hoe cropping systems in a moderately moist climate and specialty cropping systems in a dry climate. The distribution of precipitation also exerts an influence on the shaping of land use systems. Fodder cropping needs uniformly distributed precipitation. Hoe cropping also requires a fairly balanced precipitation level during the growing season and in our middle latitudes needs an especially good amount of moisture in July and August. Where the precipitation maximum is earlier and there is a pronounced spring drought, grain cropping succeeds as the lead enterprise. If the precipitation distribution becomes even more unfavorable and the climate has a decidedly dry summer, as in the Mediterranean area, bush and tree crops gain ground. Their deep root systems enable them to make much better use of the groundwater than do the annual crops, and for that reason they can better survive longer drought periods.

Fodder cropping puts its stamp on *the most extensive land use regions.* The distributional structure of fodder cropping regions is appropriately the reflection of the climatic realms, for climate, from a macrospatial standpoint, is the underlying initiator of fodder cropping in general and its modifications in particular. Three types of fodder cropping regions distinguished by climate appear in Europe. The *polar fodder cropping regions* are found in Scandinavia. Fodder cropping comes increasingly to the fore as one goes poleward, and the reasons for this are the shortening of the growing season and the unusual light and heat conditions in the realm of the midnight sun. These circumstances overwhelmingly favor fodder cropping. It becomes increasingly competitive with commercial cropping toward the pole, until finally it cimpletely displaces it.

The *maritime fodder cropping regions* embrace those of the northwest German and Dutch coastal areas, Great Britain, and the region around Cherbourg. Here it is less the amount of precipitation and more its good distribution and high humidity that favors fodder cropping. The maritime climate is better suited to grazing pasture than moweing meadow.

This is particularly the case where the winters are mild enough to allow almost year-around grazing, while on the other hand the colder winters of the German marsh belt cause a feed shortage. Types of livestock farming that make possible a seasonal adjustment of herd size to a drop in feed supplies (livestock raising and summer feeding farms) are then regarded as desirable.

A short growing season and abundant moisture combine to give their stamp to the *montane fodder cropping regions.* It stands to reason that fodder cropping will assume extreme forms in precisely the higher mountains. The largest single montane fodder cropping region is the Alpine area, including the foothills. However, fodder cropping appears at specific altitudes in almost all European highland areas, such as the Black Forest, Vosges, Hohe Venn, Sauerland, Ardennes, and French Jura and Central Plateau. The mountain climate favors pasture mowing over grazing. This has its advantages for management because the winters are long.

The farming enterprise of combined fodder cropping and livestock raising is, apart from the raising of certain specialty crops, the only one in European agriculture that permits monoculture. Today one of the most important and extreme forms of specialization taking place on fodder cropping farms of up to 30 ha is the abandoning of the remaining crop areas and a shifting to pure grassland farming. The examples in the Dutch grassland marshes or in the Württemberg Algäu show that the pure grassland farm can be operated more profitably than the mixed farm. If despite this advantage the shift to pure grassland farming is taking place only slowly, this is because of the sizable capital investments required. Even if the intensification of farming practices is now economically attractive, it also requires more livestock and more room for stalls, hay, and silage. Investment needs are also increased considerably when the grassland areas are enlarged as well as farmed more intensively.

Grain cropping farms also feature locational types that differ in cause, characteristics, and problems. The dominant grain cropping enterprise can be coupled to a companion enterprise of fodder cropping or hoe cropping, whereas it is only seldom combined with specialty crops. In the *Mediterranean countries,* a pronounced summer drought leads to the superiority of grain cropping on the large farms. So long as irrigation is impossible, preference must be given to crops that can adapt to this precipitation distribution by making good use of the winter moisture and ripening soon after the onset of the dry summer. These are the earlier ripening grain types, especially the winter grains.

Motivations are quite different in the *soil-determined grain cropping economies.* In the fertile lands of the "grain *börde"* it has been the very special loess soil with its even deposition and weak admixtures of loam that have strongly encouraged grain cropping. The physical characteristics of this soil are much more favorable for grain production than for hoe crops. The heart of this grain *börde* lies in the Westphalian border area surrounding Lippstadt, Soest, and Warburg. A northern outlier runs from Warburg into the Hanoverian Kalenberg country, a southern one into the Hessian loess zone. Everywhere in these areas, grain cropping has been widely distributed for decades.

Similar farming structures are found today on the *young cultivated marshes.* Grain yields of 40 to 55 q/ha and rape yields of 25 to 35 q/ha now make it impossible for hoe cropping, already handicapped by the fall dampness, to compete. Here it is a matter of operators of *land-rich or wage-labor family farms* being able, or even having to abandon hoe cropping and livestock raising in favor solely of crops that can be harvested by combine, thus crops whose labor productivity (net) can hardly be surpassed. Fehmarn Island is an example of where grain and rape cultivation today is strongly emphasized and uniformly developed.

The need to improve the labor setup by shifting to less intensive land use is also one that faces operators of land-rich and wage-labor family farms on other soil types, as it does all over Europe. Grain cropping is especially favored where the labor capacity of the farm is extremely limited, for it requires only minor labor expenditures because of the nature of the crops and is the most capable of being mechanized by present technological standards.

Because of the influence of place-specific motives on grain farming, which have already been noted, grain cropping farms are distributed in more isolated than agglomerated fashion, and only rarely in the form of compact regions. It is certain, though, that mechanization in the future will strengthen even more the trend to grain farming, for grain cultivation can be mechanized today far more efficiently than hoe cropping or the combined enterprise of fodder cropping-livestock raising. Fully mechanized grain cropping now requires only 20 to 25 MH/ha and has become almost impervious to wage labor costs. This contrasts with the at least 100 to 110 MH/ha needed for raising seed potatoes and the 55 to 63 MH/ha required for sugar beets. A further stimulus to grain cultivation is that a three- to four-year succession of the crop is no longer as disadvantageous as it was still two decades ago, thanks to greater efficiency and more chemical fertilization, therapeutic plant protection, and chemical weed control.

The *group of hoe cropping systems* can be subdivided two ways, by the extent of hoe cropping and by the principal hoe crop type (sugar beets, potatoes, field vegetables, and the like). Prerequisites for the development of hoe cropping farms are:
— a soil favorable for hoe cropping;
— a long growing season, which facilitates work spacing;
— only a small part of the land in grassland;
— a favorable market location for hoe crops;
— a rational distribution of fields; and
— farm sizes suitable for the most intensive cultivation, thus land-poor family farms or large scale wage-labor farms.

Hoe crop-grain crop farming is a very adaptable and well balanced system that is extensively distributed in the mid-latitude climatic zone and can still be encountered on almost all types of soil, so long as they are not too heavy. It can be found on both the sandy soils of the North German Plain and the intermediate soils of the broad loess belt deposited on the edge of the Central German Highlands, on what are often quite heavy loam soils in Hesse and Franconia, and finally even on the clay soils of the Dutch province of Zeeland and of Brittany.

In contrast, pure hoe cropping, that is with at least 35% FL in hoe crops, has much more sharply fixed soil demands. Practically no other soils but of the intermediate kind allow it to satisfy the following prerequisites:
— Several types of hoe crops must be able to be raised, since the reciprocal relationships in the system complex make it almost impossible for just one crop type to attain 35% FL. As a result,
— the soil must be heavy enough to be able to support sugar beets, but
— at the same time not so heavy that it hinders profitable potato cultivation.

With these requirements, the soil conditions for the pure hoe cropping system are already quite closely determined. These conditions are given on the darker and loamier soils of the central loess belt in the Hildesheim-Brunswick *börde,* the Cologne-Aachen Bay, parts of Flanders and Brabant, the Rhine-Main area, and the Rhine Plain in Baden, as well as on the portions of recent diluvium in the eastern parts of North Hanover around Burgdorf, Uelzen, and Lüneburg. Pure hoe cropping is more independent of soils where field vegetables are raised in the vicinity of large markets, such as West Berlin, the Vierland area near Hamburg, the *Vorgebirge,* or hill lands around Bonn, and next to the Naples-Salerno population agglomeration. Grain cropping offers only small resistance to the penetration of these locations by hoe cropping. Grain and hay harvests are so plentiful that the grain hectarage required for internal economies is at a minimum.

The developmental tendencies of hoe cropping are not the same for the two farm size classes for which it is suited. Operators on *large scale wage-labor farms,* which are extremely sensitive to wage labor costs, are now trying to parry rapidly rising wages with mechanization and other technologies. That is why they tend to restrict hoe cropping in favor of enterprises that can be more easily mechanized, the raising of grains and oil crops. In the warmer climates this tendency is not discernible because there hoe crops are available that can be fully mechanized at all stages of work. In the U.S., soybeans, peanuts, and cotton have made it possible to develop farming systems that put an extraordinary emphasis on hoe crops and yet save on labor.

In contrast, operators on the *land-poor family farms* remain true to hoe cropping in spite of mechanization difficulties, for they must use their rich and readily available family labor supply that would otherwise be unexploited and increase by all means available their gross income with the help of higher gross returns. However, in climatic zones that are warmer, sunnier, and free of late frost, the same need applies to the rainsing of *bush and tree crops of a specialty type,* which are generally adopted where a suitable climate combines with a lack of land and a high rural settlement density.

Specialty crop farms are even more climatically conditioned than fodder cropping farms, and even more in need of full labor utilization and a production increase than hoe cropping farms. The map shows them to be in southern Italy, in Sicily and Corsica, on the Riviera, in Provence and Languedoc and far into the Rhone Valley, also in the Garonne, in the Upper Rhine and Mosel valleys, as well as other places. Still this map does not do justice to the distribution of specialty crop farms, since bush and tree crops frequently do double duty in cultivated fields and in meadows. One need only think of the extensive areas of the Po Plain, which in the statistics appear as cultivated land but actually comprise grain fields interrupted by rows of fruit trees, among which are strung a third crop, grapevines.

1. The Agricultural Geography of Western Europe

Bush and tree crops of a specialty type appear in Western Europe in the forms of:

1. fruit growing with modest heat demands (*Altes Land* near Hamburg, Lake Constance area);
2. viticulture;
3. olive cultivation; and
4. fruit growing with exacting heat demands (citrus fruit, figs, almonds, cherries, apricots).

Climatic demands increase from 1 to 4. Grapevines already need a mean annual temperature of at least +9.5°C and come into conflict with hoe cropping in the dry-summer Mediterranean climate. Olives and citrus fruit are confined to Spain and central and southern Italy. Monoculture occurs in specialty crop farming just as it does on fodder crop farms. This is especially true for viticulture, which can employ family labor productively throughout the year. Citrus fruit raising also shows a strong tendency toward monoculture. In Europe, though, little use is made of the specialized farm. The standard is the mixed farm, since it has the advantages of reducing risk, providing fertilizer balance, and spacing work.

The question now is what crop groups can specialty crops associate with in the farm unit. Naturally either feed or grain crops must be preeminent, for the cultivation of specialty crops requires large amounts of manure, which in turn demand companion crops of land use enterprises that supply it. There is also the fact, however, that hoe cropping does not fit in with specialty crop farming in the economic aspects of labor, since both create high labor peaks in the fall. Hoe cropping thus competes with the specialty crops for the labor and manuring capacity of the farm, so that as a rule specialty crop farms must do without a large quota of hoe crops.

Fig. 54 shows that actually specialty crops are usually combined with fodder cropping in all mountainous areas. In the mountain lands, the different climatic requirements of both crop groups can be satisfied. Let one recall, for example, the steep slopes of the Mosel or Ruwer valleys, which are covered with vines that find ideal environmental conditions with southern exposure. The valley bottoms and the summits of the Eifel and Hunsrück, in contrast, have a larger share of fodder crops. In this way it is possible to combine manure-producing fodder crops with manure-demanding specialty crops. This is also a successful solution from the standpoint of labor demand, since viticulture requires predominantly summer work while livestock raising requires more winter work.

But where there is only level land and thus little climatic contrast, and where the climate moreover is dry, bush and tree crop farming combines with a larger share of grain cropping. Grain cropping also fits in quite well with bush and tree crops from the standpoint of labor spacing. It also has a complementary relationship with respect to manuring needs, for it furnishes raw material for humus in the form of straw.

Finally, we also come upon a combination of bush and tree crops with more emphatic hoe cropping. That such farming types can occur, though we have earlier established that hoe crops and specialty crops compete for the labor and manuring capacity of the farm, can be explained as follows:

- Farm size favors this combination in the *Upper Rhine Valley*. The small peasant farmers must increase their organizational intensity by all available means. The family labor situation makes it possible to deal with the fall labor peaks, which are caused by the grape and hoe crop harvests.
- The explanation for the specialty crop-hoe crop farms in *Italy* is quite different. The useful effect of barn manure in the Mediterranean and subtropical climate is only minor. At the same time, though, production costs for barn manure are high when year-around grazing is possible, in that animals must be stabled and fodder crops harvested simply for the sake of producing barn manure. Farming based on manure production is therefore largely dispensed with.

For these reasons, there is little competition between specialty and hoe crops in the subtropics for manure. As to the need for work spacing, we must bear in mind that hoe cropping in the Mediterranean area is mostly a winter operation. But if this is the case, the hoe crop and specialty crop harvests no longer make heavy demands on labor during the same period. The orange harvest in Sicily lasts from November to April and does not interrupt the later potato and sugar beet harvests. Thus with the shift of hoe crops to the winter, and consequently their harvests to the spring, the labor-demand relationships between specialty crops and hoe crops are turned around. Hoe cropping now not only ceases to compete with specialty crop farming, it complements it. This is why the combination of hoe cropping and specialty crop farming prevails so widely in Italy, whereas it is hardly imaginable for farm management in most of the middle latitudes. Thus we have another illustration of the fact that there are simply no models in agriculture that have validity everywhere and at all times.

e) Livestock Regions

Next we shall show, with the aid of Fig. 55, the regional structure and zonal arrangement of the livestock raising systems in the European Community. The delimitation of the system was made in the same way as was done for the land use systems (Ch. VIII-1-b). The four most important livestock enterprises selected were those emphasizing dairy cows, young cattle, small ruminants, and swine. Poultry were not considered.

Only a brief look at Fig. 55 is enough to convince one of the outstanding importance of milk production in European livestock raising. *Pure*

1. The Agricultural Geography of Western Europe

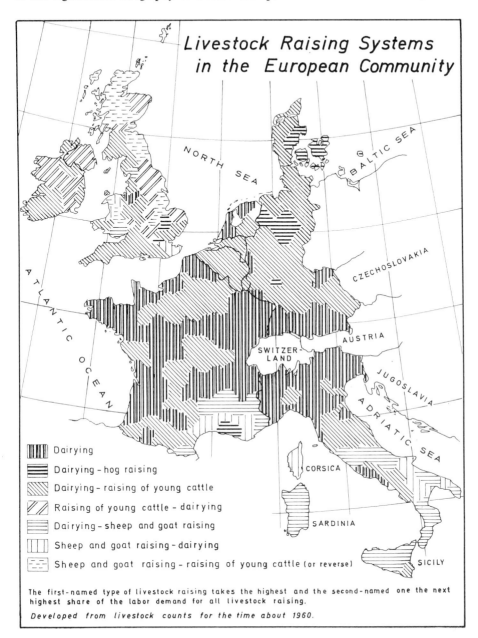

Figure 55

dairying, i.e. a system in which dairying makes up at least two thirds of the total livestock weight, is found only in more northerly latitudes, beginning with the Po Plain and Gascony. Only in these areas are the amount and seasonal distribution of precipitation favorable enough so that dairy farming systems, so highly dependent on seasonal feed balance, can strongly dominate. But, for dairy farming systems to flourish, it is not only necessary to have a good feed base and favorable labor situation, but also to have all other enterprises with ruminants impeded in their development. That is why dairying systems occur more frequently

- in *fodder cropping* than in other land use systems, since this system usually provides no appropriate feed sources for small ruminants. This is true, of course, only for the plains;
- on *family farms* than on large scale wage-labor farms, since the larger units put more emphasis on cattle raising and fattening to save on labor;
- on *plains* than in mountains, since mountainous areas may provide a summer feed surplus for raising young cattle (Norway, Alps) and are often so afflicted by drought, sterile soils, steep slopes, and rocky surfaces that they are fit only for grazing by small ruminants (Scotland, Sardinia, Apennines); and
- in the *East Bloc countries* than in the Western countries, since the economic situation for food production in the East Bloc countries favors the more efficient nutrient transformer, the dairy cow, over the far less efficient transforming agent, the feeder animal. A dairy cow with five years of useful economic life and 4,500 kg annual milk production converts 100 kg of plant protein into 38.9 kg of animal protein, whereas a feeder animal (young feeder with 500 kg final weight) yields only 12.7 kg animal protein. For carbohydrates and fats, the comparative yield figures are 25.2 versus 5.2%. Thus only the wealthier peoples can raise the nutrient-expensive feeder on a large scale.

Regions of combined dairying and raising of young cattle occur where the above-named qualifications for dairying are not quite so favorable. The raising of young cattle acquires the character of a companion livestock enterprise, either through a considerable emphasis on breeding (large parts of West Germany) or through provision of a greater role for fattening (England, France). In the latter case it may even become the lead enterprise, which as a rule requires that pure cattle fattening farms process more beef animal types (eastern England, Ireland).

Fig. 56 shows more clearly the variations that the dairy cow-young cattle combination is capable of, and how the individual cattle farming types are associated with the agricultural zones for the Common Market of the Nine (see also Fig. 57). Less common is the *association of dairying with sheep and goat raising.* These combinations are found

1. The Agricultural Geography of Western Europe

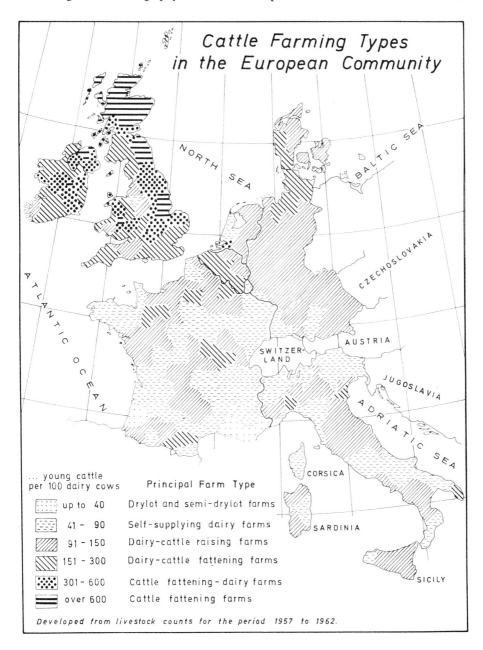

Figure 56

— *in the mountains of western Norway,* where the sterile, steep, and rocky surfaces are fit for no other agricultural use but sheep and goat grazing, and where the milking of goats for human use is a feature of the many small part-time farms; and

— *in the mountainous parts of the Mediterranean area* (see Fig. 55), where besides the already-noted inducements for the raising of small ruminants, there is the long summer drought that calls for ruminants that can range widely enough to obtain adequate feed, and in more critical situations, survive on brush (goats). Small ruminants also can, because of their great talent for climbing, find nourishment in high mountainous areas that are less affected by drought but cannot be reached by cattle.

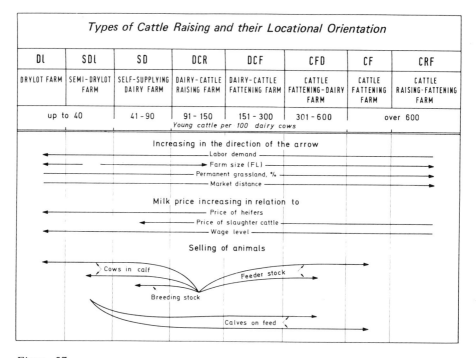

Figure 57

Where conditions for dairying are still relatively favorable, sheep and goat raising occur only as a companion enterprise (Norway, Abruzzi, Sicily). In contrast, where the place-specific factors are especially extreme, sheep and goat raising become such powerful competitors of dairying for the most efficient use of feed that they become the lead livestock enterprise. This reduction of dairying, from its normally dominating position almost everywhere in Europe to a companion role, occurs most prominently along parts of the French Mediterranean coast and in Corsica, Sardinia, and the southern Italian massif (Fig. 55).

In other regions, dairying finally drops out of the system classification altogether, as the raising of young cattle now become the principal companion livestock enterprise. This can be explained by severe summer drought or high mountain locations, combined with a variety of economic inducements for cattle feeding (Wales, Scotland). *Systems combining the raising of sheep, goats, and young cattle* become the accepted livestock raising systems.

Hog raising usually occurs as a companion enterprise of dairying and here serves as, among other things, a superior processor of skim milk. In Europe, four regions of hog raising with four different feed bases stand out:

— the *acorn hog-fattening area* of the Iberian Peninsula;
— the *grain corn hog-fattening area* in Hungary, Vojvodina, and the Banat (Mangalitza breed);
— the *potato hog-fattening area* embracing the light soils of the North European Plain (including Farther Pomerania), where in the industrial countries for the last two decades farmers have been rapidly switching over to grain feeding; and
— the old *barley hog-fattening area,* with the agricultural-export country of Denmark as the prototype. Barley feeding also dominates in West-Germany, but is being increasingly supplemented by corn.

Although hog raising is frequently a lead livestock enterprise on individual farms, it is seldom important enough to characterize large regional units.

f) Complex Agricultural Regions

The methods of delimiting land use and livestock raising systems (Figs. 54 and 55) have been selected so that systems are now statistically compatible and therefore can be combined to represent the total, or farming system. Eight farm enterprises (four land use and four livestock enterprises) and a farming weight, calculated by summing the total crop and livestock weights, are the principal elements in the computation of the farming system. Farming systems can then be characterized by the three most important farm enterprises, the lead, companion, and supplementary enterprises. They are the farm enterprises that claim the highest, second highest, and third highest share of the labor demand of the farm and thus are particularly prominent in the farming weight.

Theoretically, there are no less than 520 farming systems possible using the classification applied here. The outward manifestations of the agricultural operations on our continent are thus so diverse and complex that only a multi-patterned map (and then only in simplified form) makes it possible to obtain an overview of the geographic distribution and quantitative significance of the farming systems in the European area. There now only remains the task of combining the representations in Figs. 54 and 55 into regional models of farming systems.

In light of the farm size structure and high degree of industrialization in Europe, it is only natural that intensive farm interprises leave their stamp on family farms almost everywhere. They are predominantly:
- dairying in the oceanic climate of the northwestern part of the continent;
- hoe cropping in the transitional climate of Central Europe; and
- bush and tree crop farming in the Mediterranean climate of Southern Europe.

Our computations have supported this threefold regional subdivision. In one region the lead enterprise is dairying, in the second it is hoe cropping, and in the third it is the raising of bush and tree crops. It is also clear that these large agricultural zones are associated with the major climatic areas. Thirdly, we can now observe that the farming systems change from northern to southern Europe in such a way that a dominating secondary production gradually gives way to a dominating and finally exclusive primary production. This is illustrated with the aid of Fig. 58.

This map shows the total farming weights broken down into their crop and livestock components. It gives information on the stages of *labor intensity in land use*. One can observe that labor intensity in primary production in Europe increases progressively from north to south. Labor intensity in land use must be increased to the same extent that fodder cropping is replaced by hoe cropping, and it, in turn, by the cultivation of bush and tree crops. That the map does not show these relations very clearly on a large areal scale can be explained in part by the highly varying farm sizes within small areas, a common situation for Europe.

Further, one can see that the gradation of *labor intensity in livestock raising* is completely different. In secondary production, the labor intensity of the Northwest European countries with their heavy emphasis on dairying and hog raising puts them at the top of scale, whereas that of the Mediterranean area with its minor livestock role puts it at the lower end. In Northwestern Europe, labor on the farm is used mainly for livestock raising, whereas production stressing fodder cropping requires less work. In contrast, in Southern Europe the labor potential of the farm is used chiefly for specialty crops, while livestock raising is given little space. The overall labor intensity of the farm thus presents a more balanced picture than the gradations revealed by the isolated examinations of primary production on the one hand and secondary production on the other.

1. The Agricultural Geography of Western Europe

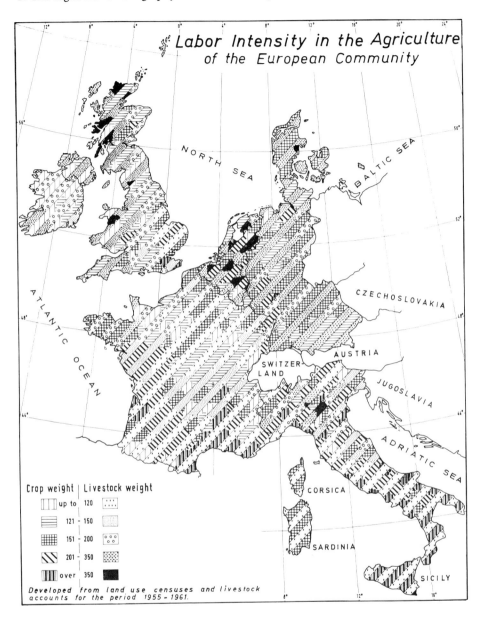

Figure 58

In a large free-trading area, every production orientation has, according to the iron laws of spatial economic equilibrium, the tendency to move to the locations where its products can be most cheaply produced and brought to consumers. The spatial arrangement of the production enterprises of agriculture then becomes the result of competition. Where goods can flow freely across political boundaries, this competition is not so much that of overall national agricultural economies as it is of similar farming systems that compete across frontiers. This is especially the case in the vicinities of the internal boundaries of the European Community. Thus, for example, the fodder cropping farmers of the German North Sea coastal areas, who cannot do away with extensive and large scale stall feeding, have come into competition with the fodder cropping farmers of the French Atlantic coast, who with year-around grazing can produce much more cheaply. Farmers in the Cologne-Aachen Bay and certain Belgian districts compete to supply the sugar refineries in the region. Also drawn into competition are the vintners of the Rhine, Mosel, Saar, and Ruwer areas with those in the climatically favored. French vineyards. Operators in the fruit farming regions in the *Altes Land* or next to Lake Constance must defend themselves from the competition of those in the climatically-favored South Tyrol, not to mention the growing challenge, in turn, to all the Central European fruit farmers by the producers of tropical and subtropical fruits. Potato cropping farmers of the Ems country and the Münster Bay contest the potato farmer of the Dutch *Geest;* vegetable growers in the Dutch province of Zeeland struggle with those of the Bonn *Vorgebirge* to satisfy consumer demand in the Rhine industrial area. French and West German farmers raising slaughter cattle compete for the Italian market. All of these examples are but a few of those illustrating that in the long run it is not the national agricultural economies as a whole that compete across international boundaries, but rather similar agricultural regions and farm types.

This interregional competition is governed by a widely varying areal range of influence, the extent varying from product to product. To cite a few examples:

– When fruit and vegetables flow from Holland to the Rhine-Ruhr area today, this is only partly because of the objective locational advantages of our neighboring country. There is also the subjective factor, the greater "know how" of the Dutch gained through much longer experience. Together they lead to such a better input-output ratio that the costs of transport into the Federal Republic are justified.

– When early potatoes come into the German markets from Sicily during middle and late March and from Campania in April, we have a situation in which there is no competition. These crops are also still competitive at the end of May, since they are ready for harvest, while the West German early potatoes usually still are not. The climatically-determined early harvest dates for southern Italy and southern France are so advantageous for prices that the long rail journey of perishable products is still worthwhile. For the West German producer of early potatoes, this means an almost ruinous competition.

— Many large farmers in the Federal Republic today would like to use residual pasture and beet leaf for raising calves, but the profit position of the enterprise is unsatisfactory. In large parts of England and France, however, the raising of calves is one of the traditional farm enterprises. Cheaper land and a milder winter make it possible for the farmer to operate with less expense for land use, buildings, winter feed, and labor. Production costs in parts of England and France are so small that the farmer can also profit from calf raising, an enterprise that is by nature weak in output but in any case still more profitable here than in West Germany.

In all three of these examples German agriculture has, through no fault of its own, locational handicaps and thus competitive disadvantages. These in turn reduce income, so long as variations in production costs cannot be sufficiently equalized through interregional differentiations of agricultural prices.

It is quite obvious from all this that:

— labor-intensive farm enterprises, other things being equal, are favored in countries with still relatively low wage levels;
— crops whose raising can be most profitably mechanized do not belong in rugged mountain lands;
— individual farm size classes have their different spheres of interest in the rest of Europe just as they do in West Germany; and
— quality is playing an increasingly decisive role for fruit, wine, malting barley, and the like.

Thus, for example, the West German farmer raising malting barley need not fear the competition of partner countries in the European Community, since domestic malting barley production has a clear advantage in quality and transport costs which the German maltman and brewer cannot do without. The qualitative advantage of our malting barleys, noticeable practically every year, results in good market opportunities in the European Community for the German farmer.

These few observations on the interregional competition of farm enterprises, farm systems, and agricultural regions along the internal boundaries of Europe are still further evidence of how indispensable a structural analysis of the European area, from an agricultural-geographic and farm management standpoint, is to an equitable and balanced regional development policy for the Community of the Nine.

g) Summary

There is no economic enterprise that is so spatially bound as agriculture. For the scientific regional analysis of large areas like that of the European Community, climate is the most important differentiating locational factor for agricultural production.

Climates in the area of the Common Market of the Nine range from a Central European transitional type to a pronounced oceanic type, and from mountain varieties to the Mediterranean and subtropical types. The area thus offers the most varied conditions for agricultural production, as well as an extraordinary diversity of farming types. Their regional distribution was presented in a series of maps, based on a newly developed classification.

The agricultural regions that were delimited almost completely blot out the international boundaries. This is an indication of the dominance of the ecological factors over the economic qualifications of production. Consequently it is not primarily the overall national agricultural economies of the European Community that compete across the frontiers, but rather the farming types and agricultural regions that are conditioned by the same physical elements.

2. The Agricultural Geography of North America

The treatment of the agricultural geography of North America will be essentially confined to the United States. The total national area will be considered, though the western part (approximately west of the 100th meridian) is classed with the dry areas and therefore was partly discussed in part in Chapter VII.

The large Canadian area will only occasionally be referred to since the agricultural geography of its southern provinces resembles that of northern U.S. (see Fig. 13, p. 61), while the agricultural-geographic conditions in northern Canada duplicate those of northern Scandinavia and northern Russia: forest regions with scattered fodder cropping farms concentrating on ley farming, followed by forest and shrub regions, and finally transition to the tundra.

a) Place-Specific Characteristics of the United States

The United States encompasses within its boundaries many agricultural regions of the mid-latitude climatic zones, and in addition some important agricultural regions of the dry climates. The reasons for this great variety lie in the sharp variations of the physical and economic bases of production.

Without a doubt, the largest industrial agglomeration of the U.S. lies in the Northeastern states. The Northeast, however, can no longer be considered as the only economic focal point in the U.S. As early as the year 1818, Goethe had expressed a belief to his friend Eckermann that

2. The Agricultural Geography of North America

some day one would establish a navigable link between the Mediterranean and Red seas. He also predicted that a canal would be cut through Central America, and that when completed, the center of gravity in the United States would gradually shift from the Northeast to the Pacific Coast. With the realization of the first two predictions by Goethe, it now appears that the third will also come true. During the 1940—50 decade, the population of California grew by 53%. The economic development of the San Francisco Bay Area and Los Angeles-San Bernardino-San Diego region has been grandiose, and obviously has also had an influence on the agriculture of the hinterlands. Today a third economic focal point is developing in the Texas oil area, though of course for ecological reasons it has had less effect on agriculture.

Insofar as physical conditions are concerned, the principal mountains and rivers generally run in a north-south direction and in so doing divide the U.S. into five major natural areas, namely

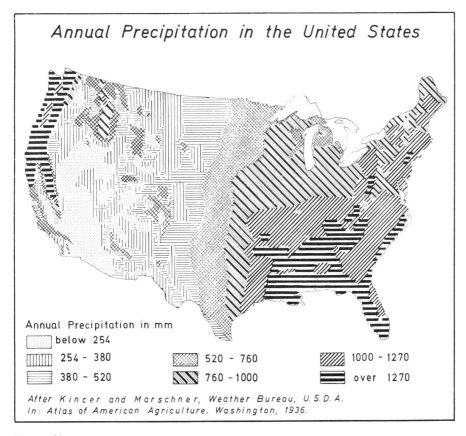

Figure 59

1. the *Atlantic Coastal Plain,* which extends from the New England states to Florida and finds its westerly limit in the Appalachian Highlands;
2. the *Appalachian Highlands* themselves, with their broad masses and deeply incised transverse valleys;
3. the *Mississippi and Missouri basins,* with their extensive, fertile, and centrally-located plains, overlain in large sections by chernozem and prairie soils;
4. the immense *highland area* between the Rocky Mountains in the east and the Sierra Nevada and Cascade mountains in the west; and
5. *the Pacific Coastal Lowlands,* with their orographic and topographic peculiarities.

This physiospatial arrangement influences the variation in precipitation conditions decisively. Viewed macrospatially, one can say that annual precipitation decreases from east to west, from 1,200 to 1,500 mm in the humid Atlantic coastal area to less than 200 mm in the arid Western Mountain states (see Fig. 59, above).

The east-west arrangement of precipitation zones is countered by a north-south differentiation in the growing season (Fig. 60). It decreases from

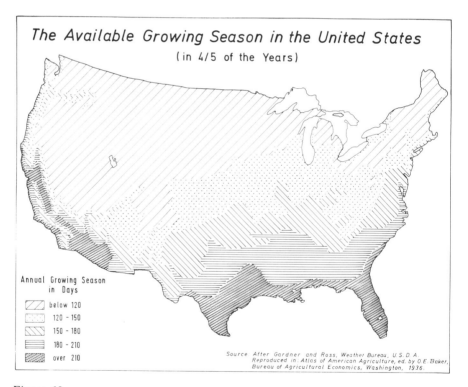

Figure 60

2. The Agricultural Geography of North America

more than 240 days on the Gulf Coast and in the states bordering Mexico to 90 to 120 days in the Great Lakes area along the Canadian border. Because the gradation patterns of precipitation and growing season meet at about right angles to each other, individual agricultural areas in the U.S. are endowed with highly varying combinations of moisture, heat, and light. Almost all kinds of transitions are found among such widely contrasting conditions as the moist and cool environment of the extreme Northeast, which recalls the Algäu; moist and hot subtropical Florida; the dry Northern Plains states with their hot summers and cold winters, a climate similar to that of the Ukraine; the desert climate and meager plant cover of Arizona; and the dry-summer Mediterranean climate of California.

A soils map of the U.S. completes our physicospatial picture (see Fig. 61). As in Europe, topographic and pedologic differences in the U.S. give rise to frequent changes in farming systems even over extremely small distances. But on a macrospatial scale, the influence of climate on farming types appears more marked. In fact, we find that the climatic

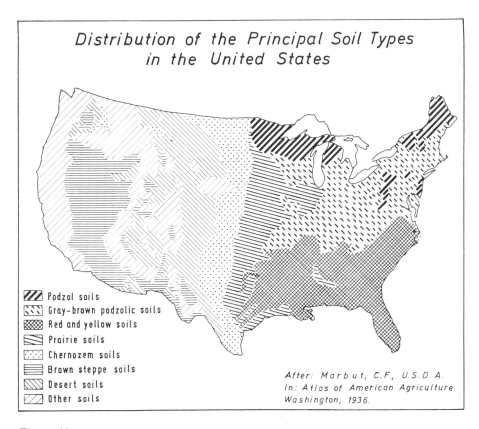

Figure 61

differences, as shown by the maps, have given areas of agricultural production a well-defined orientation. The following map (Fig. 62), which presents the agricultural zones of the U.S., is in large part a mirror of the

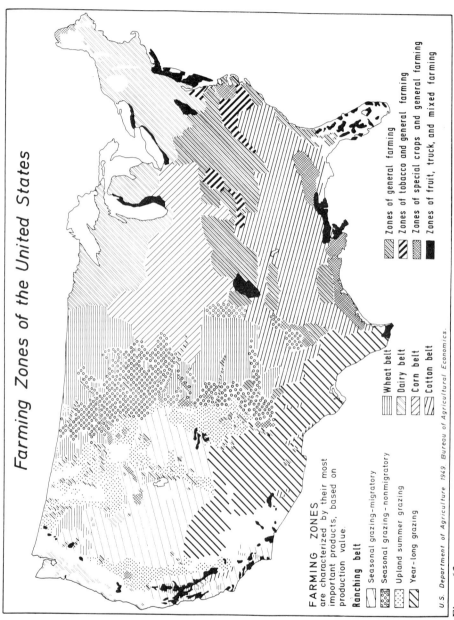

Figure 62

individual climatic areas. Even a cursory glance at this map is enough to inform one that specialization in specific land use enterprises is greater in the various agricultural areas of the U.S. than it is in European areas. This is because all U.S. agricultural areas, with their widely varying physical natures, are united in one large economic structure, one that is not fragmented by tariff barriers and other obstacles to free trade. This has allowed stronger interregional competition between production orientations, with the result that every product is produced where production and marketing costs are the lowest. This tendency has also been strengthened by the fact that all energy sources in the U.S. have, at least until recently, been relatively cheap, with low freight rates making it possible to exchange goods over great distances. In Europe, barriers to this trade are still much more formidable because of higher freight charges, and even more, international customs barriers. Further, there are the efforts at autarky by the European states, which have lead to the production in the majority of countries of products that are little justified environmentally and could be produced more cheaply in other countries. The economic integration of the Western European states is reducing these disadvantages. Whether the currently-increasing costs of oil will eventually be enough to seriously affect the course of economic integration in large areas such as Western Europe and North America remains to be seen.

b) The Dairy Zone

In the cool and moist Northeast, a large area of permanent pastures and cultivated crops stretches from the New England coast, through the Great Lakes area, and beyond the northern Mississippi. This is the dairy zone, or "Dairy Belt" as it is usually labeled in America. It correlates almost exactly with the continental cool-summer climatic zone, and hence also extends up into southeastern Canada where this climate prevails. Thus Ottawa and Winnipeg are still in the middle of this dairy region.

The most important grain in the U.S., corn, generally cannot ripen in this zone because its minimum growth period can barely be attained. Almost all of it is raised for silage. Fodder cropping, on both a permanent and rotation basis, comes to the fore, since it is favored on the production side by the cold, humid climate and the partly poorer soils, as well as by the hilly areas which discourage mechanization. On the market side, a heavy production of milk, butter, and cheese is stimulated by proximity to the industrial and urban agglomerations of the extreme Northeast and smaller urban concentrations elsewhere such as in southeastern Wisconsin-northeastern Illinois and Michigan (See Overview 19).

Overview 19: Crop Rotations in the New England States

| Valley Locations ←——————————————→ Highest Locations | | | | |
I	II	III	IV	V
1.-2. Alfalfa-grass ley 3.-4. Grain corn 5. Vegetables 6. Oats 7. W. Wheat	1.-3. Alfalfa-grass ley 4. Grain corn 5. Corn silage 6. Oats, w. Wheat	1.-3. Clover-grass ley 4. Corn silage 5. Oats	1.-5. Clover-grass ley 6. Corn silage 7. Oats	1.-8.-10. Clover-grass ley 11. Green oats
──── Decline of vegetable and wheat cropping ————→ ──── Decrease in cropping of grain corn, later also of corn silage ————→ ──── Changing from alfalfa-grass ley to clover-grass ley ————→ ──── Increasing economic life of leys ————→ Conentrates sold —→ Produced only for own needs —→ All concentrates purchased 0–60% ←———— % of sales proceeds from dairying ———— 100% ──── Increasing specialization ————→				

c) The Corn Zone

The middle and western parts of the dairy region are bordered on the south by the heart of American agriculture, the grain corn zone, the famed Corn Belt. It is roughly bounded on the south by the great curves of the Ohio River and the middle and lower Missouri. The center of grain corn cultivation lies in the continental warm-summer climatic zone.

The farming systems of the Corn Belt are not as standardized as those of the Dairy Belt. To be sure, grain corn cultivation stands supreme in the land use scheme, seldom taking less than a third of the cultivable land, usually a half, and occasionally even 100%; but the way it is used varies according to the soil, terrain, and distance to market. Corn can be sold directly, fed to dairy cows, used to fatten ("finish") hogs and cattle, or used simply to raise them. Fig. 63 offers a schematic sketch to illustrate this.

The major marketing center of the Corn Belt is Chicago, the third largest city on the North American continent. From here, agricultural products fan out in all directions, but particularly to the industrial centers of the Northeast. With increasing distance from Chicago in a southerly, westerly, and easterly direction, not only does distance from the market increase, but soil quality diminishes and terrain becomes more rolling. Since all three of these elements influence farm organization in the same direction, there has developed, when viewed schematically, a combina-

2. The Agricultural Geography of North America 239

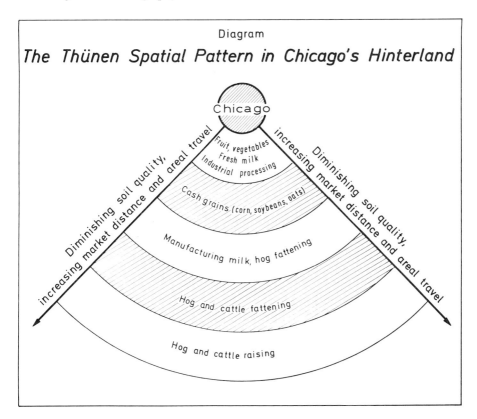

Figure 63

tion ring-sector arrangement of agricultural production zones south of the city, which brings to mind the Thünian spatial picture.

On the outskirts of Chicago and farther south into central Illinois, where the plains are fertile and ideal for machinery, and grassland is minor, an economy without livestock has formed. Corn, soybeans, and oats are sold directly to the market close at hand. Next, and farther south, comes a ring-like zone in which corn is ultimately coverted into manufacturing milk or pork. In the next and very broad ring, hog and cattle fattening dominate. The area in permanent grassland has naturally increased, and since the hilly surfaces also increase the threat of erosion, it becomes necessary to insert grass-clover leys into the rotation, perhaps even for several years duration. Livestock must then be kept to use the fodder growth. They are then brought up, through intensive feeding of corn, to prime slaughter weight.

Finally, in a fourth zone, the proportion of FL in plowed land declines still further. Increasing dryness toward the west causes corn yields to fall faster than fodder crop yields, just as increasing precipitation to-

Overview 20: Crop Rotations in the Corn Belt

I	II	III	IV	V	VI
1. Corn	1. Soy-beans	1. Soy-beans	1. Clover-grass ley	1. Clover-grass ley	1.-2. Clover-grass ley
2. Corn	2.-4. Corn	2.-3. Corn	2.-3. Corn	2. Corn	3. Corn
3. Corn	5. Oats	4. Oats	4. Oats	3. Oats	4. Oats
in % of cultivable land					
100	60	50	Corn 50	33	25
–	20	25	Soybeans –	–	–
–	20	25	Oats 25	33	25
–	–	–	Clover-grass ley 25	33	50
Monocultural farms	←		Specialized farms	→	Mixed farms

ward the east does. Also, the greater mechanization difficulties encountered farther south and east have so restricted corn cropping relative to fodder cropping that the operator can no longer support intensive feeding of the now quite large cattle herds with only his own production. Consequently farmers have adjusted to a system stressing the raising of hogs and cattle for the adjacent fattening zone.

Overview 20 shows some typical crop rotations of this agricultural region that come under the banner of grain corn cropping.

d) Agricultural Regions with Mixed (General) Farming

Adjacent to the Corn Belt on the south is an area that is already subtropical in its warm summer, a zone of mixed, or general farming. Here it is already too warm for optimal corn production, while for optimal cotton production it is still not warm enough. Moreover, since a large part of this zone is extensively traversed by the broad masses of the Appalachian Highlands, growing conditions vary on even a microspatial scale. Thus the range of crops is broad and includes corn, winter wheat, oats, clover-grass ley, soybeans, fruit, and vegetables.

Most farms also raise livestock since large areas of permanent grassland and the threat of erosion call for clover-grass ley farming and the age and poverty of the soils make a manuring economy advisable. A special feature of this zone is the cultivation of tobacco, which concentrates especially in Virginia, North Carolina, and central Kentucky. Although it holds a central position in the overall farm operation, its high labor peaks and great capacity for using barn manure create strong pressures for mixed farming. Diversified production is the distinguishing characteristic of this zone. Farming is more diversified here than in any other large agricultural region of the United States. The integration of several enterprises on the same farm has clearly left its stamp on farm management.

e) The Cotton Zone

The zone bordering the mixed farming strip on the south, the Cotton Belt, can only be described with qualifications. Its subtropical warm-summer climate still helps to make this area the third largest cotton region in the world, but the crop has been increasingly retreating to just a few specially suited sections, particularly the lower Mississippi Valley. The distribution is somewhat more belt-like if only farm income or sales source is considered, though even this pattern is no longer quite so extensive as indicated on Fig. 62. With these spatial contractions, the traditional monoculture has largely disappeared. Still typical, though, and particularly among many of the smaller farms, is a specialized operation that avoids livestock and makes cotton the principal money crop, accompanied by only one to three companion crops such as corn, peanuts, soybeans, sweet potatoes, or sorghum.

Livestock raising in the South has notably expanded since the last world war, but it is still insufficiently developed relative to regional potentials. The best croplands are still reserved almost exclusively for cash crops, and here cotton raising produces no feed byproduct other than cottonseed. Also, few pure feed crops have a place in the more intensive cropping systems. Even where livestock farming has made important advances, and lands have been sown to grass and heavily fertilized, there have been limits. The required large initial investments and the high incidence of risk have restricted the greatest benefits of the highly profitable beef cattle raising to areas with the largest farms. The minor scale of grain cropping in the South, and made even more so by the recent decline of corn in favor of soybeans, is also posing a barrier to continued livestock expansion. Finally, the lesser availability of grain for feed combines with the opportunity for year-around grazing to make manuring uneconomic.

In the last two decades, cotton production in the U.S. has been rapidly shifting from the Southeastern to the dry Southwestern states. The costs of the irrigation that is necessary there are easily compensated by the more mechanized large scale operations and the dependability of the dry harvest weather.

With that, we have described in coarse outline the most important agricultural regions east of the 100th meridian, which corresponds roughly to the 500 mm-precipitation line. However, at least three more agricultural regions may be mentioned. Two encompass the dry and much less productive western half of the United States, excluding the widely dispersed but highly intensive irrigated oases, as well as the irrigated cotton areas already noted. The third region is actually a collection of widely scattered coastal areas, all of which share a reputation for highly intensive fruit and vegetable farming, and in some cases livestock farming that suggests factory operations.

f) The Wheat Regions

The zone of wheat cultivation forms the transition to the semiarid western areas. It extends from northern Texas, through the states of Kansas, Nebraska, North and South Dakota, and far up into Canada. It also has an exclave in the Columbia Basin, which takes in parts of the states of Washington, Oregon, and Idaho. Fertile grassland soils and gentle terrain combine with dryness to make dry farming indispensable to this area, but particularly to the western part of the zone. Here the fully mechanized grain farm is at home, a unit on which is still frequently practiced the purest form of monoculture simply because there is no cultivated plant other than wheat that has sufficient drought resistance, though sorghum has now often become a major companion in certain warmer parts of the zone. This heavy emphasis on wheat has given rise to the well-known suitcase farmer. The southern winter wheat zone (Winter Wheat Belt), with its core in Kansas and its up to 200 growing days, is to be differentiated from the northern spring wheat zone (Spring Wheat Belt), which in part has only 100 growth days.

To the west, precipitation falls from 750 to less than 400 mm. The lower precipitation limit for dry farming of wheat (taking into account evaporation, solar radiation, and precipitation regime) is 375 mm in Texas, 300 mm in South Dakota, and only 230 mm in the Canadian province of Saskatchewan. The more recent extensive cultivation gains made with large machine aggregates have strongly reduced production costs of wheat and have made it more competitive with extensive grassland farming. Farm data and types of crop rotations for a variety of locations in the wheat region can be found in Table 12 and Overview 14 of Chapter VII-2-a, pp. 179 f.

Unfortunately, the coincidence of high wheat prices with relatively high precipitation during and after both world wars encouraged the expansion of wheat farming too far into the drier grasslands, which become ever more so toward the west. Here the frequent drought years pose great risk and also threaten farms with erosion. The mistake, to be sure, has been recognized, but the process of returning the critical areas to grazing is not only technically difficult, because of the difficulty of reestablishing a grass cover in semiarid areas, but also economically problematical, since the now denser settlement would require considerable thinning. The minimum support for a farm family that shifts from wheat farming to extensive grassland farming must increase from 400–700 ha up to 1,800–2,200 ha, this within a span of 250 to 520 mm/year. Thus the capacity of a specific region for wheat farms is about four times as high as that for ranches.

g) Regions of Extensive Grassland Farming (Ranching)

The last large agricultural region of the United States to be named is that of extensive grassland farming, or ranching. It embraces almost all of the Intermountain states and includes extensive highland areas that lie between the Rocky Mountains and the Sierra Nevada, dry steppes, semideserts, and mountains. Minimal rainfall or rugged terrain, or more rarely, great remoteness from transportation, even excludes the dry farming system, or grain-fallow economy, and generally permits only the most extensive form of agricultural land use, the grazing of large areas of sparse natural vegetation by feeder stock. In the hot-summer state of Arizona, in extreme southwestern U.S., livestock raising like this can be carried on with as little as 250 mm of annual precipitation, and extensive sheep raising can be practiced with as little as 150 mm. Naturally, then, quite extensive pasture areas per animal are needed (see Fig. 64 and Fig. 42, p. 168).

Subzones must also be distinguished within the grassland farming region. A southern area, with year-around grazing and only occasional emergency feeding in drought periods, is joined on the north by a zone in which regular winter feeding is mandatory. In the vicinity of the high mountain

Figure 64

chains, herds are driven, as in the high Alps, in long seasonal migrations. Cattle cover up to 80 to 110 km in their trek from the valley ranch, where winter grazing goes on in the open, to the foothill pastures and eventually the summer pastures in the mountains. Sheep travel even as much as 160 to 200 km.

h) Coastal Regions with Fruit and Vegetable Growing Dominant

Most fruits and vegetables are distinguished by high hectare yields, low value per unit weight, and high perishability. Thus they are sensitive to transport and are located as close as possible to markets. But since major population centers have risen as transport nodes in certain coastal areas, the largest and most important fruit and vegetable regions are also to be found in those very same coastal strips, especially since they are favored climatically. They include the lower Great Lakes littorals; the Boston, New York, Philadelphia, and Washington areas; large sections of Florida; the Gulf Coast in the vicinity of Mobile, New Orleans, and Baton Rouge; the area at the mouth of the Rio Grande; and the longitudinal valleys of California.

The following comments will be restricted to the irrigated valleys in California with their Mediterranean climate, and will concentrate on the specialized production programs of individual farms and regions. In the last three decades, California has experienced an unexpected economic upswing and registered huge population growth rates. This especially applies to the Los Angeles, San Bernardino, and San Diego areas. The result has been rapidly rising wage and income demands, relatively favorable prices for capital, and increasing cost of land. The altered relations between factor costs have thus forced a recombination of the production factors. Much greater emphasis has been put on capital investment, labor inputs have been sharply curbed, and farming has been intensified. In return, extremely high capital investment has brought with it highly advanced stages of specialization that conflict with the basic principles of mixed farming, especially as concerns the need for spacing work, rotating crops, and spreading risk (see Ch. IV-1).

Briefly described below, and without any claims for completeness, are a few interesting examples of farm management types that have resulted from this opposition of economic forces:

aa) Specialization has been especially successful with *bush and tree crops*. Three-, two-, and even one-crop farms are common. There are farms whose only support is a crop such as grapes, citrus fruit, dates, walnuts, or apples. Harvests, then, bring substantial labor peaks, which cannot be controlled through internal economies.

Monoculture of this kind is possible only because Filipino or Mexican migrant labor crews stand on call, for they seek to space their own work by migrating so as to utilize the ecological distinctiveness of the California area. Indeed, the same fruit types ripen not only in different locations, but at different times, and here not only according to latitude, but to altitude and distance from the ocean as well. A corollary of this flexibility is that California accomodates almost all fruits of the mid-latitude and subtropical climatic zones with the most varied seasonal labor demands. Requirements associated with crop rotation do not need to be observed with the perennial crops, and with irrigation the production risks are also smaller, so that monoculture under these conditions is relatively easy.

bb) Production risks are even smaller with *greenhouse crops,* the production factors of which — moisture, heat, air, nutrients — can be regulated by the farmer. A horticultural farm in the vicinity of San Francisco grows nothing but chrysanthemums on 1.4 ha under glass, and on a continuing basis. Flowers are cut every two weeks, and 24 hours later are standing in the vases of customers in Chicago or New York (aerial transport). Market risks are reduced by the grower taking advantage of the laws of photoperiodism. The chrysanthemum, as a short-day plant, needs a shorter period of daylight to produce blooms. Depending on the market situation, the farmer will either slow bloom formation with nocturnal lighting during the vegetative phase or accelerate it by darkening the greenhouse for a few hours in the morning and evening. Thus in addition to the production factors mentioned above, the farmer has the period of light under his control.

cc) *Vegetable growing,* with its high degree of specialization, encounters substantial difficulties. Almost all vegetables can be raised in a short growing season, so that in California often three different types can follow one another in the same year. The next year other types should be grown for reasons of plant hygiene. Rotations thus bring advantages, but they also require diversified production. Specialization in vegetable growing is also hindered not just by high market risks, but by the annual labor program. Specialization can nevertheless be occasionally successful, as shown by a farm operator who has 40 ha FL in irrigation, employs six tractors, purchases considerable manure, and draws 70% of his sales revenues from celery. Two years of celery are followed by only one year of lettuce, spinach, and dry beans, this to obtain such crop rotation benefits as loosening of the soil and addition of humus. This three-year quasi-rotation, on a farm geared for maximum production, is then repeated.

dd) In the Salinas Valley and other places, complete vegetable rotations are made compatible with a high degree of specialization through *lease rotation*. Since many farmers wish to specialize in one or two vegetables, but cannot do so because of the diseases, pests, and weeds associated with such continuous cropping, they rotate leases so that they can continue to grow the same vegetables for short periods on diverse soils. Thus in place of the alternation of crops on the same land (crop rotation), we have the alternation of land with the same crops (lease rotation). The canneries are anxious for supplies and therefore offer inviting prices to the growers so that they can put up with the inconveniently short leases and the annually fluctuating distances to the fields.

i) Industrialized Animal Production on Specialized Farms

Strong specialization tendencies also make themselves felt in animal production, where the pressures for drastic labor economies call for uniformly extensive capital investments. Cost-effectiveness requires that these investments be exploited to the maximum, which can only be done through specialization. Thus the same locational forces that have stimulated California fruit and vegetable growing also have lead to an extensive industrialization of production in livestock raising.

aa) Included here is the fattening ("finishing" in the parlance of American stockmen) in *feedlots* of cattle raised on the range. One feedlot I visited in the Sacramento Valley holds 3,000 two- to three-year old animals in its corrals. A feeding period of three months permits four turnovers of stock annually, so that this operation sends 12,000 slaughter cattle to market every year. Another feedlot even supplies the market with 40,000 feeders annually. No feeds of any kind are produced; everything is purchased. Feeds include alfalfa green meal, barley, dried molasses, and cottonseed cake. They are prepared in large mixing machines and transported to the feeding troughs by trucks equipped with mechanical unloaders that operate laterally. An overseer on horseback completes this picture of large scale feeding operations.

bb) Operators of *drylot dairy farms* in the Riverside-San Bernardino area will, in extreme cases, buy alfalfa hay from the El Centro area, over 230 km away, and deliver the milk over the same 230 km back to El Centro. Milk cannot be economically produced around El Centro because of the hot desert climate, and alfalfa cannot be profitably raised around Riverside because of the high cost of land. The herds of milking cows average 200 to 500 head. Milking goes on 20 hours a day.

2. The Agricultural Geography of North America

cc) Most advanced of all is the *specialization in poultry raising*. In Germany, a pure poultry farm is already designated as highly specialized if operations are organized into the following divisions:

1. breeding for hatching eggs;
2. incubation;
3. brooding;
4. raising layers;
5. fattening fryers;
6. processing market eggs; and
7. processing manure.

In California, however, specialization is often even much more advanced in the way many operators participate in the production process, each taking care of only one of the above functions.

Industrialized livestock and poultry farms obey the same laws that industries observe in their locational orientation. Based on the character of the particular animal production, they are

— *either supply-oriented:* near feed suppliers (sugar refineries, breweries, wineries, orange juice plants, oil mills, etc.)
— *or market-oriented:* near population agglomerations, packing plants, dairies, egg-processing plants, etc.

j) Regional Differences in the Factor Combination

Just as production orientations sharply vary from place to place in the U.S., so also are the comparative availabilities and consequently the princes of the production factors quite variable. Thus the optimal productive combination for individual areas must also be a variable one. Of labor may be considered typical of all the U.S. For the individual agricultural areas, though, this is not always true, or if so, only with marked cultural areas, though, this is not always true, or if so, only with marked gradations. This can be understood more clearly if we divide the U.S. into three parts, first by drawing a north-south line that follows the western boundaries of the states of North and South Dakota, Nebraska, Kansas, Oklahoma, and Texas, and then by drawing an east-west line that corresponds with the northern boundaries of the states of Delaware, Maryland, West Virginia, Kentucky, Arkansas, and Oklahoma. We thus arrive at three broad regional complexes that the Bureau of the Census has used for statistical purposes: North, South, and West (see Fig. 65, p. 248).

Land in the sparsely settled, isolated, and *little-developed West* is in surplus almost everywhere and therefore is cheap, especially since the climate bestows upon it only minor natural productivity. On the other

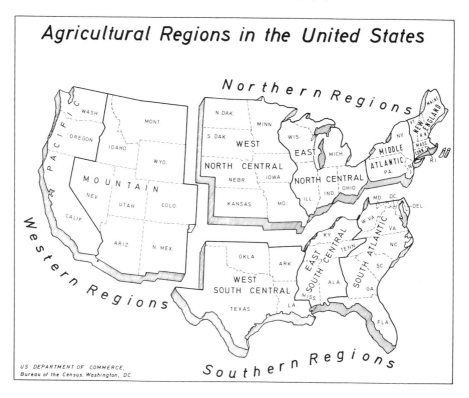

Figure 65

hand, labor is extremely scarce and expensive, since life on the isolated farms offers little attraction for the wage laborer. It is therefore advisable to farm as large an area as possible with the fewest possible workers. This goal comes ever closer to realization the more the radius of human action is expanded by large scale use of livestock or machinery. Thus large inventories are accumulated per worker. In relation to the area farmed, however, inventories remain as small as labor investments, so that one must speak of *extremely extensive* economic practices.

The situation is completely different in the *South*. It still has a relatively high rural population density and is the least industrially developed of the three regions relative to population. The lesser competition of industry for labor and the system of tenancy ensure that labor in the Southern states is still available to agriculture relatively cheaply, and thus its application need not be as frugal as in the West. Land prices, though, are much higher than in the West, not only because of the high value of output, but because of the brisk demand for cropland. The fullest possible utilization of the productive reserves of the soil must therefore be the first economic objective. Since livestock raising is still encountering obstacles and mechanization is hampered by many small farms, while labor is still relatively cheap, operators economize on inventories and greatly increase labor expenditures per areal unit, and thus are *labor-intensive* in their practices.

Finally, in the *industrial North,* land prices are still higher than in the South and wages and income demands of the farmer are about as high as in the West. Here neither land nor labor can be invested in rich amounts, but instead the highest possible use must be made of them with the most economical application. This can only be done by applying large inventories of every kind (livestock, machinery, fertilizers, feeds) on both a per-hectare and per-worker basis. These measures are facilitated by the relatively low costs of industrially-produced farm inputs at the farm gate, due to their nearness to the production centers. Operations in the North are therefore *markedly capital-intensive* and overall even more intensive than in the South.

Land prices increase from West to South to North, wages increase from South to North to West, and prices of capital goods increase from West to North to South. The proportions of the investment in the production factors vary inversely with their prices to a degree hardly attained in any other country, largely because hardly anywhere else is agriculture organized so much on the profit principle as it is in the United States. The combination of these patterns in the three agricultural areas are schematized in Overview 21.

Overview 21: Proportional Investment of Production Factors in Maximizing Income in U.S. Agriculture

	West	South	South
Land	much	little	little
Labor	little	much	little
Working Capital	much	little	much
Consumptive Capital	little	moderate	much
Management	moderate	little	much

Indicative of the contrasting combinations of production factors in these three large regions is that more than half of the land in the U.S. is in the West, more than half of the employed are in the South, and more than half of the value of agricultural production is in the North.

3. The Agricultural Geography of the East Bloc Countries

a) Place-Specific Characteristics

Because of the major influence of socialization policies, the extensive territory of the East Bloc can be considered as essentially one agricultural area. It extends from the eastern Harz and the Bohemian Forest to the Okhotsk and China seas, and from the Arctic Ocean to the Himalayas.

This land mass is broadly divided into the following *climatic zones* (see also Ch. II):

1. The *Arctic climate belt* begins somewhat north of Archangel, runs up to the mouth of the Ob near the Arctic Circle, and then extends, in varying width, to the Bering Straits. The tundra vegetation generally permits only seasonal grazing by the reindeer of seminomads, such as the Turkish Yakuts.

2. Next comes the widely distributed *Subarctic climate belt,* stretching from the White Sea to Kamchatka. Its southern boundary runs approximately from Karelia, through Sverdlovsk and Ulan Bator, to Khabarovsk.

3. Then follows, still farther south, a *continental cool-summer climate bloc,* which differs from the Subarctic climate belt in areal trend by starting out with a wide western part and gradually narrowing to the east, finally disappearing completely in Mongolia. The delimitation of this climatic wedge is given approximately by the three points of Leningrad, Bucharest, and Novosibirsk. An exclave is found north of the Sea of Japan.

4. There is a *continental warm-summer climate,* thus a pronounced grain-corn climate, in the East Bloc in two relatively small regions. One prevails in northeastern China, on the southern edge of the exclave just mentioned. Peking and Seoul are located in this area. The second area of this climatic type is the largest part of the Balkan Peninsula, i.e. the Danube Basin south of Budapest and the bordering mountain regions.

5. Still farther south are the bordering zones with a *subtropical warm-summer climate.* They include the area on the Adriatic as well as parts of the Black Sea coast. Above all, however, a major part of the People's Republic of China, the entire southeast, lies in this climatic zone.

6. Finally there are still other climatic types in the East Bloc, but their agricultural influence will not be considered here because they are not to be classed as middle latitude types. In mind here are, on the one hand, the enormous inner-Asian areas with dry climates (arid and

3. The Agricultural Geography of the East Bloc Countries

semiarid), and on the other, the Indochinese areas with tropical rainy climates (wet, wet-and-dry).

The extreme importance of the climatic factor of precipitation amount for agriculture in the East Bloc can also be seen in Fig. 5, p. 35.

The soil conditions that help to make for agricultural differentiations on a microspatial scale are of only minor significance in the macrospatial view of agricultural zones as used here. Still the following can be said (Krische 1933, p. 22, citing Glinka 1927):

- By far the largest parts of Poland and the USSR consist of *podzolized (leached) soils,* which prevail in both Subarctic and continental cool-summer climatic zones. Although easily worked, they are also poor in nutrients. They are farmed especially for potatoes, rye, and leys.
- Adjacent to this podzolized area on the south is an extremely wide and fertile belt of Chernozem (blackerths) and Chernozemic soils that are superbly qualified for the cultivation of sugar beets and wheat. This band surrounds the Black Sea to the north and runs in an east-northeast direction to China.
- Adjoining this fertile area on the south are the chestnut-brown, brown, and gray soils, which are distributed in the Danube area.
- Brown-and-yellow soils dominate in central Germany, Jugoslavia, and eastern China. Their often considerable resistance to cultivation is becoming even more disadvantageous with mechanization, while the advantage of a naturally rich supply of nutrients is being increasingly lost as chemical fertilizers become cheaper.
- Still farther south on the southeast coast of China, starting from about Shanghai, and in the southern half of Albania, we are already encountering the red soils of the subtropical climate.

Another important place-specific factor for the agriculture of the East Bloc is the *density of the transport net.* In 1960, the railroad density in central Silesia amounted to over 200 km of lines per 1,000 km^2; in the Leningrad and Moscow areas, 50 to 100 km; in the greatest part of the Ukraine, only 5 to 20 km; and in Uzbekistan, as little as 0 to 5 km of lines. Enormous Siberia has only one commercially significant east-west rail line, a situation made even more serious by the fact that most of the navigable rivers flow from south to north, thus perpendicular to the rail route. Moreover, they are handicapped by ice for much of the year.

The variations in the density of the transport net are further exaggerated by the increasing sparsity of roads in Russia in both a northerly and easterly direction. Road construction in the northern part of the country is expensive, since forests must be cleared and hard winter freezing requires large amounts of gravel fill. When this problem is combined with the decreasing population density to the north and east, one can easily see why transport construction costs per inhabitant progressively increase, and why even wealthy countries with the same conditions must be content with only a loose transport network (cf. northern Sweden). The effects of conditions like these on the character of agricultural regions is by no means unimportant.

b) Large Socialist Farms as a Regionalizing Feature

The extensive physicogeographic differences in the East Bloc are not reflected in the farming types, agricultural regions, agricultural areas, and agricultural zones as strongly as one might assume. The reason for this is that the socialization of agriculture has led to a definite leveling of the farm size structure, so that farm size is not capable of exerting the decisive influence on agricultural differentiation that it does in large parts of the Western world. The large socialist farms (APC or Kolkhoz and PE or Sovkhoz) are widely established in the agriculture of the East Bloc, and thus their influence on the agricultural region is clearly visible in the macrospatial aspects of the economy.

c) Problems of an Areally-Suitable Agricultural Production

The physical agents in agriculture, such as climate, soil, and relief, do not always make their full weight felt in the centrally-planned economies of the socialist countries. In the Western free market economy, production is largely controlled by the market and after harvest takes place. In contrast, production in the Eastern centrally-planned economy is mainly controlled by the Plan, which is drawn up well before the growing season. Macrospatial production plans, however, can only cope to a certain extent with highly-varying local conditions compared with the effectiveness of the production decisions of independent peasants, farmers, ranchers, or planters, who are directly involved in the operations and are motivated by material rewards.

If allowance is made for this involuntary weakness in adaption in socialist agriculture, then there is still to be considered what in certain cases is a deliberate arbitrariness in locating agricultural production. Since prices must not be fixed by the market, a specific type of production also need not be completely areally suitable, but can be forced into less favorable locations by the priorities of politics. The hopyards along the autobahn between Helmstedt and the Elbe are certainly not locationally justified, and a large part of Krushchev's land cultivation program in Kazakhstan was no less unqualified. The first situation of course represents an effort to save currency, while the latter was an action designed to reduce the grain deficit in Russia. Thus what would be absurd in the private economy may be economically sensible from the viewpoint of the national economy.

d) Stages of Socialization

Because socialism bears so heavily on the absolute and relative effectiveness of the many formative agents in agriculture, it is necessary to consider briefly collectivization itself. Table 15 shows, according to the latest census reports at this writing, that for all the countries listed, except Poland and Jugoslavia, less than 10% AL was in the private sector and for many countries the figure approached the zero mark. This land was split up among the hectarages of the independent peasant farms and the private parcels of collective farmers and state employees. Also to be noted in Table 15 is that

1. socialization in all countries is clearly increasing with time;
2. within the socialist sector the collective farms generally dominate by far, while the state farms are retreating (exceptions: Cuba, USSR, Jugoslavia, Romania);
3. in the Soviet Union a clear shift of emphasis in favor of the Sovkhoz has taken place; and
4. Polish agriculture has maintained, with 84.3% AL, by far the largest areal share in the private sector.

Table 15: Areal Share of the Socialist Sector in the Agriculture of the East Bloc

State	Year	Socialist Sector, % AL			Private Sector, % AL
		Collective Farms	State Farms	Total	
Albania	1959	73.5	21.6	95.1	4.9
Bulgaria	1972	...	...	99.7	0.3
China (PR)	1956	96.0	4.0	100.0	...
Czechoslovakia	1972	...	...	91.6	8.4
GDR	1960	...	...	92.5	7.5
	1968	85.0	7.0	92.0	8.0
	1972	...	...	94.4	5.6
Jugoslavia	1971	...	...	15.0 [1]	85.0 [1]
Cuba	1963	...	65.0	...	...
North Korea	1957	93.7	...	...	...
Poland	1969	1.2	14.5	15.7	84.3
Romania	1959	44.1	28.6	72.7	27.3
	1972	...	...	90.7	9.3
USSR	1928	1.2	1.2	2.4	97.6
	1940	78.6	8.9	87.5	12.5
	1962	53.0	43.9	96.9	3.1
Hungary	1960	...	...	77.8	22.2
	1972	...	...	97.2	2.8

[1] % of cultivated land.
Source: Beyme, K. von: Ökonomie und Politik im Sozialismus. Munich, Zurich 1975, pp. 55 f.

In general, the reapportionment of property and land use rights among the agricultural production factors is accomplished in the following four phases of collectivization (Beyme 1975, p. 56):

Phase 1: expropriation of the lands of the large landowner;
Phase 2: beginning of actual collectivization;
Phase 3: transition to more socialized forms of collective farming:
Phase 4: elimination of the coexistence of collective farms and state farms (future plans).

In the Soviet Union, the first phase took place in the time from 1917 to 1921, the second phase occurred between 1929 and 1933, and the third phase began in 1952. In the German Democratic Republic, land reform was initiated in 1945 and the first APC was established in 1952. By 1960, about two-thirds of the APCs had been converted to Type II status. In the People's Republic of China, the first three phases were initiated in the years 1950, 1953, and 1958 (Beyme 1975, p. 57).

Only by recognizing the primacy of politics over economy can one understand why the large socialist farm has been exalted as the prototype for the agriculture of all East Bloc countries (except Poland) regardless of such things as the stage of national economic development, the educational level of the population, and the ecological realities. In the preceding chapters of this book, attention has been repeatedly directed to the fact that under various ecological conditions there are also various farm sizes that can be regarded as optimal. In Chapter IX-2 it will be shown that during the course of national economic development the farm size structure in agriculture must change. All the countries for whom an agricultural census has been available for many decades substantiate this. All the more astonishing then, is that in the East Bloc, under the most varied conditions, the same concept is extolled: the superiority of the large socialist farm.

If it is imperative for national economic interests that an agriculture involving millions of illiterate small farmers be raised to a much higher level in a short time, then the large socialist farm may have a place. It offers a much greater radius of action for a small elite of proficient, flexible, and progressive farm operators, and thus provides much greater efficiency. This need does not apply to highly developed industrial states, though, since almost every farmer has enough of the entrepreneurial qualities whose application is to the national interest. Capitalism motivates the individual through material inducements, whereas socialism motivates him through social consciousness. In Central and Eastern Europe this has not worked out. Also, a detailed system of worker remuneration based on amount and quality of work (work units) on the large socialist farms has not been able to evoke the same will to work displayed by the independent West European farmers in exploiting their opportunities and avoiding their risks. As a result, the large socialist farms, in spite of their good provisions for mechanization, are by our concepts oversupplied with labor.

Table 16 provides some details on the Agricultural Producers' Cooperatives of the German Democratic Republic. The observable tendencies toward fewer and larger units and toward greater socialization of the production factors could be considered representative of the greatest part of the East Bloc.

Table 16: Agricultural Producers' Cooperatives in the GDR

Characteristic	Type I APC	Type II APC	Type III APC	APC Total
For operations in common in the APC	only the cultivable land	field crops, grassland, perennial crops, draft animals, machinery	total farm unit	–
Computation of income by work units, by land share	at least 40 to 60%	at least 30 to 70%	at least 20 to 80%	–
Shift in the number of APCs, by types 1960 1968	12,976 1,185 (1973)		6,353 5,759	19,329 11,513
Number of APCs in 1968, by ha AL: up to 200 ha 500 to 1,000 ha 2,000 & more ha	3,372 400 –		313 2,410 91	3,585 2,810 91
Source: Statistisches Jahrbuch der DDR 1963, p. 230; 1964, p. 246; 1966, p. 264; 1969, p. 184.				

As to how much these political goals conflict with the increase that is needed in production is shown by Table 17, the data being extracted from the report presented by Gomulka at the Eighth Plenum of the Central Committee of the Plish United Workers' Party. Here it should be noted that Polish state farms stem almost entirely from former German estates, whose field patterns and buildings were taken over and whose place-specific advantages over farms in the Congress Poland area are often considerable. Also, even if one takes into account the difficulties associated with the initiation of collectivization in the middle 1950s, the deficiency in production remains, which may be an overall reflection of the weaknesses in areal adjustment and performance enthusiasm that have been mentioned earlier.

Table 17: Effects of Socialization on the Agricultural Production of Poland

Characteristic	Unit	Private Farms	Collective Farms	State Farms
Share of AL	%	78.7	8.6	12.6
Share of agricultural production	%	83.9	7.7	8.4
Gross returns	zloty/ha	621.1	517.3	393.7
Share of gross returns	%	100.0	83.3	62.8

After Zotschew, T.: Die Entwicklungsprobleme der polnischen Wirtschaft. Kiel 1957, pp. 3f.

e) Agricultural Zones in the Baltic-Adriatic Area

The area from the Baltic Sea to the Adriatic, as shown in Fig. 66, includes all the types of agricultural regions that are generally found in Eastern Europe. The agricultural regions on the map have been delimited on the basis of official statistics for primary administrative units, and are organized according to the five most important farm enterprises or farm enterprise groups: grain cropping, fodder cropping, hoe cropping, specialty crops (grapes, fruit, and other bush and tree crops), and dairying. The secondary enterprises were assigned to these main enterprises, so far as possible, according to the closeness of their farm enterprise relationship. The scale of the land use enterprises was expressed in % AL, and that of dairying in cows per 100 ha AL. Cropland values and herd densities were then weighted by their corresponding mean labor expenditures. This was accomplished for the land use enterprises by multiplying the cropland value in % AL by the weighted index, and for dairying by multiplying the herd density per 100 ha AL by the weighted index. The agricultural regions were then named after the three farm enterprises having the highest, second highest, and third highest weights.

The area being discussed is composed of the following nine types of agricultural regions (see Fig. 66):

1. The *hoe crops-dairying-grain region* forms the largest unbroken block of territory. Eastern Mecklenburg, Brandenburg March, Pomerania, Silesia, and the largest part of Congress Poland belong to it. A warm summer with a July/August maximum, a sufficiently long fall, largely level terrain and light soils, as well as the dominance of small peasant farms in Poland, are all factors that have strongly favored hoe cropping and especially the cultivation of potatoes. Because potatoes can be used in various ways, their cultivation can be quite well adapted to the needs of different farm sizes and varying market distances. Labor expenditures per hectare climb in the sequence of industrial potatoes – table potatoes – seed potatoes – feed potatoes for fattening hogs. Transport costs per

3. The Agricultural Geography of the East Bloc Countries 257

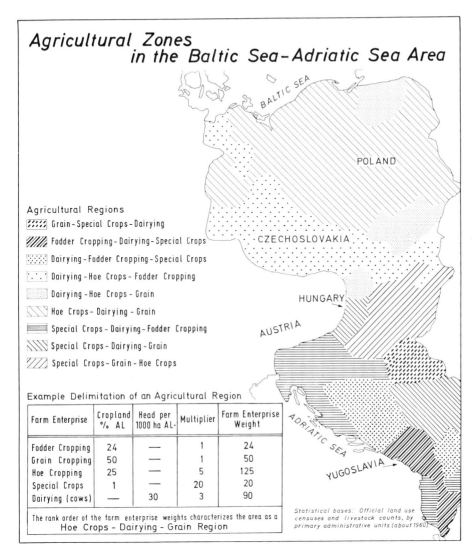

Figure 66

hectare fall in the sequence of seed potatoes — table potatoes — hog fattening — alcohol. Accordingly, operators

— on large socialist farms in the vicinity of cities such as Szczecin (Stettin), Berlin, Poznań (Posen), and Warsaw will raise table potatoes.
— on large socialist farms farther from markets will produce seed potatoes.
— on large socialist farms at a very great distance from markets, e.g. in Farther Pomerania, will cultivate industrial potatoes and make use of a distillery or starch factory.

– on land-poor family farms in Congress Poland will combine the cultivation of feed potatoes and fattening of hogs so as to provide productive work and to save on transport costs.

No more than second place can be given to dairying in the designation of the agricultural region, since in most cases only a little permanent grassland is available and the light soils are not adequate for clover. In addition, intercropping must be held within strict limits because of the dry climate and a growing season that is shorter than in West Germany. Only on the better soils where sugar beet cropping is favored over potatoes is the feed base for dairying broadened, with fresh beet leaf, beet leaf silage, and dry beet chips as the products. This is especially true of the Magdeburg Börde, the Ucker March, and the Wroclaw (Breslau) Platform.

Crop rotations for this large area range

– from the pure three-field economy, occurring in a broad belt from Poznań to the Russian border,
– through many intermediate stages in the Oder-Vistula area,
– to the pure crop rotation economy in the German Democratic Republic and Farther Pomerania.

2. The roles of the dominating farm enterprises are reversed in the *dairying-hoe crops-grain regions.* They are areas where heavy soils hinder hoe cropping and encourage fodder cropping, or where groundwater conditions make it necessary to leave a larger portion of the land in permanent grassland. Warmia (Ermland) and Masuria are as much examples of these conditions as are Galicia or the Bratislava Lowland.

3. A small modification is to be found in the *dairying-hoe crops-fodder cropping regions* that include western Mecklenburg and by far the largest part of Czechoslovakia. The prevailing cropping system in the CSSR is a clover-grass economy, with a two-year useful economic life for fodder cropping. In the surrounding Bohemian Mountains, a three-year period is also common. Fodder crops take up 20 to 40% CL (Andreae 1964, p. 80).

4. The agricultural regions of the continental steppe climate are of quite a different character. A large *special crops-grain-hoe crops region* stands out in Hungary because viticulture consumes the largest part of the farm labor potential, even though it also claims only a relatively small quota of the agricultural area. Since the locationally-favored viticulture competes with hoe cropping for manure and for labor in the fall, hoe cropping can only be given third place in farm enterprise weight.

5. *Grain-special crops-dairying regions* show up on the map only in the Vojvodina, in the Banat, and in eastern Slovenia. Here extremely large state farms reign, having taken over the traditionally heavy emphasis on grain cropping because of the compelling ecological conditions. Emphatic

grain landscapes like those here are no longer found anywhere else in Europe except in parts of the Ile-de-France.

6. South of this region, in Serbia, is a *special crops-dairying-grain region.* In hilly country, this enterprise combination is a good one for making the most of the varying place-specific conditions on the farm. Cool northern slopes and high plateaus can be allocated to grassland, southern slopes can be given over to viticulture, and valley bottoms can be used for large scale mechanized operations. Fodder cropping/dairying and viticulture are complementary in their labor and manuring needs. Crop farming is of service in great part because of the raising of grain corn, so typical of the Balkans and a critical source of not only food but supplementary feed. Thus crop rotations are governed by nurse crops from almost beginning to end, or in other words, the leafy crops take up more than half of the cultivated land.

7. *Special crops-dairying-fodder cropping regions* dominate Slovenia and Croatia. Thus they are similar to the agricultural region described in 6, except for the difference that fodder cropping now attains a greater farm management weight than that of grain cropping.

8. The agricultural regions of Herzegovina and Montenegro are again areas of stronger concentration on fodder cropping and dairying. A special characteristic of these areas is migratory grazing, without which the *dairying-fodder cropping-hoe crops region* and *fodder cropping-dairying-hoe crops region* would hardly have been able to develop and which, on their part, prescribe a rugged terrain. The Mediterranean climate is not conducive to dairying, first, because plant growth ceases in the summer, and second, because taurine cattle are less resistant to heat. On the other hand, well-defined grassland areas in the Dinaric and North Macedonian mountains would hardly be of use to farms because under difficult winter conditions, large amounts of winter feed would have to be procured. Migratory grazing now provides the feed balance. When, by the end of May, the pastures in the lowlands are exhausted and the hot season has set in, the migrations of the herds to the cool mountain pastures begin. In this way, from the beginning of June to the end of September, feed is obtained for the livestock. After that, the lowlands again provide feed sufficient to the next May. The individual migratory trails run to 45 to 85 km, so that the herds travel from 90 to 170 km per year (Beuermann 1967). The longer the migrations and the scantier the pasture, the more the sheep and the goat displace the cow. For this information we are indebted to A. Beuermann and his penetrating study of migratory grazing in Southeastern Europe.

f) Agricultural Zones of the Soviet Union

Table 18 gives an overview of the development of Soviet agriculture as a whole from 1940 to 1967, and further reinforces what was said earlier about the general trend. Here, however, the spatial aspect is of greater interest.

A large land block like the Soviet Union, with a significant north-south dimension, offers the opportunity of working out the relationships between the physical-geographic zones and agricultural zones. The situation is especially favorable because the great zonal structure in the western part of the country is hardly disturbed by mountains. From the Kola Peninsula to Tashkent, zones of tundra, coniferous forest, mixed forest, forest steppe, steppe, semidesert, and desert form a pattern of succession, the sequence finally terminating at the edge of the large mountainous area near Tashkent (Fig. 67, p. 263). The agricultural zones associated with these natural zones are characterized in Overview 22, p. 262, by some examples of crop rotations (Könnecke 1967, pp. 301 ff.).

1. The *coniferous forest zone* is not favorable for agricultural production. Podzol soils, precipitation that declines in an easterly direction from about 600 to 250 mm, severe winters, and above all, a short growing sea-

Table 18: Number, Size, and Inventories of Large Socialist Farms in the Soviet Union, 1940 to 1967

Characteristic	Unit	1940	1950	1960	1967
A. *Kolkhozes*					
Number of kolkhozes	in 000s	235.5	121.4	44.0	36.2
Size of kolkhozes	ha AL	1,429	3,061	6,446	6,000
Cultivable land per kolkhoz	ha	614	1,221	2,962	3,000
Labor force	MY/kolkhoz	...	212	395	...
Cattle	head/kolkhoz	85	224	807	1,092
Swine	head/kolkhoz	35	98	609	599
Tractors	no./kolkhoz	1.8	3.9	14.3	23.6
Combines	no./kolkhoz	0.6	1.4	6.1	7.0
B. *Sovkhozes*					
Number of sovkhozes	in 000s	4,159	4,988	7,375	12,783
Size of sovkhozes	ha AL	12,200	12,900	26,200	22,800
Labor force	MY/sovkhoz	330	334	745	617
Cattle	head/sovkhoz	592	562	1,957	2,017
Swine	head/sovkhoz	459	500	1,715	916
Tractors	no./sovkhoz	17.7	14.8	54.6	54.8
Combines	no./sovkhoz	6.4	6.6	27.9	20.7

Source: Leončarević, I.: Die landwirtschaftlichen Betriebsgrößen in der Sowjetunion in Statistik und Theorie. Wiesbaden 1969, pp. 21 and 26.

son, are little conducive to diversified and flexible land use. The northern boundary of the compact cultivated area runs approximately along the 60th parallel. In European Russia, potatoes, spring barley, vegetables, and fodder crops are cultivated up to the Arctic coast. Individual useful plants also penetrate the river valleys far to the north in Asiatic Russia. Crop rotations A to C evidence the short growing season in several ways:
— first, in the perennial ley cropping, the only system capable of utilizing the growth period from the first day to the last since the soil must be worked in the spring and in the fall;
— also, in the extremely minor role of winter grain cropping, which is possible only after full fallow or after a crop that permits bastard fallow (ley cropping);
— further, in the frequent occurrence of full fallow, which ensures a thorough cultivation of the soil; and
— finally, in the possibility of planting potatoes after potatoes since in this climate there is no need to be concerned about a nematode infestation.

In the east Siberian coniferous forest zone, small cultivated areas are scattered over the extensive forest and grass areas in much the same way as in higher locations among the European sub-Alpine mountains. With a growing period of but 90 to 110 days and precipitation falling to as little as 220 mm, as in Yakutia, crop rotation C is quite understandable. Only the pressure of self-subsistence in these isolated areas can induce cropping under such difficult conditions.

2. The *mixed forest zone* can be viewed as being far more favorable agriculturally. Crop rotations show this clearly: a rich complex of crops, no more full fallow, hardly any perennial fodder cropping, and a larger share of winter grains. The opportunities for winter grain cropping are especially evident in the Byelorussian crop rotation F, in which winter rye can be planted right after the potato harvest. Annual precipitation lies in a range quite favorable for crop farming, 500 to 750 mm. Milk, potatoes, and pork are the main products, the result of relatively prolific forage growth in the wetter areas, an average of about 40% AL in permanent grassland, and a relatively dense and prosperous population in the Moscow-Kiev-Minsk-Leningrad area. Vegetables are also raised near the consuming centers, and sugar beets are increasingly cultivated to the south.

The centrally-planned economy has produced a Thünian spatial pattern in the near vicinity of Moscow. Highly specialized farming zones are found in concentric circles around the capital city, their ordering caused by diminishing transportability of the principal products in an outward direction. Thus fresh milk and vegetable production are found in the innermost ring. Farther from the market center is a ring with emphasis on potatoes, with up to 50% CL in the crop. Manufacturing milk belongs

Overview 22: Crop Rotations in the Agricultural Zones of the Soviet Union (Könnecke 1967)

	Coniferous Forest Zone	
A. Leningrad Oblast	B. Yaroslavl Oblast	C. Siberia
1.-2. Leys 3. Potatoes 4. Potatoes, root crops 5. Peas, oats	1.-2. Leys 3. Flax (for seed), s. wheat 4. Potatoes, legumes 5. Oats 6. Fallow 7. W. wheat, w. rye 8. Spring grains	1. Full fallow 2. Spring grains 3. Spring grains 4. Full fallow, hoe crops 5. Spring grains
	Mixed Forest Zone	
D. Moscow Oblast	E. Grodno Oblast	F. Byelorussia
1.-2. Leys 3. Flax (for seed) 4. Early potatoes 5. W. wheat 6. Potatoes 7. Oats	1. Field forage 2. Flax (for seed) 3. Potatoes 4. Feed corn 5. Field forage 6. Winter grains	1. Potatoes 2. Winter rye-lupine stubble seed (on very thin soils)
	Forest Steppe Zone	
G. Southern Forest Steppe	H. Central Russia	J. Siberia
1.-3. Leys 4. Corn, w. grains-stubble seed 5. Vetch-oats	1. Flax (for fiber) 2. Sugar beets 3. Flax (for fiber) 4. Potatoes 5. Flax (for fiber) 6. Field beans	1. S. beets, potatoes, field beans 2. S. wheat 3. S. Wheat
	Steppe Zone	
K. Kuban Area (irrigation)	L. Virgin Lands	M. Virgin Lands (irrigation)
1.-2. Field forage 3.-4. Rice 5. Vegetables, hoe crops 6.-7. Rice	1. Corn 2. Corn, carrots, s. beets 3. Green forage	1.- 2. Alfalfa 3.- 5. Cotton 6. Legumes, corn 7.-10. Cotton

3. The Agricultural Geography of the East Bloc Countries 263

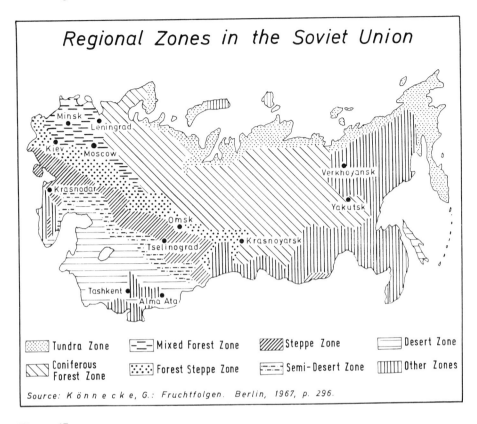

Figure 67

to one of the outermost rings, and is supported by annual fodder crops like feed corn and mixed legumes. Flax raised for flaxseed also earns a place in these crop rotations since it is fairly independent of market distance.

3. The extraordinarily agriculturally valuable *forest steppe zone* is a belt that extends for 4,500 km along the southern edge of the European mixed forest zone and the Siberian coniferous forest zone. A humid climate and thick, humus-rich black soils favor the cultivation of grains and sugar beets. Annual precipitation and growing season steadily diminish to the east, the latter from 220 days in the Ukraine to 160 days in western Siberia. This means a declining importance of winter wheat and an increasing significance of spring wheat. Situations are frequently encountered in the Ukraine where less than 10% AL is in permanent grassland, about 40% CL is in leafy crops (sugar beets, corn, sunflowers, legumes, and potatoes), and 60% CL is in grain. The diversity of cropping is not nearly so great on the other side of the Urals because of the severe winter temperatures and lack of snow and the unfavorable soil conditions.

4. South of the forest steppe zone is the *steppe zone.* Farmers here can rely on 500 mm annual precipitation, and in some places the figure is as low as 250 mm. The efficiency of these amounts for plant physiology is only minor because of the high rate of evaporation. Water and wind erosion are also problems. But the soils are naturally productive, so that their cultivation has become profitable with irrigation. The situation has become even more attractive when the climatic warmth can be used for crops that can thrive in only a few parts of the USSR: rice in the Kuban area or cotton. The potential for further irrigation lies in the lower courses of the great rivers of the Dnieper, Donets, Don, Volga, and Ural.

5. The agricultural handicaps of dryness, very high evaporation, saline soils, and wind erosion are intensified in the *semidesert and desert zones.* Here irrigation is the absolutely indispensable condition for cultivation, since natural rainfall is usually less than 200 mm. Alfalfa/cotton rotations, as well as sugar beets and grain corn, then match vegetables and tobacco in gaining the upper hand. Where irrigation cannot be practiced – and this is the case for by far the largest part of this region, extensive grazing dominates the landscape.

In the last few decades, the Soviet Union has made great efforts to increase its agricultural production. Of the measures that have drastically changed certain agricultural regions, three above all should be named:

1. *Expansion of irrigation systems:* Primitive forms of irrigation were replaced by more effective types and additional irrigation installations were constructed. An additional 800,000 ha of cultivated land were created by irrigation in the area of the Volga-Don Canal. Eventually the Volga will be used to irrigate 15 millon ha in the eastern Volga area and along the Caspian Sea (see Fig. 68). For the more distant future, water reclamation plans of the Soviet Union call for the conversion of more than 25 million ha into productive irrigated land.

2. *Pushing back of the polar boundaries of cropping:* Biological technological advances in plant breeding have made it possible to raise spring barley, potatoes, and winter rye increasingly north of the Arctic Circle.

3. *Cultivation of the virgin lands:* The most renowned and significant part of the virgin lands that has been cultivated is the Siberian steppe zone, which embraces approximately the northern third of Kazakhstan. Here, from 1953 to 1960, 20 million ha of arable land were reclaimed from natural grassland. However, just as in the western U.S., part of this expansion by dry farming has taken place in areas that are much too dry, where the risk of drought is great. To return these lands to extensive grazing is difficult because problems such as wind erosion hinder the regeneration of the natural steppe grass.

A million hectares of virgin lands and old fallow were also reclaimed for cultivation in the Soviet Far East, from 1954 to 1960. Soybeans, grains,

3. The Agricultural Geography of the East Bloc Countries 265

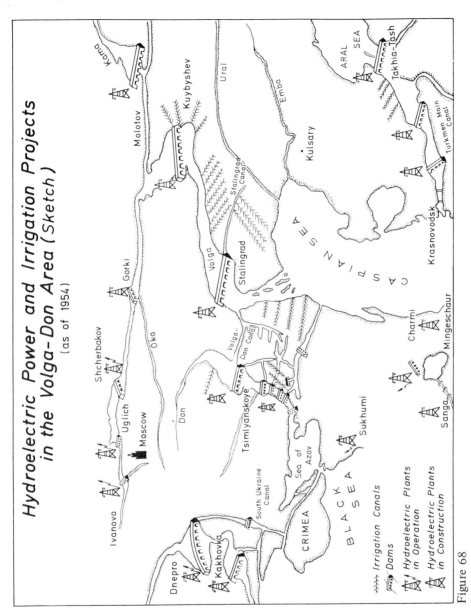

Figure 68

and livestock dominate, with vegetables, potatoes, and dairying concentrating near the cities (Könnecke 1967, p. 308). The major reserves here of virgin lands suitable for cultivation lie in the Amur area, the granary of the Russian Far East, which forms a transition to the agricultural zones of North China.

g) Agricultural Zones of the People's Republic of China

The discussion of the agricultural zones of China must first be prefixed by a general overview of the agricultural economy of the country. Collectivization has been carried out in the following stages (Rochlin and Hagemann 1971, p. 87):

before 1950: *individual peasant farms,* with an average farm size of 1.62 ha (1949); the average holding of the "large landowner" came to 18.2 ha, for large-scale farmers, 4.4 ha, for medium-scale farmers, 1.1 ha, and for small farmers, 0.4 ha (Böttcher 1971, p. 16).

1950 to 1955: *mutual aid teams;* private property remained intact, though draft animals and equipment were used in common by six to fifteen households.

1952 to 1956: *agricultural producers' cooperatives* of 10 to 50 households, with an average holding of 37.5 ha; ownership in common of land, large livestock, and equipment; also private farmland.

1953 to 1958: *advanced agricultural producers' cooperatives;* 100 to 160 households; average holding of 153 ha in 1958; private property restricted to gardens and small animals.

1958 to 1959: *people's communes;* agricultural production, forestry, fishing, local industry, large agrotechnical projects; over 5,000 households, subdivided into production brigades and production teams; average holding of 4,565 ha in 1959.

after 1959: *revised people's communes,* i.e. reduction to as little as 1,600 households and an average of 1,360 ha; stronger emphasis on agriculture; extensive intensification through more irrigation and use of commercial fertilizer; mechanization and cultivation of marginal soils. In 1974, the approximately 50,000 people's communes in the PRC were almost the only organizational form of agriculture.

In Table 19 some economic data on Chinese agriculture are compared with those of several important and widely contrasting countries in order to illustrate the difficult situation of the People's Republic. The two most populous countries on earth, China and India, which take in 36% of the world population, are still densely settled agrarian states. Poverty of capital and lack of industrialization reinforce each other in such countries and contribute to a situation in which agricultural inputs of industrial origin, such as the labor-saving tractor and yield-increasing chemical fertilizers, are still extraordinarily scarce and expensive. Small harvest returns are the result, and this forces a sharp reduction in the consumption of animal products so as to avoid the food losses inherent

Table 19: Economic Data on the Agriculture of the People's Republic of China in Comparison with Some Other Countries, 1974

Characteristic	Unit	PR China	India	Japan	FRG	USA
Population [1,2]	million	740	548	105	61	203
Population density [3]	people/km [2,5]	85	175	291	249	22
Agricultural land, [1,4]	mill. ha	327	178	5.7	13.3	435
of which is irrigated [1,4]	%	23.3	17.7	45.4	2.0	3.7
Cropland: grassland [1,4]	1: ...	1.6	0.08	0.07	0.65	1.3
Chemical fertilizer expenditures [3]	kg pure nutrient/ha AL	16.4	15.6	404	243	40.2
Tractors [3]	per 1,000 ha AL	0.5	1.0	49.6	97.0	10.1
Yields:						
Wheat	q/ha	12.8	11.6	28.0	47.6	18.4
Rice	q/ha	32.7	16.4	58.4	–	49.8
Milk	kg/cow/yr.	544	486	900	3,880	4,666
Food production per head:						
Grains	kg	311	198	157	372	1,000
Meat	kg	11.2	1.1	13.2	58.2	84.0
1974 index	1961/65 = 100	110	96	109	113	112

[1] Latest census. – [2] Mostly 1970. – [3] 1973. – [4] Mostly 1972. – [5] Total area.
Sources: UNO: Statistical Yearbook 1974. New York 1975. – FAO: Production Yearbook 1974. Rome 1975. – Statistisches Jahrbuch über Ernährung, Landwirtschaft und Forsten der BRD 1975. Hamburg and Berlin 1975.

in the processing of feed by the animal. In 1969 only 14.6% of the gross production of Chinese agriculture consisted of animal products, compared to 80.5% of the food production of the German Federal Republic. Nevertheless, not even grain production per head in the People's Republic of China attained the same levels as those in the USA and the FRG, though grain is a much more basic food item in China and India (rice) than it is in the FRG or the USA (wheat, rye).

Because of the large north-south extent of the PRC, the agricultural geography of the country reflects a zonal structure that is just as clear and interesting as that of Soviet Russia. The characteristics of the two countries beautifully complement each other since China lies closer to the equator. The agricultural zones are differentiated in Fig. 69 as follows, from north to south (see Biehl 1973, pp. 14 ff.):

1. First is the *spring wheat-millet-sorghum zone,* with only one harvest per year. It includes the Manchurian Plain in the north and runs along the southern edge of Inner Mongolia. As is the case in the Russian coniferous forest zone, the northern part of the US grain belts, or the wide reaches of Canada, this is an area where the hard winters scarcely permit

any forms of winter grain cropping. The desert-and-oasis province of Sinkiang can be similarly evaluated, for Siberian climatic characteristics are also found there.

Figure 69

2. The *winter wheat-cotton zone,* which begins south of Peking and encompasses the largest part of the great Yellow Plain with the provinces of Hopeh, Shantung, Honan, and Shansi, can be considered as already much more favorable agriculturally. The very reference to cotton in the zonal designation points to a much warmer climate. Here is the one thing that is typical of the greater part of China and of most warm countries: more than one harvest per year. In this zone, it is possible to raise winter wheat that ripens soon enough to allow grain corn to be planted after the harvest and then brought to maturity, all in the same

3. The Agricultural Geography of the East Bloc Countries

year. Cotton follows in the second year, and it of course takes up the entire growing season. Thus the outcome of this crop rotation is three harvests in two years, something that may be viewed as a typical and not isolated occurrence.

3. Two harvests every year can be obtained in the adjoining and still more southerly zone, the *mixed zone*. Rice, which thrives in the heat, is often raised in the summer, while wheat, which has little resistance to heat, is raised in the winter. Exactly the same rotation within a calendar year can frequently be found under similar ecological conditions in other countries: in the constantly wet subtropics, as in parts of the La Plata states; in the dry-summer subtropics, as in Egypt or Turkey in the form of summer irrigation of rice; or in semiarid areas, as in parts of Chile where there is year-around irrigation of both crops.

In other cases, one can find the following rotations in the mixed zone:

1st year winter: winter wheat
summer: grain corn
2nd year winter: legumes
summer: cotton.

Thus here also two harvests are obtained every year.

4. The *rice-tea-silk zone* begins somewhat north of the Yangtze line. The production program here is even more diversified. Now as much as two and a half harvests can be taken from the rice fields per year in a pattern somewhat like the following:

1st year winter: early wheat or green manures (legumes)
summer: two successive rice crops
2nd year winter: late, high-yielding winter crop
summer: more demanding rice.

Since the rice planted is usually wet rice, leveled fields are required. Hills and slopes are consequently used differently. Tea bushes and mulberry trees are cultivated there. Tung oil and citrus fruits are also produced on those sites.

5. The southernmost agricultural zone of the country is probably best designated as the *rice-tropical crops zone*. While paddies are worked as they are in the northern adjoining zone, bush and tree crops that find their physiological optimum only in the tropics gain acceptance as well: oil palms, coconut palms, coffee bushes, silk-cotton trees, sisal agave, pineapples, and the like, Here one cannot help but feel that concessions are being made to a foreign trade economy. With a shortage of currency, strivings for autarky in agriculture lead to a production that is little justified locationally, which in turn will diminish the overall value of that productivity.

That today double cropping is possible on about half of the arable land of China is due not only to the geographic position of the country, but to the expansion of irrigation, from 16% of the cultivated land in 1949 to already 33% of it in 1969 (FAO). Droughts as well as floods must be combatted. Thus efforts have first concentrated on building dikes in central and southern China in order to protect large areas from floods. China has recognized that large flood control projects must provide for industrialization as well as irrigation. Moreover, it is in the fortunate position of being able to capitalize on a half a century of experience by other countries in the modern technology of dike construction and stream regulation. Experience has taught that only if a river is fully harnessed up to its source with a step-like chain of reservoirs, can it be regulated for maximum efficiency and used to the fullest for irrigation, navigation, and hydroelectric power. The projects in the three large river basins of the Yangtze, Hwai, and Hwang are being carried out with especially this concept in mind. By the middle of the fifties, the People's Republic of China had already about 70 million ha of land under irrigation, and had plans for expanding this to 93 million by 1967. How extremely vital these questions of water control are to the country may be seen in a report of the Peking *People's Daily* that in 1960 more than half of the Chinese cultivated area was afflicted by floods or droughts.

IX. Structural Changes in World Agricultural Space with Economic Growth

We shall conclude this book by treating some of the evolutionary and theoretical aspects involved in the tendencies of farms to shift location and to change form, insofar as these changes help to differentiate agricultural space.

1. Forces for Development

The processes of national economic growth give rise to significant evolutions and adaptions among farms because — as explained in Chapter IV-3 — they generate technological advances and these in turn decisively influence the level and structure of agricultural prices. Abel (1967, p. 355) has investigated the long-term development of prices of production factors in Germany, beginning with the end of the Middle Ages (see Fig. 70).

From the 15th century to the first half of the 19th century, i.e. in the agrarian period, land prices rose by sixteen times, prices of capital goods (here represented by iron prices) by about 80%, and wages by only 9%. Hence, in these centuries, there was steadily increasing pressure for intensification, and above all for increasing labor intensity. At the beginning of the industrial period, however, these price trends either changed their course or their rate of increase. From the first half of the 19th century to the end of the thirties in our century, land prices only doubled, while wages increased almost five times and prices of capital goods became cheaper and decidedly so. The industrial era was thus marked by a moderation of labor intensity and an increase in capital intensity. The development of prices in West Germany during the postwar period has continued these trends. On the whole, these trends appear to have, at least for densely settled countries, a definite universality, and are of basic importance for the following discussion.

Technological advances, price and wage trends, and institutional and political changes, are the engines of agricultural development. The consequences are the following:

a) Industrialization raises mass income, demand for food, and with that, agricultural prices. Other conditions being equal, the purchasing power (exchange value) of agricultural products for wages and farm inputs must rise. But because the prices of manufactured inputs are also fall-

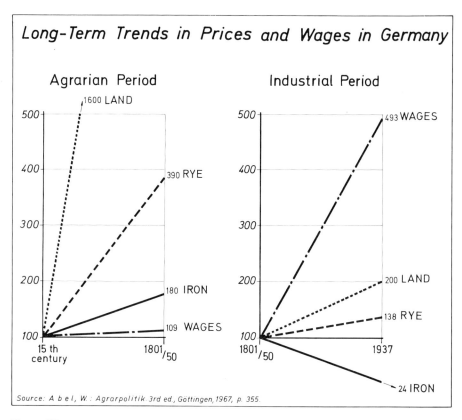

Figure 70

ing with industrialization and wages are increasing because of the competition for labor, the following differentiations in farming operations must occur:

aa) Farming intensity is increased by expanding the intensive enterprises at the expense of the extensive enterprises, and by restricting fallowing;

bb) Capital is substituted for labor, in sparsely settled countries with labor-saving capital goods (machines for working the soil, planting, harvesting, transporting, processing) and in densely settled countries more with yield-increasing inputs (irrigation, chemical fertilization, veterinary hygiene, possibly feed concentrates);

cc) Whether specialization intensity is raised overall, i.e. whether the reduction of labor expenditures in the individual enterprises is more than compensated by the additional application of yield-increasing inputs, depends on the ratio of the wage increase to the decline in capital costs. In densely settled developed countries this question is always answered in the affirmative; in sparsely settled ones, it is not always so.

b) As general affluence increases, and with that, changes in consumption habits of the population as well, prices of animal products definitely increase faster than those of plant products. Livestock raising is therefore stimulated in its present locations and its expansion is encouraged elsewhere. Farms on which there is no livestock become mixed operations, as land use is integrated with livestock raising and the two are mutually promoted through fodder cropping and a manuring economy. In this way a close symbiosis has developed between land use and livestock raising, one that has, among other things, helped Central European agriculture to achieve its very high productivity.

c) Improvements in infrastructure and agricultural structure reduce costs, and thus improve the input-output ratio, and with that increase the intensity of agriculture as well. Expansion of markets and consolidation of the transport net exert a leveling influence on agriculture. Regional differences that were caused by varying distances to market become blurred, as transport costs diminish overall and thus become less variable among the farms. But the more this happens, and the effect of isolation from the market is reduced, the more agriculture adapts to the physical conditions affecting production.

d) At the highest stage of national economic development, wages and income expectations in agriculture rise much more quickly than the prices of agricultural products and the costs of land use and industrially-produced farm inputs. The purchasing power of agricultural products for hand labor falls drastically, and the purchasing power of hand labor for every input rises. The result is that every worker in agriculture must be provided with much more land and capital goods. In the present stage of technology in the FRG, this means that with the great efforts to enlarge the farm, there must also be a reduction in farming intensity by continuing to increase specialization intensity (e.g. less hoe cropping and dairying, but higher productivity per hectare and per animal).

2. Changes in the Factor Combination

If one takes a good look at the progress of economic development, one can roughly distinguish four development stages:
- aa) agrarian countries,
- bb) agrarian-industrial countries,
- cc) industrial-agrarian countries,
- dd) industrial countries.

a) Factor Costs and Factor Combination in Sparsely Settled Countries

Industrialization makes land and labor more expensive, but capital cheaper. The combination mix of production factors is a product of reciprocal actions based on cost relationships. The more the marginal productivity of a production factor falls, the more this factor must be expended relative to the others. One must always strive to expand inputs of each production factor just enough to ensure that its marginal productivity covers its costs. The minimum cost combination is achieved when the marginal yields of all three production factors are proportional to their marginal costs. Consequently expensive production factors, other things being equal, must be applied sparingly, whereas the cheapest production factor must be accorded quantitative superiority in the production process, since as its cost is small its marginal productivity can also be low.

Thus it is possible to deduce the optimum combination of production factors for the different stages of national economic development directly from the theory of marginal productivity. To demonstrate this, we shall turn to a table elaborated by Herlemann (1954, pp. 355 ff.), whose basic ideas we shall be largely following here.

In Fig. 71, which shows the various combination possibilities of the land, labor, and capital factors in agricultural production, the shortage of a factor relative to the other two is indicated by one or two minus signs and the predominance of a factor relative to the others is indicated by one or two plus signs.

The agrarian states represent developing countries in which no differentiation of practical significance for the national economy has so far taken place. The remaining stages depict the result of increasing levels of industrialization, which is usually a characteristic of economic development. But the process of industrialization can take two fundamentally different courses, depending on whether it takes place in sparsely settled countries or in countries with a high population density (see Fig. 72). First, let us deal with the changes in the minimum cost combination arising in the course of national economic development in *sparsely settled agrarian countries* with fewer than 60 inhabitants per 100 ha FL. The greatest part of Central and South America and many African countries can be included in this group. Land is still fairly abundant in these areas, so it is also cheap. Labor is less cheap, while all capital goods purchased from industry are exceedingly expensive. Therefore it is possible to do without high land and labor productivity, whereas in contrast heavy stress must be put on achieving high capital productivity. As a result, land is used generously and capital goods are used sparingly wherever the desired effects can be achieved by increasing labor input. The agricultural system

2. Changes in the Factor Combination

Diagram
Factor Costs and Factor Combinations in the Course of National Economic Development

Typical Examples	Sparsely Settled Countries (Land / Labor / Capital)			Development Progression	Densely Settled Countries (Land / Labor / Capital)			Typical Examples
				A. Cost Relationships				
Zaire, Ethiopia	− −	+	+ +	Agrarian Country	+ +	− −	+ +	Egypt, Ghana, Thailand
Mexico, Zambia, Kenya	−	+	+	Agrarian-Industrial Country	+ +	−	+	Colombia, Iran, Ecuador
Argentina, Uruguay	+	+ +	−	Industrial-Agrarian Country	+ +	+	−	Chile, Finland
U.S.A., Australia	+	+ +	− −	Industrial Country	+	+ +	− −	Great Britain, FRG
				B. Input Combinations				
see above	+ +	+	− −	Agrarian Country	− −	+ +	− −	see above
	+ +	+	−	Agrarian-Industrial Country	− −	+	−	
	+	−	+	Industrial-Agrarian Country	−	−	+	
	−	− −	+ +	Industrial Country	−	− −	+ +	

Costs and inputs of production factors are − − very low   − low   + high   + + very high
Depending on Herlemann, H.-H.: Technisierungsstufen der Landwirtschaft „Ber. üb. Landw.", Hamburg and Berlin, NF, Vol. XXXII (1954), p. 335 ff

Figure 71

as a whole must be regarded as extensive. Farming types typical of this stage are the forest-burning system of shifting cultivation in the humid tropics and the grain-fallow and extensive grassland farming systems in the dry climates.

The next stage is characterized by the countries in which industrialization is beginning to make itself felt in the agricultural structure, the *agrarian-industrial countries.* Although land is still abundantly available, the labor potential in agriculture is being reduced because of competition from industry. Capital goods are less scarce. The effect of all this on the scarcity ratios of the production factors is that land and labor are becoming more expensive, whereas capital is becoming cheaper. In contrast to the stage represented by the agrarian counries, labor productivity must now be increased and less stress need be put on capital productivity. The labor input is therefore reduced through an increase in the capital input. The agricultural system as a whole must still be regarded as extensive.

With the third stage of development, and further industrialization, areas of concern become the *industrial-agrarian countries.* Land is somewhat scarcer but still relatively cheap. Labor, though, has become significantly

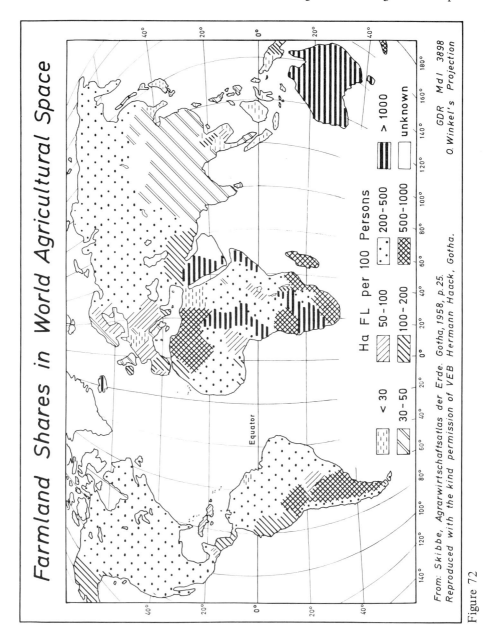

Figure 72

more expensive in the course of the industrialization process, while the cost of capital goods and credit has decreased. As a result labor productivity must again be greatly increased, while capital productivity may be

2. Changes in the Factor Combination

allowed to relax even further. Agricultural operations are now labor-extensive but capital-intensive.

The changes in the minimum cost combination for the agricultural systems sketched so far have been directed toward raising the efficiency of human labor with increasing use of labor-saving machinery and vehicles (mechanization). With the help of this equipment it is possible to open up new areas to cultivation. The bearer of progress in the development of new cultivation techniques is the agricultural machinery industry, whose progress is in turn favored by the expansion of areas under cultivation and the increasing specialization of farms. Output per unit of area of labor-extensive agriculture is still relatively low. Agriculture is at the mechanization stage (Herlemann 1954).

The United States is an example of the *dominantly industrial countries* of stage 4. Because of the very high level of industrialization land and labor have become even more scarce and expensive, while capital goods have become still cheaper. It is therefore now urgent to combine high labor productivity with high land productivity, even if this can only be bought with an exceptionally high capital input and the resulting drop in capital productivity. The agricultural system must be intensive, particularly capital-intensive.

But the increase in capital input takes different forms at this stage. The improvement of the purchasing power ratio between agricultural products and industrially produced inputs now makes it possible to increase the use of yield-increasing inputs in combination with better quality seed, improved crop rotations, better tillage, seed cleaning and seed dressing equipment, and the like. The agricultura machinery industry, which is becoming increasingly concerned with the manufacture of labor-saving machines and equipment (seed drills, mechanical hoes, fertilizer spreaders, cultivators, etc.), and the chemical industry (fertilizers, weed killers, seed dressings, vaccines, etc.) are becoming increasingly important for agricultural production. As a result of the rising cost of land, the use of land-saving inputs has come to take its place beside the use of labor-saving capital goods (Herlemann 1954).

b) Factor Costs and Factor Combination in Densely Settled Countries

The progression of capital investment is reversed in the agriculture of densely populated areas, with more than 60 inhabitants per 100 ha FL. Here land-saving capital goods are most important at first, while labor-saving equipment is introduced only later. The intensification phase here thus precedes the mechanization phase (see Fig. 71, p. 275, right half).

The starting point for development is now the *overpopulated agrarian state.* Countries such as India, Taiwan, and Java fit this category. Land is scarce and therefore expensive, labor is abundant and consequently cheap, and capital goods are expensive because of the lack of industrialization. The result is that while one can be content with low labor productivity, land and capital productivity must be high. These economic goals are reached with production methods that are as capital-extensive as they are labor-intensive. Appropriate farming types are wet-rice cultivation, combined with substantial cropping of roots and tubers, or labor-intensive bush and tree crops.

Economic conditions in these densely settled developing countries are usually unfavorable because industrialization is too slow relative to population growth, the result being an increasing overpopulation of the rural areas with all its adverse effects on labor productivity and living standards. In spite of a high intensity of labor and a predominantly vegetarian diet, the domestic agricultural output guarantees only a minimum caloric supply for the population. An excessively high birthrate, large fluctuations in crop yields, and an inadequate importing capacity increase the difficulties of supply which find their expression in periodic famines and epidemics. Since birth control promises little success and possibilities for emigration are limited, the ultimate solution lies in a well-planned program of industrialization which, in view of the dearth of domestic capital resources, will have to be supported by foreign capital (Herlemann 1954).

The beginning of the industrialization process gives rise at first to the *agrarian-industrial countries.* Land prices are still high, wages are on the increase, but the prices of capital goods are lower. To allow for these cost conditions high land productivity must be combined with growing labor productivity, even at the price of diminishing capital productivity. Thus one farms with increasing capital intensity and decreasing labor intensity.

The prevailing scarcity of land coupled with a still adequate labor supply at first appears to justify all expenditures that serve to increase the productivity of land. The intensity of agriculture, which is based on relatively high labor expenditures and increasing employment of yield-increasing inputs, is expressed in increasing yields per unit of area, though with a still comparatively low labor productivity. Only with the increasing shortage and expense of farm labor that come with *advanced industrialization* are the farmers in the densely settled states finally forced to improve labor productivity with heavy mechanization.

Thus the objective of economic development — whether in an overpopulated or sparsely settled agricultural country — is at all times the *industrial state,* which is characterized by a high input of capital designed to obtain a high productivity of land and labor. A high degree of industrialization makes for a high level of wages and a shortage of labor, so that efforts must concentrate on obtaining high labor productivity. A high population density in relation to agricultural land also makes necessary high land productivity. This is feasible as soon as the price levels of agricultural products increase with the growing purchasing power of the consumer. Thus land and labor productivity must be increased simultaneously. This can only happen, however, because the costs of capital inputs now are low, which make possible a decisive increase in labor productivity through mechanization and in land productivity through intensification. Human labor is replaced by machines and land is replaced by yield-increasing inputs. The ultimate combination of production factors is thus characterized by the economical use of land and labor and a very generous use of capital in every form. Agriculture now is capital-intensive.

c) Differences in the Overall Trend of Change in the Factor Combination

Let us conclude by summarizing. Three elementary production factors in agriculture must be differentiated: land, labor, and capital. Technically they do not have to be combined according to a specific set of ratios. Indeed, within wide limits they are interchangeable. From the economic point of view, though, there is for each stage of economic development only one minimum cost combination of these three production factors, one derived from the relationship between their prices and which alone ensures that production is at its cheapest.

Basic to the attainment of the most economic factor combination is the recognition that, everything else being equal, the higher the price of a production factor the more sparing its input must be. This is so because first, on a farm run on the profit principle the marginal yield (measured in money) should cover the cost of the input of the factor in question, and second, the marginal productivity of the factor sinks if its input is progressively increased in relation to the other two. Meanwhile, the cheapest production factor should be accorded the quantitative predominance in the production process.

Progress in national economic development, which is generally achieved through increasing industrialization, leads to a situation in which capital goods gradually become cheaper while labor becomes more and more expensive. One of the results of this is that human labor is increasingly supplemented by capital goods. Which capital goods will come to the fore in the course of economic development will depend on the availability of land. In sparsely settled agrarian countries the initial concern is the replacement of manual labor. Here then, the mechanization phase precedes the intensification phase. Not until later, when agricultural land begins to grow scarce, will yield-increasing inputs be increasingly expended.

The situation in overpopulated agrarian countries is entirely different. Here land becomes scarce more quickly than labor, so that yield-increasing inputs must be given priority because they make it possible to use land sparingly. Mechanization comes in only later, when labor has become increasingly scarce and expensive with industrial progress.

In short, it is clear that the changes in the minimal cost combinaton in the course of national economic development can vary greatly from one country to another, depending on whether the countries were initially sparsely or densely populated. It also follows that in developing countries with varying economic structures, technological progress will take different forms. Mechanical technological advances will be of primary benefit to the sparsely settled agrarian countries, whereas organic technological advances will particularly lighten the future path of overpopulated agrarian countries.

d) Developmental Tendencies in Farm Size

The development of farm size during economic growth can also be inferred from what has been said (see Fig. 73):

Fig. 73 allows us to conclude that, among other things,

1. changes in farm size in sparsely and densely settled countries show quite different trends.
2. farms in densely settled countries are considerably smaller than those in sparsely settled countries at all stages of national economic development.
3. farm size differences among farms with varying density of settlement are the smallest at the beginning of national economic development because farms are of the subsistence type, with farmers striving to provide all of the food needed by the family.
4. farm size differences are the greatest in the middle stages of economic development. Farms must contract in densely settled countries because of population growth, and the substitution of yield-increasing capital goods for land makes this possible. In sparsely settled countries the substitution of machine capital for labor forces the farmer to cultivate larger areas.
5. farm sizes for the two country groups begin to approach each other at the highest stage of national economic development, though naturally not as closely as at the beginning of development. In the densely settled countries, industrialization now permits rural out-migration

2. Changes in the Factor Combination

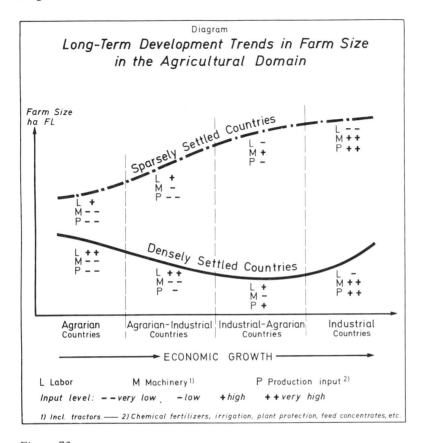

Figure 73

and the establishment of large mechanized farms. In the sparsely settled countries, it is becoming increasingly difficult to obtain agricultural workers, so that the operator must fall back on as advanced a technology and specialization intensity as possible.

Fig. 73 of course shows only the typical long-term trends in farm size in the agricultural domain, and cannot be taken to mean that development will always take place in this way and no other. Above all, one should not get the impression that for each of the stages of economic development there is only one optimum farm size in agriculture. More likely, it can be generally said that a mixture of different farm sizes is the most favorable, no matter what the time or place and regardless of whether farming is of the free enterprise or socialist type.

Fig. 74 shows how profoundly the chain reaction in the current adaption process in West European agriculture has influenced the farm size structure. Of particular note is that

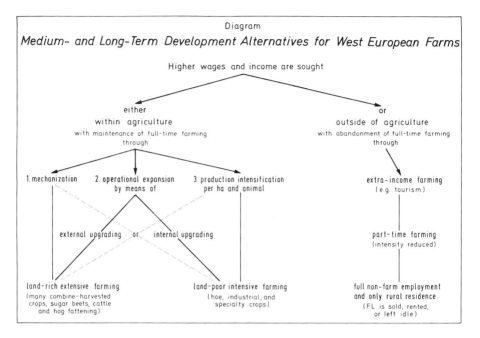

Figure 74

1. above all, wage and income expectations are rising, which raises the question of whether the full-time farm can and should be maintained, or whether agriculture can continue as a part-time operation, with the farmer taking an off-farm job.
2. a greater share of the farmers who wish to remain full-time operators are under considerable pressure to expand their operations;
3. external upgrading with further mechanization leads to land-rich extensive farms, whereas
4. internal upgrading with a major increase in productivity has as its consequence land-poor intensive farms. The conditions for their survival are usually more difficult than are those for the farms named under 3.

Fig. 75 lists six different farm size classes for West German agriculture today, among which there are still transitional types. This figure also shows that the adaption procedures of the six main classes to changing economic conditions are highly variable. Farms of varying size have very contrasting economic and sociological potentialities and limits, as already described in Chapter IV-2-c.

The information in Fig. 75 is not of agricultural-geographic interest in itself. It is of use, though, in providing the necessary comparisons referred to in this chapter on changes in the factor combination, and in

2. Changes in the Factor Combination

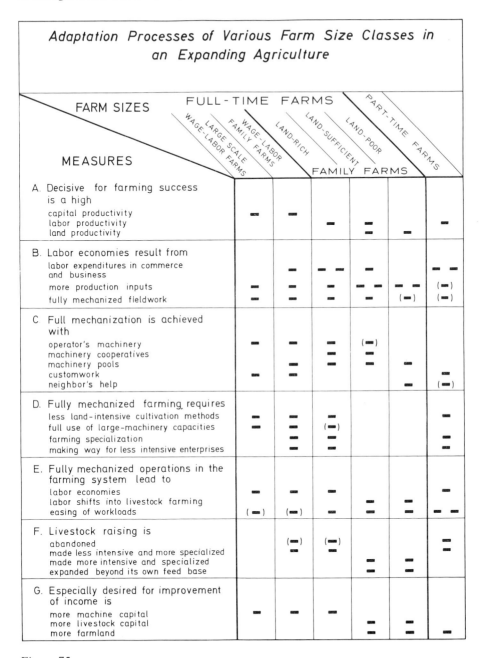

Figure 75

explaining those of the next section on the development patterns of the production program in economic growth.

3. Deversification and Specialization of the Production Program

In the following we present an outline of the historical development of farms as seen from the viewpoint of degree of diversification. Without wishing to ignore the flowing transitions of economic actuality, we shall emphasize only three stages of farming diversity (see Overview 23).

Overview 23: Stages in Farming Diversity

Characteristic	Single-Product Farms	Specialized Farms	Mixed Farms
Number of potential cash crops	one	several	many
Share of sales returns obtained by the principal farm enterprises	> 90%	50–90%	< 50%
Farm organization	one-sided	severalfold	many-sided
Developmental trends	——— diversification ———→		
	←——— specialization ———		

a) One-sided Farms at the Beginning of Development

At the beginning of national economic development, low costs of land use, meager settlement, and high prices for capital goods lead to an extensive form of operations. From 1947 to 1965 Brazilian agricultural production rose annually by 4.5%, whereas yields of the 24 most important crops increased by only 0.5% per year. In a country with only 10 inhabitants per square kilometer it is definitely more economic to farm large areas extensively than to farm small areas intensively.

But extensive farms are frequently organized quite one-sidedly, since
- livestock production and fodder cropping are still frequently lacking where low living standards prevail;
- the principles of crop rotation hardly need to be followed because it is still possible to exchange land instead;
- the requirements of work spacing make for fewer problems because of lower costs; and
- the risk associated with more extensive operations is less than that accompanying more intensive farming because of a smaller input-output ratio.

3. Diversification and Specialization of the Production Program

Examples of extensive one-sided farms at lower stages of national economic development are offered by steppe shifting cultivation, dry farming, dry (hill or upland) rice farms, and grassland farming in the shrub savannas.

Overview 24 shows that the Kikuyu farms in Kenya were still hardly diversified at about 1860. Present-day industrial countries have in most cases also passed through this extensive one-sided stage. Examples are:

Overview 24: Diversification on Typical Small Farms in the Kikuyu District, Kenya

Farm Enterprise	ca. 1860	ca. 1920	ca. 1950	ca. 1960	ca. 1980 [1]
A. Annual crops:					
Corn	X ⎫ M[2]	X ⎫	X	X	X
Beans	X ⎭	X ⎬ M[2]	X	X	X
Sweet potatoes		X ⎭	X	X	X
Potatoes				X	X
Field forage				X	X
Vegetables				X	X
B. Perennial crops:					
Bananas			X	X	
Coffee				X	X
Tea					X
C. Livestock enterprises:					
Goats [3]	X	X	X	X	
Slaughter cattle	X	X	X		
Dairy cattle				X	X
Swine					X
Poultry					X
Number of farm enterprises	4	5	6	10	11

[1] Preliminary estimate. — [2] M = Mixed cropping, usually in shifting cultivation. — [3] On roadside pasture.
Source: Ruthenberg, H.: Farming Systems in the Tropics. Oxford 1971, p. 250.

— pure grain cropping with insignificant livestock raising in Germany, before the introduction of clover and hoe crop cultivation;
— pure grain cropping with shifting cultivation on the Don Steppe, still at the beginning of our century; and
— corn monoculture in parts of the Transvaal, up to the present.

b) Diversification Tendencies in the Pre-industrial Era

National economic development at first stimulates both intensification and diversification. Economic and population growth in the pre-industrial period cause the costs of land use to rise faster than labor costs. This in turn forces a progressive increase in labor intensity. The diversification process is initiated by the growing influence of the integrating forces of farm organization, a consequence of increasing labor intensity.

When the farmer combines farm enterprises into an organic whole, three economic benefits are gained. Production costs per areal and product unit are reduced because of the constant farm labor charge; yields per hectare are increased because soil fertility is more completely exploited; and the cash value of the farm output also frequently increases as well because of the better possibilities for disposal (Brinkmann 1922). A few examples may be cited to illustrate these processes:

The principles of crop rotation gain ever greater importance with increasing labor intensity. The greater the share of cultivated land in intensive crops that deplete the soil, the more a compensation must be sought in soil-building crops with narrow row spacing, more coverage of the soil, a longer growing season, and good rooting capabilities (legumes, catch crops).

So long as chemical fertilizers continue to be expensive, the land use system must also provide nutrient balance (nitrogen from legumes), which requires ever greater diversity the more the intensive nutrient-demanding crops come to the fore. Intensive crops also show a greater need for organic fertilizer substances than do extensive crops. With increasing intensity, this situation also encourages diversification on farms.

Intensive forms of livestock farming, such as dairying, place greater demands on feed balance (nutrient concentration, protein ratio, more uniform fodder yield through the year) than do extensive forms (raising wool sheep or oxen), and thus contribute to diversified fodder cropping.

So long as market contacts are lacking, every effort must be made to spread costs, which leads to diversification of farm operations.

Intensive farm enterprises are usually riskier than extensive ones since
— the input-output ratio is smaller;
— the products of many intensive farm enterprises (potatoes, beets, yams, sweet potatoes, manioc) are less easily transported and stored, so that harvest fluctuations cause wide price swings; and
— intensive farm enterprises are in general more sensitive than extensive enterprises to damage wreaked by diseases and pests.

The more intensively the farmer farms, the more necessary it is for him to avoid the greater threat of failure by diversifiying.

3. Diversification and Specialization of the Production Program

Finally, considerations of work spacing also increase the necessity of diversification with growing labor intensity. Intensive crops are generally more onerous than extensive crops in their demand on labor at specific times. Thus, for example, the planting and harvesting of the
- extensive crops of millet and oil palms require 32 and 42% of the total labor expenditure, respectively, whereas the same operations for the
- moderately intensive crops of sweet potatoes, peanuts, and rubber demand 71, 81, and 83% of the total labor expenditure, respectively.

Hence the more intensive the farming, the more reliance must be put on combinations of enterprises having complementary labor demands. Labor in the pre-industrial period is provided by human and animal force, which by their nature give rise to fixed costs and of course potential joint costs, the latter costs being an expression of the ability of man and animals to be used in a variety of enterprises. These can and must be spread over several enterprises. Also the modest amount of machinery and equipment that is available in this type of agriculture still entails, in great part, joint costs. This materiel can be employed in many up to all enterprises.

Looking back, one can see that intensive farm enterprises are far more dependent on diversification than extensive enterprises are. With weak national economic development, they usually can constitute only one part of the farm operation and require supplementation with complementary branches of production. There are exceptions of course (plantations specializing in cotton, sisal, tea, sugarcane, or coffee), but as a rule monocultural operations are possible only with extensive enterprises and intensification above all compels diversification.

The development of typical farms in the Kikuyu country (Overview 24, p. 285) exemplifies these relationships in a more concrete way. From 1860 to 1960 the number of farm enterprises rose from four to ten.

Our considerations up to now, under the assumption of only a weak involvement of farms in the national economy, have left us with the simple function:

  Rising land values → increasing intensity → greater diversification.

c) Specialization Tendencies in the Industrial Era

Once, though, a particular stage in the involvement of agriculture in the national economy is attained, then the relations between intensity and diversification change. Specialization occurs.

From 1954/55 to 1964/65 the average number of enterprises for 128 farms in Hanover (over 50 ha AL) fell from 14.1 to 11.1 Today farmers often have only two to three enterprises. The causes of this phenomenon lie in the fact that the influences of all integrating forces lessen with advanced national economic development. Let us stay at first with work spacing. With weak involvement in the national economy, human and animal labor as well as the machinery and equipment already on hand all press for diversification. This occurs because all of them give rise to potential joint costs, thus not only permitting but requiring the insertion of complementary enterprises so as to repay the costs of their excess capacities. With more involvement in the national economy, however, labor becomes so scarce and expensive and machines so comparatively cheap that harvesting must also be mechanized. The required machinery usually creates absolute special costs. They are good for only one purpose. Their fixed costs cannot be spread over several enterprises, but instead must be met by providing sufficient scope for the enterprise that the respective machine serves.

The costs that are incurred in the employment of harvesting machinery and are thrusting into the concentration of forces in the farming economy are significant. One need only think of the capital investments and costs for such things as cotton or corn picking machines; self-propelled combines for grain, corn, millet, sorghum, or soybeans; and sugarcane, potato, and beet harvesters. Farm operations are organized around these machine aggregates in such a way that only on the large farm is it still possible to practice diversification with a fully mechanized technology. For the mass of the family farms, there remains no other choice but to specialize, since it is impossible to mechanize fully six or eight enterprises with the farm area available. Motorized traction power also facilitates farm specialization, since tractors can cut costs because they need to be used far fewer hours than are oxen or horses.

If one concentrates on the core of these facts, then it can be generally said that:
— Labor intensity promotes diversification, since human and animal labor give rise to potential joint costs.
— Capital intensity promotes specialization, since at the very least, harvesting machinery allows absolute special costs to accrue.

Thus, in the transition from labor- to capital-intensive agriculture, labor considerations force an elimination of diversification and a switch from a joint to a specialized operation. A series of technological aids, which industry offers agriculture at favorable prices at this stage, also permits this specialization. To give only a few examples of how these and other agents have encouraged specialization:

Land exploitation as an integrating factor in the transformation of the farm is losing its force with greater involvement of agriculture in the national economy. Chemical fertilizers, therapeutic plant protection, and herbicides are loosening the bonds of crop rotations.

3. Diversification and Specialization of the Production Program

The conquering of place-specific barriers through expansion of the infrastructure eliminates most self-subsistent operations on the farm, since it makes possible the purchase of food.

Also, the pressure for a many-sided farm organization so as to avoid losses is moderating, for society is more and more relieving the farmer of risk. Insurance companies cover harvest risk. The state, though, is increasingly assuming the market risk as it develops ever better agricultural instruments for moderating competitive and seasonal price fluctuations.

Finally, farm specialization is being promoted as the increasing scientific penetration of agriculture places ever greater demands on the knowledge and abilities of the farmer, no matter what the enterprise. It is becoming impossible for the individual to acquire all the special knowledge and experience required by many farm enterprises. The farmer is becoming a specialist.

In summary, one can say: Strong integration into the national economy leads to specialization on farms and plantations, to the formation of stress points in farm organization.

d) Stages in Farming Diversification and Specialization in the March Toward Total Economic Integration

Agriculture that is little developed is labor-intensive, whereas advanced agriculture is capital-intensive. Labor intensity, however, leads to farm diversification, while capital intensity culminates in farm specialization. With low land values, farming is carried on extensively rather than intensively. One-sided farms must be developed. If land values, and with them, intensity, rise, then farms must become more diversified so long as they are little integrated into the national economy. With increased integration of agriculture into the national economy, farming operations again become more specialized, but without attaining the one-sidedness that was associated with low land values. The processes are schematized in Fig. 76.

As shown, usually there is a correspondence between

1. monocultural farms and extremely extensive agriculture;
2. mixed farms and labor-intensive agriculture; and
3. specialized farms and capital-intensive agriculture.

But since agricultural intensity changes in the course of national economic development through a sequence of

　　　　extensiveness → labor-intensiveness → capital-intensiveness,

the following change in farming diversity must be expected:

　　　　monoculture → diversification → specialization.

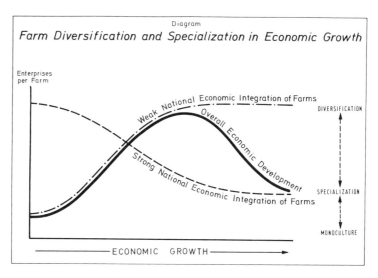

Figure 76

4. Changes in Farming Systems in Selected Climatic Zones

In the course of economic development land becomes increasingly scarce, so that it must be farmed ever more productively, i.e. used more intensively. With the change from a sparsely to a densely settled agrarian state, there is at first an increasingly large labor force available, while all forms of capital goods continue to be very expensive. Hence the necessary agricultural intensification in this phase of development must be effected by increasing labor expenditures. Only when considerable industrialization develops does the labor force decline and, as with land, become more expensive while commercially-produced capital goods become cheaper. In this latter phase of development, further intensification must be strived for by increasing capital investment.

In the course of economic development, the productivity of land must first be increased, and later also that of labor. However, the triad of intensity stages that satisfies this demand follows a sequence of:

extensiveness → *labor-intensiveness* → *capital-intensiveness.*

This resume of the last section will help us to explain and to understand the following developmental sequences. It already explains the old cultural-historical developmental theories:

Richard Krzymowski's three-stage theory: hunting and sishing → nomadism → crop farming.

Eduard Hahn's three sequential forms:
      1. gathering → hunting and fishing (only in particular locations);
or   2. gathering → hoe culture → garden culture;
or   3. gathering → hoe culture → plow culture → nomadism.

Here, though, developmental sequences peculiar to a few selected climates will be listed and briefly explained (Andreae 1972).

a) Tropical Rainforest Climate

Farming systems in the constantly wet tropical rainforest climate will always show a marked tendency to remain land use (i.e. cropping) systems and will in the course of economic development change according to the following sequence:

Stage 1: haphazard forest-burning system of shifting cultivation;
Stage 2: organized forest-burning system of shifting cultivation with natural forest fallow;
Stage 3: organized forest-burning system of shifting cultivation with man-induced forest fallow;
Stage 4: organized forest-burning system of shifting cultivation with plow cultivation;
Stage 5: legume fallowing;
Stage 6: grass fallowing;
Stage 7: ley farming; and
Stage 8: bush and tree crop farming.

This progression of stages is consistent with the need to raise land productivity first and then labor productivity. It is therefore the usual pattern, though this does not exclude the possibility of some stages being skipped where there is a very rapid and dynamic economic development. In some cases, also, the sequence can change. The haphazard forest-burning system of shifting cultivation was replaced directly by oil-palm growing in parts of Nigeria and by cocoa production in southern Ghana, because the geographical position of these countries on an ocean allowed them to benefit from high prices in the world market. The combination of man-induced forest fallow with plow cultivation presents problems and is skipped in many countries. Foreign trade often makes bush and tree cultivation possible at a time when the home market still cannot fully absorb the animal products of ley farming.

In general, however, the sequence of farming systems given here will prevail during economic development. The order is realistic because an expansion of the proportion of productive land with a simultaneous. increase in hectare yields is at first achieved with increased labor expenditures. The change from the haphazard to the organized forest-burning system of shifting cultivation simply increases the proportion of cropped land, i.e. simply increases the amount of labor used. The change from natural to artificial fallow does require some capital in the form of plant stock, but this kind of capital can be created by labor on the farm itself. Only with the change from hoe to plow culture do capital expenditures, in the form of implements and draft animals, increase substantially. When forest fallowing is replaced by legume fallowing, additional capital is required for seeds, and still more for fertilizers when legume fallowing is replaced by grass fallowing. The highest form of land use, ley farming or bush and tree crop farming, necessitates considerable additional capital investment in the respective form of livestock or permanent plant stock.

b) Humid Savanna Climate

In the humid savanna the emphasis in farming will probably change in the course of economic development as follows:

Stage 1: forest-burning system of shifting cultivation;
Stage 2: bush-fallowing system of shifting cultivation;
Stage 3: grass-fallowing system of shifting cultivation;
Stage 4: rainfed farming without fallowing and livestock raising;
Stage 5: rainfed farming without fallowing but with livestock raising; and
Stage 6: irrigation farming without fallowing but with livestock raising.

Here, too, development is at first directed toward an increase in the proportion of productive land through a continuing reduction in the number of fallow years, the fallow vegetation consequently changing from forest via bush to grass. Later an attempt is made to dispense with fallow altogether in the interest of land productivity, and it soon becomes obvious that livestock are a prerequisite of permanent rainfed farming in the interests of fertilizer production. The increased demand for animal products now makes this possible.

For its part livestock raising, with its demands for a seasonal feed balance, helps to bring about the last and decisive step to irrigation farming. The humid savanna is perfectly suited for this, since it has a relatively small water requirement but abundant supplies of groundwater. A rational system of dams and reservoirs is now also possible because rivers flow throughout the year. As long as land productivity is the first concern and labor still cheap, various methods of irrigation can be used. Beginning with a specific wage level, sprinkler irrigation will become progressively more economic, since it requires much less labor and capital equipment becomes cheaper in the course of development. Thus sprinkler irrigation, together with chemical fertilization, plant protection, and the like, ensures high labor productivity while maintaining high land productivity.

c) Dry Savanna and Steppe Climates

In spite of the greater need for irrigation in the dry savanna, irrigation farming finds fewer opportunities because of the shortage of water. In this area, the decisive sequence of changes in farming systems in the course of national economic development can be described thusly (see Fig. 77, p. 294):

- *Stage 1:* extensive grassland farming;
- *Stage 2:* savanna shifting cultivation;
- *Stage 3:* grain-fallow farming with extensive grassland farming (ranching);
- *Stage 4:* crop rotations organized to include fodder and legume cropping, in conjunction with extensive grassland farming; and
- *Stage 5:* penetration of dryland crop farming by moderately intensive grassland farming.

At first the fodder cropping in stage 4 serves primarily the aims of soil biology rather than feed production. Planted grass fallow is at first introduced primarily in support of wheat and barley cropping (steppes) or of millet, sorghum, or corn cropping (dry savannas), and not in support of livestock farming during times of feed shortage. This does not exclude, of course, the possibility of using fodder crops for grazing during the dry season or for producing hay. In this way they form the link between cattle fattening and corn, millet-sorghum, wheat, or barley cropping.

Fodder cropping allows rotations to be used for grain cropping and broadens the feed base for cattle fattening in the dry season. It supplies root humus for the cultivated crops and winter feed for the grazing animals; the animals transform part of this feed into manure which in turn benefits the grain crops. In this way the two farm enterprises of slaughter cattle production and grain cropping, which once existed side by side in isolation, now give rise to an association, an integrated whole, a farming system.

In the course of further development, more productive field crops are also cultivated, such as garbanzos, peanuts, sunflowers, sesame, and eventually even crops like cotton. This crop farming as a whole continues to encroach upon the natural pasture areas, until it finally occupies

IX. Structural Changes in World Agricultural Space

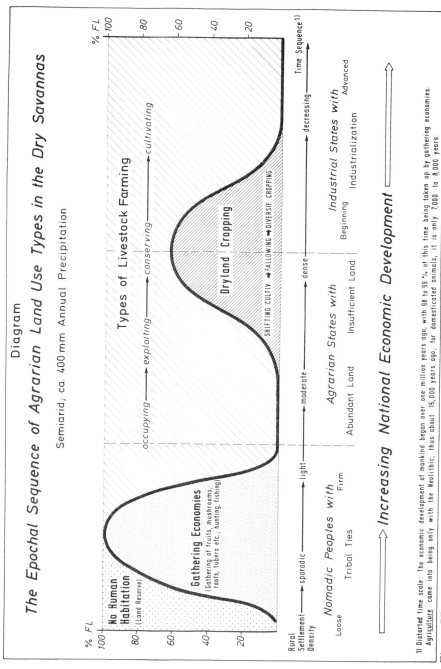

Figure 77

4. Changes in Farming Systems in Selected Climatic Zones

all the farmland where conditions are suitable for cultivation (gradient, depth of topsoil, groundwater table, etc.). The penetration of dryland crop farming by moderately intensive ranching, in stage 5, is due ultimately to the fact that even with the most modern technology, the yields of the field crops in these dry locations are not sufficient to satisfy the income demands typical of the developmental stages of a highly advanced national economy. Ranching is clearly superior to dryland crop farming in labor productivity.

However, in the shrub savannas and dry steppes, which lie outside of the potential rainfed farming zone, the evolution can occur only within the ranching or extensive grassland farming system. According to the principle of

extensiveness → labor-intensiveness → capital-intensiveness,

the sequence of changes in the course of national economic development will be as follows (see Fig. 43, p. 172):

Stage 1: nomadic grazing;

Stage 2: sedentary grassland farming with seasonal migrations;

Stage 3: bridging of the dry season with the farm's own feed reserves and by drawing on the animals' fat deposits;

Stage 4: increasing the number of watering places;

Stage 5: increasing pasture subdivision;

Stage 6: purchasing additional feed; and

Stage 7: farms producing on the farm all feed for the dry season.

This gradual advancement of capital investment is set in motion both by an improvement in the market exchange value of animal products for farming inputs and by an improvement in market access for the farm. The improvement of the feed balance also contributes, in accordance with physical conditions, to the increase in productivity of livestock farming types.

d) Marine Cool-Summer Climate

An example of the changes in farming systems in a mild and humid middle latitude climate is provided by the genesis of the crop rotations of central England (see Fig. 78). As in many other climatic zones, cropping here had its beginnings in shifting cultivation. This was subsequently replaced by primitive ley farming and later, beginning with the eighth and ninth centuries AD, by the three-course rotation with fallow. Toward the end of the eighteenth century, the early industrialization of England had already caused agricultural prices to rise so much that the consequent intensification had lead to the Norfolk rotation with no less than 25% CL in hoe cropping.

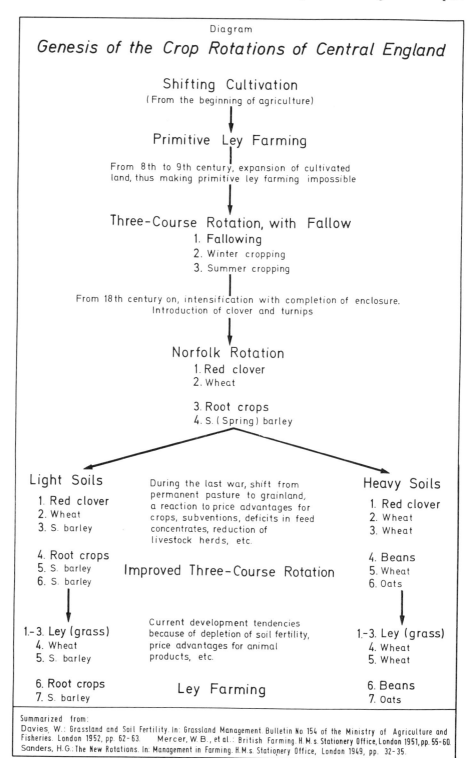

Figure 78

4. Changes in Farming Systems in Selected Climatic Zones

This positive development of English farming came to an abrupt end in the 1870s, when England adopted the free trade doctrine and sacrificed its agriculture for industrialization. From 1873 to 1896 wheat prices in England fell by about half, barley prices by a third, and potato prices by more than half; meanwhile wages rose by a third. During the 1870–1937 period, England converted 40% of its cultivated land to permanent grassland. With the convulsions of the world economic crisis in the 1930s and the second world war, this trend was reversed: 2.7 million ha of permanent pasture were plowed up from 1937 to 1944 and incorporated into rotations. While at first the emphasis was on crop production within the framework of an improved three-course rotation because of the need for food, a shift was made after the war back to the climatically-favored livestock raising, though now less on the basis of permanent than on rotation pasture. The ley farming system is an intensive form of grassland farming.

e) Continental Cool-Summer Climate

The development of English agriculture is not too typical for a middle latitude climate since it has been subjected to drastic foreign trade and other political measures.

Aereboe (1923, p. 320) characterizes the main line of development of farming types under the influence of changing economic conditions as follows:

Stage 1: extensive grassland farming;
Stage 2: one-field farming with shifting on pastureland;
Stage 3: two-field farming with shifting on pastureland;
Stage 4: three-field farming with shifting on pastureland;
Stage 5: three-field and fallow farming without shifting;
Stage 6: improved three-field farming, fallow partly grazed;
Stage 7: improved three-field farming, fallow completely grazed;
Stage 8: rotation farming with intensive legume cropping;
Stage 9: rotation farming with intensive fodder cropping (clover);
Stage 10: rotation farming with intensive hoe cropping and green manuring;
Stage 11: pure farming of hoe crops, grains, and green manure crops; and
Stage 12: market gardening forms of cropping.

Aereboe's conception of the evolution of farming systems was strongly conditioned by his knowledge of steppe agriculture in the Government-General of Voronezh. The situation in the Voronezh district suggested to him that all agricultural development was preceded by extensive grassland farming, which later evolved into grain farming. Still later fodder crops, as well as legumes and oil crops, would be cultivated. Hoe cropping would then come in, and gradually it would displace the earlier systems, first the cropping of legumes and than fodder cropping.

Outlook: The Agricultural Evolution Theory of Friedrich Aereboe in the Light of this Agricultural Geography

Many of the highly absorbing chapters in Aereboe's *Allgemeinen landwirtschaftlichen Betriebslehre (General Farm Management)* that deal with price relationships in their influence on farm organization focus on the evolution of farming systems. It generally begins with extensive grassland farming as the first stage. But extensive grassland farming is not the first stage of agricultural development in all climatic realms. To be sure, Aereboe recognized that extensive grassland farming was not at the beginning of historical development everywhere, for man to some extent had cultivated land before he had succeeded in domesticating productive animals. However, Aereboe at first did not perceive that even in the geographic coexistence of the various economic development stages of today, extensive grassland farming does not always make up the lowest stage.

An extensive grazing operation in the tropical rainforest belt is in no way possible, since pasture would be taken over by bushes and forest in short order. In addition, the nagana disease that is carried by the tsetse fly still makes it impossible to raise cattle or sheep in wide areas. It is the hunter, fisherman, and gatherer, and not the herdsman, who mark the beginning of development in the constantly wet tropical rainforest climate. Nor is the beginning of any agriculture later associated with extensive livestock raising; instead, farming is one of pure plant production using digging stick and bush hoe.

Pure grain farming without shifting as a middle stage in the development of cultivation also does not exist in the tropical rainforest belt. Permanent grain cropping is possible here only in the form of wet rice cultivation, but this must be considered as hoe culture from the farm management point of view, given the requisite tillage procedures in the inner tropics which usually include the starting of the plants in seedbeds and their subsequent transplanting in the fields. Grain cropping in the strict sense can never acquire great significance in the tropical rainforest, so that the Aereboe model of farming development does not apply there. His thesis, ,,Grain farming remains the backbone of cultivation at all stages of agricultural development. . .," also requires qualification for other climatic zones. Even in the warm-temperate oceanic climate of eastern United States, up to about the Mississippi, grain cropping plays only a subordinate role despite the currently high wage levels. This is because the hot summer ensures that leafy crops such as cotton, grain corn, soybeans, and peanuts, with their amenability to full mechanization in all operational phases, are superior to grain cropping in the competition for production factors.

In other world climatic belts, of course, the thesis of Aereboe that extensive grassland farming signifies the first stage of any agriculture holds true; not, however, his conception of extensive grassland farming developing further via grain farming. On the hot and dry steppes and semi-deserts the amount of precipitation does not even allow the practice of the dry farming system, grain-fallow farming. Thus here, beyond the dry boundary of cropping in the marginal zone of the inhabited area, further farm development occurs in one great leap, from extensive grassland farming directly into intensive irrigation farming. Examples of this are offered by such places as the American Intermountain states, the South African Karroo, and the semideserts of the Russian area.

In later years, Aereboe himself recognized the limitations of his evolution theory with respect to climatic zones. Five years after the last edition of his *Betriebslehre,* in 1928, Aereboe came out with his *Agrarpolitik (Agrarian Policy),* in which he again devoted an extensive chapter to farming systems and their dependence on price relationships and the physical qualifications of agriculture. Toward the end of this section, he said (Aereboe 1928, p. 116):

"With that, the essential points concerning the economic systems of agriculture in the middle latitude zones have been covered. Space prevents a more careful condideration of the same items for the subtropics and tropical zone. If that were to be done, then for the subtropics the various systems of irrigation farming would demand special attention, and for the tropics the forms of plantation farming there would have to be discussed. With regard to tropical plantation farming, it is to be noted that in all rainy areas of the tropics weed growth is so prolific that not only is the cleared land quickly covered by large and small weeds, it is also very rapidly taken over by bushes and forest if clearing operations are not constantly maintained. But this precaution again requires so much labor that the minimum expenditure per hectare of land must be extremely large. Thus cultivation on the scale of mid-latitude agriculture is generally not possible in tropical agriculture."

These observations by Aereboe also are in need of debate. Both of the tropical farming systems named by Aereboe force in fact a high minimum intensity; yet irrigation farming is even today not the most important farming system in the subtropics, and plantation farming was also at that time not the most widely distributed farming system in the tropics. The conclusion by Aereboe that "cultivation on the scale of mid-latitude agriculture is generally not possible in tropical agriculture" needs the qualification that Aereboe had already expressed two decades earlier: "Tropical and subtropical agriculture are from the beginning subject to pressures for a much higher farming intensity than that of the colder zones." Even if Aereboe were to have meant by "rainy areas" only the tropical rainforest zones, his conclusion would hardly hold. His reference to the prolific weed growth and the threat of bush and forest encroachment is of course quite correct. However, the forest in the humid tropics is in several ways not an enemy of crop farming, but its best friend. During the many years of forest fallow, the trees help man to fight the weeds, to provide for concentration of plant nutrients, and to guarantee a free regeneration of soil fertility. Thus even today the forest-burning system of shifting cultivation holds its own in the humid tropics, a system which up to now has been much more important than plantation farming and today is still practiced by more than 200 million people scattered over 30 million km².

This system of shifting cultivation must be considered as a thoroughly extensive form of cultivation. In extreme cases, only one year of cropping is followed by eighteen to twenty years of fallow, which makes for extremely small capital expenditures, consisting almost solely of hand tools and approaching zero value. Labor expenditures, then, when related to area, are only about 2.5 MY per 100 ha of usable land, whereas grain-fallow farming on the Spanish Meseta requires up to 3 to 4 MY/100 ha AL. Although sheep grazing on the Spanish Meseta uses only 1.5 MY/100 ha AL, attendant capital expenditures for livestock, wells, watering installations, and fences, as well as labor aids come to as much or half again as much as that of the monetary labor expenditures. There are certainly even more extensive forms of grassland farming, but the difference from the forest-burning system is less one of amount than orientation of intensity. While the forest-burning system of shifting cultivation in the tropical rainforest zone is more labor-intensive than the most extensive grassland farming, it is also even more capital-extensive and overall must be regarded as a definitely extensive form of cultivation.

The thesis of Aereboe on the tropical rainforest zone applies to the perennial crops. With respect to annual crops, however, this can only mean that extensive cropping occurs in forms different from those in the middle latitudes, and not that there is actually none. Quite the opposite is true: cultivation in this climatic zone suffers less from an overly high intensity minimum than from a deficient capacity for intensification. The shifting cultivation system can at the maximum feed only 40 to 50 people per square kilometer, and is the cause of undernourishment in many overpopulated developing countries. Only 20% of the land can be used, since for a maximum of three crop years a minimum of twelve years of forest fallow must follow. How to develop cultivation systems that allow a higher intensity stage but do not destroy soil fertility has been a cardinal problem of tropical agriculture for decades (see p. 136 ff).

Finally, the twelve-stage theory of Aereboe (see p. 297) cannot even be applied to the areas of cool middle latitude climates without restrictions, though he claimed a much greater general validity for it there. The twelve developmental stages posited by Aereboe are a succession of progressively increasing farming intensities. In actuality, however, farming intensity reaches its culmination point in highly industrialized states and on large farms, and then begins to decline. Aereboe proceeded on the assumption that the purchasing power of agricultural products for capital and labor would continually increase. But in West Germany, since about 1950, the exchange value of agricultural products has increased only for capital goods, whereas it has fallen for wage labor. The result has been a decrease in intensity in farm organization, which has led in part to the elimination of livestock and hoe crops. Aereboe's hypothesis runs to the effect: always organize more intensively and always operate more intensively. In contrast, the guiding purpose of farm development in Central Europe today is:

organize extensively — operate intensively,

i.e. less intensive farm enterprises, but higher productivity per hectare and per animal.

These considerations of the agricultural-gegraophic limits of the Aereboeian evolution theory do not in any way lessen the value of its penetraiting and logically-based insights. They only show its unavoidable limits. An all-embracing and exhaustive genetic model of farming types can never and will never come from one pen. The types of world agriculture are still too many, their zonal structure and spatial differentiation all too diverse.

Bibliography

Abel, W. 1967: *Agrarpolitik.* 3d ed. Göttingen.

Aereboe, F. 1923: *Allgemeine landwirtschaftliche Betriebslehre.* 6th ed. rev. Berlin.

Aereboe, F. 1928: *Agrarpolitik. Ein Lehrbuch.* Berlin.

Agrawal, G.D., and Bansil, P.C. 1969: *Economic Theory as Applied to Agriculture.* Delhi.

Anderson, J.R. 1970: *A Geography of Agriculture.* Dubuque, Iowa.

Anderson, J.R. 1973: *A Geography of Agriculture in the United States Southeast.* Vol. 2 of *Geography of World Agriculture.* Ed. by G. Enyedi. Budapest.

Andreae, B. 1964: *Betriebsformen in der Landwirtschaft.* Stuttgart.

Andreae, B. 1965: *Die Bodenfruchtbarkeit in den Tropen. Nutzbarmachung und Erhaltung. Betriebswirtschaftliche Überlegungen für die Arbeit in Entwicklungsländern.* Hamburg and Berlin.

Andreae, B. 1966: ,,Weidewirtschaft im südlichen Africa. Standorts- und evolutionstheoretische Studien zur Agrargeographie und Agraröknomie der Tropen und Subtropen. No. 15 of *Geographische Zeitschrift, Beihefte Erdkundliches Wissen.* Wiesbaden.

Andreae, B. 1972: *Landwirtschaftliche Betriebsformen in den Tropen. Bodennutzung und Viehhaltung im Spannungsfeld von Tradition und Fortschritt.* Hamburg and Berlin.

Andreae, B. 1973: *Strukturen deutscher Agrarlandschaft.* Vol. 199 of *Forschungen zur deutschen Landeskunde.* Bonn-Bad Godesberg.

Andreae, B. 1974: "Diversifizierung und Spezialisierung der Farmwirtschaft im Tropenraum," *Berichte über Landwirtschaft,* Vol. 52, New Series, pp. 497 ff.

Andreae, B. 1974: *Die Farmwirtschaft an den agronomischen Trockengrenzen.* No. 38 of *Geographische Zeitschrift, Beihefte Erdkundliches Wissen.* Wiesbaden.

Andreae, B. 1976: "Strukturzonen und Betriebsformen in der europäischen Landwirtschaft," *Geographische Rundschau,* Vol. 28, No. 5.

Andreae, B. 1976: "Agrarsysteme," *Handwörterbuch der Wirtschaftswissenschaft,* Vol. 1, pp. 155 ff. Göttingen, Tübingen, and Stuttgart.

Andreae, B. 1977: "Farming Regions in the Tropics: Environment, Geographical Comparisons and Socio-Economic Development."
W. Krause (ed.) *Handbook of Vegetation Science,* Vol. 2: *Agriculture.* The Hague.

Atlas der deutschen Agrarlandschaft, 1962–1972. Wiesbaden.

Atlas de la France rurale, 1968. Paris.

Aubert, H.-J. 1964: "Die Entwicklung der Plantagenwirtschaft Sri Lankas im Rahmen der Agrarwirtschaft nach der Erlangung der Unabhängigkeit." Cologne Dissertation.

Auty, R.M. 1976: "Caribbean Sugar Factory Size and Survival," *Annals of the Association of American Geographers,* Vol. 66, pp. 76 ff.

Bachman, L.L., and Christensen, R.P. 1967: "The Economics of Farm Size," In: Southworth, H.M., and Johnston, B.F. (eds.): *Agricultural Development and Economic Growth.* Ithaca, N.Y.

Bähr, J. 1968: *Kulturgeographische Wandlungen in der Farmzone Südwestafrikas.* No. 40 of *Bonner Geographische Abhandlungen.* Bonn.

Bähr, J. 1970: "Strukturwandel der Farmwirtschaft in Südwestafrika," *Zeitschrift für ausländische Landwirtschaft,* Vol. 9, pp. 147 ff.

Baker, O.E. 1926–1933: "Agricultural Regions of North America," *Economic Geography,* Vols. 2–9.

Ball, A.G., and Heady, E.O. (eds.) 1972: *Size, Structure, and Future of Farms.* Ames, Iowa.

Baren, F.A. van 1964: "Fysisch-geographische aspecten van de aride en semiaride gebieden," *Tijdschrift von Het Koninglijk Nederlansch Aardrikjkskundig Genootschap Amsterdam,* Vol. 81, pp. 182 ff.

Bartha, R. 1971: *Das Klima der Sahel-Zone.* No. 59 of *Afrika-Studien,* pp. 23 ff. Munich.

Bartlett, H.H. 1956: "Fire, Primitive Agriculture, and Grazing in the Tropics." In. Thomas, W.L., Jr. (ed.) 1956: *Man's Role in Changing the Face of the Earth,* Chicago, pp. 692 ff.

Benneh, G. 1972: "Systems of Agriculture in Tropical Africa," *Economic Geography,* Vol. 48, pp. 244 ff.

Bennett, M.K. 1960: *A World Map of Food-Crop Climates.* Vol. I, No. 3, of *Food Research Institute Studies.* Stanford, Calif.

Beuermann, A. 1967: *Fernweidewirtschaft in Südosteuropa.* Braunschweig.

Beyme, K. von, 1975: *Ökonomie und Politik im Sozialismus.* Munich and Zürich.

Biehl, M. 1973: *Die Landwirtschaft in China und Indien.* 4th ed. rev. Frankfurt/M.

Birch, J.W. 1954: "Observations on the Delimitation of Farming Type Regions, with Special Reference to the Isle of Man," *Transactions of the Institute of British Geographers,* Vol. 20, pp. 141 ff.

Birch, J.W. 1963: "Rural Land Use and Location Theory: A Review," *Economic Geography,* Vol. 39, pp. 273 ff.

Blanckenburg, P. von, and Cremer, H.-D. (eds.) 1971): *Handbuch der Landwirtschaft und Ernährung in den Entwicklungsländern.* Vol. 2. Stuttgart.

Blaut, J.M. 1967: "The Ecology of Tropical Farming Systems," *Revista Geografica,* Vol. 28, pp. 47 ff.

Blenk, J. 1971: *Die Insel Reichenau.* No. 33 of *Heidelberger Geographische Arbeiten.* Heidelberg.

Boesch, H. 1966: *Weltwirtschaftsgeographie.* Braunschweig.

Boesch, H. 1968: *Wirtschaftsgeographischer Weltatlas.* Munich.

Boesch, H. 1968: "The World Land Use Survey," *International Yearbook for Cartography 1968,* pp. 156 ff.

Bonnamour, J. 1973: *Géographie rurale – méthodes et perspectives.* Paris.

Bonuzzi, V. (ed.) 1972: *Agricultural Typology and Land Utilisation.* Proceedings of the Agricultural Typology Commission of the International Geographical Union. Center of Agricultural Geography and Institute of Agricultural Economy and Policy of the University of Verona. Verona.

Borcherdt, C. 1961: "Die Innovation als agrargeographische Regelerscheinung," *Arbeiten aus dem Geographischen Institut der Universität des Saarlandes,* Vol. VI, pp. 13 ff.

Borchert, G. 1967: *Die Wirtschaftsräume Angolas. Transportprobleme – Rentabilitätsgrenzen.* Special No. of *Hamburger Geographische Studien.* Hamburg.

Borchert, G., Oberbeck, G., and Sandner, G. (eds.) 1971: *Wirtschafts- und Kulturräume der außereuropäischen Welt – Festschrift für Albert Kolb.* Hamburg.

Borchert, G. 1972: *Die Wirtschaftsräume der Elfenbeinküste.* Vol. 13 of *Hamburger Beiträge.* Hamburg.

Boserup, E. 1965: *The Conditions of Agricultural Growth.* Chicago.

Böttcher, G. 1971: *China als kommunistisches Entwicklungsland.* Munich and Zürich.

Bowman, I. 1931: *The Pioneer Fringe.* Special Publication No. 13 of the American Geographical Society. New York.

Brinkmann, T. 1930: "Entwicklungslinien und Entwicklungsmöglichkeiten der landwirtschaftlichen Erzeugung Argentiniens," *Berichte über Landwirtschaft,* New Series, Vol. XIII, pp. 569 ff.

Brinkmann, T. 1935: *Economics of the Farm Business.* Trs. and eds. Benedict, E.T., Stippler, H.H., and Benedict, M.R. Berkeley.

Britton, D.K., and Hill, B. 1975: *Size and Efficiency in Farming.* Lexington, Mass., and Westmead, England.

Brown, D. 1971: "Agricultural Development in India's Districts," Dissertation. Cambridge, Mass.

Brunet, R. 1968: "Die Bedeutung der Sozialstruktur und der Region für die Agrargeographie," *Münchner Studien zur Sozial- und Wirtschaftsgeographie,* Vol. 4, pp. 15 ff.

Buchanan, R.O. 1959: "Some Reflections on Agricultural Geography," *Geography,* Vol. 44, pp. 1 ff.

Busch, W. 1936: *Die Landbauzonen im deutschen Lebensraum.* Stuttgart.

Cantor, L.M. 1969: *A World Geography of Irrigation.* London and Edinburgh.

Carlin, J. 1956: "Trends in Western Australian Agriculture," *Farm Policy,* Vol. 5, pp. 86 ff.

Chang, J.-H. 1968: "The Agricultural Potential of the Humid Tropics," *Geographical Review,* Vol. 58, pp. 333 ff.

Chang, J.-H. 1968: *Climate and Agriculture: An Ecological Survey.* Chicago.

Chang, J.-H. 1977: "Tropical Agriculture: Crop Diversity and Crop Yields," *Economic Geography,* Vol. 53, pp. 241 ff.

Chao, K. 1970: *Agricultural Production in Communist China: 1949–1965.* Madison, Wis.

Chisholm, M.: 1963: "Tendencies in Agricultural Specialization and Regional Concentration of Industry," *The Regional Science Association,* Vol. 10, pp. 157 ff.

Chisholm, M. 1964: "Problems in the Classification and Use of the Farming Type Region," *Transactions of the Institute of British Geographers,* Vol. 35, pp. 91 ff.

Chisholm, M. 1970: *Rural Settlement and Land Use.* Chicago.

Chou, M., et al. (eds.) 1977: *World Food Prospects and Agricultural Potential.* New York and London.

Clark, C., and Haswell, M.R. 1970: *The Economics of Subsistence Agriculture.* 4th ed. rev. London.

Clarke, J.I. 1959: "Studies of Semi-Nomadism in North Africa," *Economic Geography.* Vol. 35, pp. 95 ff.

Clarke, J.I. 1959: "Studies of Semi-Nomadism in North Africa," *Economic Geography,* Vol. 35, pp. 95 ff.

Clayton, E.S. 1964: *Agrarian Development in Peasant Economies: Some Lessons from Kenya.* Oxford.

Clout, H.D. 1972: *Rural Geography.* Oxford and New York.

Coppock, J.T. 1964: *An Agricultural Atlas of England and Wales.* London.

Coppock, J.T. 1964: "Crop-Livestock and Enterprise Combinations in England and Wales," *Economic Geography,* Vol. 40, pp. 65 ff.

Coppock, J.T. 1968: "The Geography of Agriculture," *Journal of Agricultural Economics,* Vol. XIX, pp. 153 ff.

Courtenay, P.P. 1965: *Plantation Agriculture.* New York and Washington.

Crkvencić, J., and Klemencić, V. 1967: "Arbeitsrichtungen and -ergebnisse der Agrargeographie in Jugoslawien," *Mitteilungen für Agrargeographie,* Vol. 17, pp. 201 ff.

Csáki, N. 1974: *Land Supply and International Specialization in Agriculture.* Vol. 3 of *Geography of World Agriculture.* Ed. by G. Enyedi. Budapest.

Dahlke, J. 1973: *Der Weizengürtel in Südwestaustralien. Anbau und Siedlung an der Trockengrenze.* No. 34 of *Geographische Zeitschrift, Beihefte Erdkundliches Wissen.* Wiesbaden.

Das landwirtschaftliche Betriebsgrößenproblem im Westen und Osten, 1961. Special No. 13 of *Agrarwirtschaft.*

Darby, H.C. 1956: "The Clearing of the Woodland in Europe." In: Thomas, W.L., Jr. (ed.) 1956: *Man's Role in Changing the Face of the Earth,* Chicago, pp. 183 ff.

Datoo, B.A. 1978: "Toward a Reformulation of Boserup's Theory of Agricultural Change," *Economic Geography,* Vol. 54, pp. 235 ff.

Deshler, W.W. 1963: "Cattle in Africa: Distribution, Types, and Problems," *Geographical Review,* Vol. 53, pp. 52 ff.

Duckham, A.N., and Masefield, G.B. 1970: *Farming Systems of the World.* New York.

Dumont, R. 1957: *Types of Rural Economy: Studies in World Agriculture.* Tr. D. Magnin. New York.

Dumont, R. 1964: *Sovkhoz, kolkhoz, ou le prolèmatique communisme.* Paris.

Dumont, R. 1965: *Chine surpeuplée: tiers-monde affamé.* Paris.

Dunn, E.S. 1954: *The Location of Agricultural Production.* Gainesville, Fla.

Dussart, F. (ed.) 1971: *L'habitat et les paysages ruraux d'Europe.* Vol. 58 of *Les congrès et colloques de l'université de Liège.* Liège.

Edwards, A., and Rogers, A. 1974: *Agricultural Resources.* London.

Edwards, D., and Rees, A.M.M. 1964: "The Agricultural Economist and Peasant Farming in Tropical Conditions." In: *International Explorations of Agricultural Economics.* Ames, Iowa.

Engelbrecht, T. 1930: *Die Landbauzonen der Erde.* Supplementary No. 209 of *Petermanns Geographische Mitteilungen.* Gotha/Leipzig.

Enyedi, G. 1967: "The Changing Face of Agriculture in Eastern Europe." *Geographical Review,* Vol. 57, pp. 358 ff.

Enyedi, G. 1968: "Landwirtschaftliche Bodennutzung in Ungarn," *Petermanns Geographische Mitteilungen,* No. 2, pp. 81 ff.

Enyedi, G. (ed.) 1976: *Rural Transformation in Hungary.* Budapest.

Erickson, F.C. 1948: "The Broken Cotton Belt," *Economic Geography,* Vol. 24, pp. 263 ff.

Falkner, F.R. 1939: *Beiträge zur Agrargeographie der afrikanischen Trockengebiete.* Stuttgart.

Faucher, D. 1949: *Géographic agraire.* Paris.

Feder, E. 1973: *Agrarstruktur und Unterentwicklung in Lateinamerika.* Frankfurt/M.

Fels, E. 1967: *Der wirtschaftende Mensch als Gestalter der Erde.* Vol. 5, 2d ed. rev., of *Erde und Weltwirtschaft,* ed. by R. Lütgens. Stuttgart.

Fochler-Hauke, G. (ed.) 1959: *Allgemeine Agrargeographie.* Frankfurt/M.

Food and Agriculture Organisation of the United Nations 1975: *Production Yearbook 1974.* Vol. 28. Rome.

Found, W.C. 1971: *A Theoretical Approach to Rural Land Use Patterns.* New York.

Franke, G., et al. 1975, 1967: *Nutzpflanzen der Tropen und Subtropen.* 2 vols. 1st vol. 2d ed. Leipzig.

Frenzel, K., et al. 1968: *Harms Handbuch der Erdkunde.* Vol. VII: *Australien, Ozeanien, Polargebiete, Weltmeere.* 6th ed. rev. Munich.

Fuller, A.M., and Mage, J.A. 1976: *Part-Time Farming: Problem or Resource in Rural Development.* Proceedings of the First Rural Geography Symposium, University of Guelph, Guelph, Ontario, Canada. Norwich, England.

Fussell, G.E. 1965: *Farming Technique from Prehistoric to Modern Times.* Oxford.

Garrison, W.L., and Marble, D.F. 1957: "The Spatial Structure of Agricultural Activities," *Annals of the Association of American Geographers,* Vol. 47, pp. 137 ff.

George, P. 1963: *Précis de géographie rurale.* Paris.

George, P. 1965: *Géographie agricole du Monde.* Paris.

Gerling, W. 1954: *Die Plantage.* Würzburg.

Gilbank, G. 1974: *Introduction a la géographie générale de l'agriculture.* Paris.

Glaser, G. 1967: *Der Sonderkulturanbau zu beiden Seiten des nördlichen Oberrheins zwischen Karlsruhe und Worms.* No. 18 of *Heidelberger Geographische Arbeiten.* Heidelberg.

Gleave, M.B., and White, H.P. 1969: "Population Density and Agricultural Systems in West Africa." In: Thomas, M.F., and Whittington, G.W. (eds.) 1969: *Environment and Land Use in Africa.* London.

Gnielinski, S. von, 1968: "Zuckerrohranbau in Liberia und seine wirtschaftliche Bedeutung," *Zeitschrift für ausländische Landwirtschaft,* Vol. 7, pp. 276 ff.

Gómez-Ibáñez, D.A. 1977: "Energy, Economics, and the Decline of Transhumance," *Geographical Review,* Vol. 67, pp. 284 ff.

Gourou, P. 1961: *The Tropical World: Its Social and Economic Conditions and Its Future Status.* Tr. E.D. Laborde. 3d ed. New York.

Gourou, P. 1956: "The Quality of Land Use of Tropical Cultivators," In: Thomas, W.L., Jr. (ed.) 1956: *Man's Role in Changing the Face of the Earth,* Chicago, pp. 335 ff.

Gregor, H.F. 1959: "Push to the Desert – The pressure of agriculture on California's arid land illustrates the law of diminishing returns," *Science,* Vol. 129, pp. 1329 ff.

Gregor, H.F. 1962: *Environment and Economic Life.* Princeton, N.J.

Gregor, H.F. 1963: "Industrialized Drylot Dairying: An Overview," *Economic Geography,* Vol. 39, pp. 299 ff.

Gregor, H.F. 1965: "The Changing Plantation," *Annals of the Association of American Geographers,* Vol. 55, pp. 221 ff.

Gregor, H.F. 1970: "The Large Industrialized American Crop Farm – A Mid-Latitude Plantation Variant," *Geographical Review,* Vol. 60, pp. 151 ff.

Gregor, H.F. 1970: *Geography of Agriculture: Themes in Research.* Englewood Cliffs, N.J.

Gregor, H.F. 1974: *An Agricultural Typology of California.* Vol. 4 of *Geography of World Agriculture.* Ed. by G. Enyedi. Budapest.

Gregor, H.F. 1975: "A Typology of Agriculture in Western United States – In World Perspective." In: Montresor, E., and Pecci, F. (eds.) 1975: *Agricultural Typology and Land Utilisation.* Proceedings of the Agricultural Typology Commission of the International Geographical Union. Center of Agricultural Geography and Institute of Agricultural Economy and Policy of the University of Verona. Verona.

Gregor, H.F. 1976: "Agricultural Intensity in the Pacific Southwest," *Proceedings of the Association of American Geographers,* Vol. 8, pp. 45 ff.

Gregor, H.F. 1979: "The Large Farm as a Stereotype: A Look at the Pacific Southwest," *Economic Geography,* Vol. 55, pp. 71 ff.

Gregory, S. 1969: "Rainfall Reliability." In: *Environment and Land Use in Africa.* London.

Gribaudi, D. 1964: "Geografia agraria." In: *Un sessantennio di Ricera Geografica Italiana*, pp. 301 ff. Rome.

Grigg, D.B. 1966: "The Geography of Farm Size – A Preliminary Survey," *Economic Geography*, Vol. 42, pp. 204 ff.

Grigg, D.B. 1969: "The Agricultural Regions of the World: Review and Reflections," *Economic Geography*, Vol. 45, pp. 95 ff.

Grigg, D.B. 1970: *The Harsh Lands*. London.

Grigg, D.B. 1974: *The Agricultural Systems of the World*.

Groeneveld, S., and Meliczek, H. (eds.) 1978: *Rurale Entwicklung zur Überwindung von Massenarmut – Hans Wilbrandt zum 75. Geburtstag*. Saarbrücken.

Grotewold, A. 1959: "Von Thünen in Retrospect," *Economic Geography*, Vol. 35, pp. 346 ff.

Hahn, E. 1882: "Die Wirtschaftsformen der Erde," *Petermanns Mitteilungen*, Vol. 38, pp. 8 ff.

Hall, P. (ed.), and Wartenberg, C.M. (tr.) 1966: *Von Thünen's Isolated State*. Oxford.

Hambloch, H. 1966: *Der Höhengrenzsaum der Ökumene*. No. 18 of Westfälische *Geographische Studien*. Münster.

Hart, J.F. (ed.) 1972: *Regions of the United States*. New York.

Hart, J.F. 1975: *The Look of the Land*. Englewood Cliffs, N.J.

Harvey, D.W. 1966: "Theoretical Concepts and the Analysis of Agricultural Land Use Pattern in Geography," *Annals of the Association of American Geographers*, Vol. 56, pp. 361 ff.

Hase, H.J. von, 1964: *Die Auswirkungen der Dürrejahre in Südwestafrika und ihre Überwindung*. Vol. 65 of *Der deutsche Tropenlandwirt*. Witzenhausen.

Haystead, L., and Fite, G.C. 1955: *The Agricultural Regions of the United States*. Norman, Okla.

Heady, E.O. 1952: *Economics of Agricultural Production and Resource Use*. New York.

Heady, E.O. 1966: *Agricultural Problems and Policies of Developed Countries*. Oslo.

Herlemann, H.-H. 1954: "Technisierungsstufen der Landwirtschaft," *Berichte über Landwirtschaft*, New Series, Vol. XXXII, pp. 335 ff.

Hewes, L. 1956: "Risk in the Central Great Plains – Geographical Patterns of Wheat Failure in Nebraska, 1931–1952," *Geographical Review*, Vol. 46, pp. 375 ff.

Hewes, L. 1958: "Wheat Failure in Western Nebraska, 1931–1954," *Annals of the Association of American Geographers*, Vol. 48, pp. 375 ff.

Higbee, E. 1958: *American Agriculture: Geography, Resources, Conservation.* New York.

Highsmith, R.M. 1965: "Irrigated Lands of the World," *Geographical Review,* Vol. 55, pp. 328 ff. and Pl. 1.

Hodder, B.W. 1968: *Economic Development in the Tropics.* London.

Hoffmann, E., Ewert, H., and Güther, A. 1954: "Die Abgrenzung von Bodennutzungs- und Betriebssystemen," *Agrarwirtschaft,* Vol. 3, pp. 263 ff.

Hofmeister, B. 1961: "Wesen und Erscheinungsform der Transhumance," *Erdkunde,* Vol. 2, pp. 121 ff.

Hofmeister, B. 1974: "Die quantitative Grundlage einer Weltkarte der Agrartypen." In: *Festschrift für Georg Jensch aus Anlaß seines 65. Geburtstages.* Berlin.

Hohnholz, J. 1975: "Agrargeographische Beobachtungen im Norden Thailands," Naturwissenschaftliche Rundschau, Vol. 28, pp. 311 ff.

Hunter, J.M. 1967: "Population Pressure in a Part of the West African Savanna: A Study of Nangodi, Northeast Ghana," *Annals of the Association of American Geographers,* Vol. 57, pp. 101 ff.

Igbozurike, M.U. 1971: "Ecological Balance in Tropical Agriculture," *Geographical Review,* Vol. 61, pp. 519 ff.

Ilesić, S. 1968: "Für eine komplexe Geographie des ländlichen Raumes und der ländlichen Landschaft als Nachfolgerin der reinen 'Agrargeographie,'" *MSSW,* Vol. 4, pp. 67 ff.

Instituto Geografico de Agostini, Novara, and International Association of Agricultural Economists (eds.) 1969–1970: *World Atlas of Agriculture.* Novara.

Jackson, W.A.D. 1956: "Virgin and Idle Lands of Western Siberia and Northern Kazakhstan: A Geographic Appraisal," *Geographical Review,* Vol. 46, pp. 1 ff.

Jäger, F. 1946: *Die klimatischen Grenzen des Ackerbaus.* Zürich.

Jäger, H. 1963: "Zur Geschichte der deutschen Kulturlandschaft," *Geographische Zeitschrift,* Vol. 2, pp. 90 ff.

Jaron, B., Danfors, E., and Vandie, J. (eds.) 1973: *Arid Zone Irrigation.* Vol. 5 of *Ecological Studies.* New York.

Jätzold, R. 1970: *Die wirtschaftsgeographische Struktur von Südtanzania.* No. 36 of Tübinger Geographische Studien. Tübingen.

Jensch, G. 1969: *Klima-Globus.* Berlin.

Jentsch, E.-G. 1965: "Die Struktur der Nahrungsversorgung und der landwirtschaftlichen Produktion Tunesiens in Vergangenheit, Gegenwart und Zukunft." Berlin Dissertation.

Joerg, W.L.G. 1932: *Pioneer Settlement.* Special Publication No. 14 of the American Geographical Society. New York.

Johnson, E.A.J. 1970: *The Organization of Space in Developing Countries.* Cambridge, Mass.

Johnston, B.F. and Kilby, P. 1975: *Agriculture & Structural Transformation.* New York and London.

Juelich, V. 1975: *Die Agrarkolonisation im Regenwald des mittleren Rio Huallaga (Peru).* No. 63 of *Marburger Geographische Schriften.* Marburg/L.

Jurion, F., and Henry, J. 1969: *Can Primitive Farming be Modernised?* Kisangani, Zaire.

Kayser, B. 1963: *Economies et sociétés rurales dans les régions tropicales.* Paris.

Kellerman, A. 1977: "The Pertinence of the Macro-Thünian Analysis," *Economic Geography,* pp. 255 ff.

Keuning, H.J. 1964: "Agrarische Geografie: Doelstelling, Ontwikkeling, Methoden," *Tijdschrift van het koninklijk Nederlandsch Aardrijkskundig Genootschap,* Vol. 81, pp. 10 ff.

Klages, E. 1949: *Ecological Crop Geography,* London.

Kollmorgen, W.M., and Jenks, G.F. 1958: "Suitcase Farming in Sully County, South Dakota," *Annals of the Association of American Geographers,* Vol. 48, pp. 27 ff.

Kollmorgen, W.M., and Jenks, G.F. 1958: "Sidewalk Farming in Toole County, Montana, and Traill County, North Dakota," *Annals of the Association of American Geographers,* Vol. 48, pp. 209 ff.

Könnecke, G. 1967: *Fruchtfolgen,* Berlin.

Köppen, W. 1923: *Die Klimate der Erde.* Berlin and Leipzig.

Kostrowicki, J. 1970: "Land Use Studies as a Basis of Agricultural Typology of East-Central Europe," *Geographia Polonica,* Vol. 19, pp. 263 ff.

Kostrowicki, J., and Tyszkiewicz, W. (eds.) 1970: *Essays on Agricultural Typology and Land Utilization.* Vol. 19 of *Geographia Polonica.* Warsaw.

Kostrowicki, J. 1970: *Agricultural Typology. Selected Methodological Materials.* Warsaw.

Kostrowicki, J., and Szczesny, R. 1972: *Polish Agriculture.* Vol. 1 of *Geography of World Agriculture.* Ed. by G. Enyedi. Budapest.

Kostrowicki, J., and Tyszkiewicz, W. (eds.) 1979: *Agricultural Typology* – Proceedings of the Eighth Meeting of the Commission on Agricultural Typology, International Geographical Union, Odessa, USSR, 20–26 July 1976. Vol. 40 of *Geographia Polonica.* Warsaw.

Kramer, F.L. 1967: "Eduard Hahn and the End of the 'Three Stages of Man,'" *Geographical Review,* Vol. 57, pp. 73 ff.

Krische, P. 1933: *Landwirtschaftliche Karten als Unterlagen wirtschaftlicher, wirtschaftsgeographischer und kulturgeschichtlicher Untersuchungen.* Berlin.

Ländliche Problemgebiete. Beiträge zur Geographie der Agrarwirtschaft in Europa. No. 13 of *Bochumer Geographische Arbeiten.* Paderborn 1972.

Laur, E. 1930: *Einführung in die Wirtschaftslehre des Landbaues.* 2ed ed. rev. Berlin.

Laut, P. 1968: *Agricultural Geography.* 2 vols. Melbourne and London.

Leaman, J.H., and Conkling, E.C. 1975: "Transport Change and Agricultural Specialization," *Annals of the Association of American Geographers,* Vol. 65, pp. 425 ff.

Lebeau, R. 1969: *Les grandes types de structures agraires dans le monde.* Paris.

Leoncarević, I. 1969: *Die landwirtschaftlichen Betriebsgrößen in der Sowjetunion in Statistik und Theorie.* Wiesbaden.

Leser, H. 1971: *Landschaftsökologische Studien im Kalaharirandgebiet um Auob und Nossob (Östl. Südwestafrika).* Vol. 3 of *Erdwissenschaftliche Forschung.* Wiesbaden.

Lewis, R.A. 1962: "The Irrigation Potential of Soviet Central Asia," *Annals of the Association of American Geographers,* Vol. 52, pp. 99 ff.

Löhr, L. 1971: *Bergbauernwirtschaft im Alpenraum.* Graz und Stuttgart.

Manshard, W. 1962: "Agrargeographische Entwicklungen in Ghana," *Westfälische Geographische Studien,* No. 15, pp. 81 ff.

Manshard, W. 1962: *Beiträge zur Geographie tropischer und subtropischer Entwicklungsländer. Indien–Westafrika–Mexiko.* No. 2 of *Gießener Geographische Schriften.* Gießen.

Manshard, W. 1968: *Einführung in die Agrargeographie der Tropen.* Mannheim.

Manshard, W. 1970: *Afrika – südlich der Sahara.* Vol. 5 of *Fischer Länderkunde.* Frankfurt/M.

Martin, A. 1958: *Economics and Agriculture.* London.

Masefield, G.B. 1948: "The Life of Perennial Crops," *East African Agricultural Journal of Kenya.* Nairobi.

Matzke, O. 1975: "Der Hunger wartet nicht," *Der Förderungsdienst,* Vol. 23, pp. 362 ff.

Mayhew, A. 1970: "Structural Reform and the Future of West German Agriculture," *Geographical Review,* Vol. 60, pp. 54 ff.

McCarty, H.H. 1954: "Agricultural Geography." In: James, P.E., and Jones, C.F. (eds.) 1954: *American Geography: Inventory and Prospect,* Syracuse, N.Y., pp. 258 ff.

McPherson, W.W. (ed.) 1968: *Economic Development of Tropical Agriculture.* Gainesville, Fla.

Montresor, E., and Pecci, F. (eds.) 1975: *Agricultural Typology and Land Utilisation.* Proceedings of the Agricultural Typology Commission of the International Geographical Union. Center of Agricultural Geography and Institute of Agricultural Economy and Policy, University of Verona. Verona.

Morgan, W.B., and Munton, R.J.C. 1971: *Agricultural Geography.* London.

Muller, P.O. 1973: "Trend Surfaces of American Agricultural Patterns: A Macro-Thünian Analysis," *Economic Geography,* Vol. 49, pp. 228 ff.

Niederstucke, H. 1970: "Bodennutzungsformen in tropischen Höhenlagen," *Landwirt im Ausland,* Vol. 4, pp. 74 ff.

Nye, P.H., and Greenland, D.J. 1960: *The Soil under Shifting Cultivation.* Commonwealth Bureau of Soils. Technical Communication. No. 51. Bucks.

Obst, E. 1965: *Allgemeine Wirtschafts- und Verkehrsgeographie.* 3d. ed. rev. Berlin.

Olsen, K.H. 1963: "Die geographische Bedingtheit agrarischer Wirtschaftsformen," *Zeitschrift für ausländische Landwirtschaft,* Vol. 2, pp. 1 ff.

Otremba, E. 1960: *Allgemeine Agrar- und Industriegeographie.* Vol. 3 of *Erde und Weltwirtschaft,* ed. by R. Lütgens. 2d ed. rev. Stuttgart.

Otremba, E. 1969: *Der Wirtschaftsraum – seine geographischen Grundlagen und Probleme.* Vol. 1 of *Erde und Weltwirtschaft,* ed. by R. Lütgens. 2d ed. rev. of *Die geographischen Grundlagen und Probleme des Wirtschaftslebens,* by R. Lütgens. Stuttgart.

Otremba, E. 1976: *Die Güterproduktion im Weltwirtschaftsraum.* Vols. 2 and 3 of *Erde und Weltwirtschaft.* 3d ed. rev. Stuttgart.

Otremba, E., and Kessler, M. 1965: *Die Stellung der Viehwirtschaft im Agrarraum der Erde.* No. 10 of *Erdkundliches Wissen.* Wiesbaden.

Paffen, K.H. 1953: *Die natürliche Landschaft und ihre räumliche Gliederung (Mittel- und Niederrheinlande).* Vol. 68 of *Forschungen zur deutschen Landeskunde.* Remagen.

Papadakis, J. 1966: *Climates of the World and their Agricultural Potentialities.* Buenos Aires.

Peattie, R. 1931: "Height Limits of Mountain Economies: A Preliminary Survey of Contributing Factors," *Geographical Review,* Vol. 21, pp. 415 ff.

Peet, J.R. 1969: "The Spatial Expansion of Commercial Agriculture in the Nineteenth Century: A von Thünen Interpretation," *Economic Geography,* Vol. 45, pp. 283 ff.

Perpillou, A. 1961: *L'agriculture dans les régions climatiques du globe.* Paris.

Perpillou, A. 1965: *Géographie rurale.* Paris.

Pfeifer, G. 1956: "The Quality of Peasant Living in Central Europe." In: Thomas, W.L., Jr. (ed.) 1956: *Man's Role in Changing the Face of the Earth,* Chicago, pp. 240 ff.

Pfeifer, G. (ed.) 1971: *Symposium zur Agrargeographie – anläßlich des 80. Geburtstages von Leo Waibel am 22. Februar 1968.* No. 36 of *Heidelberger Geographische Arbeiten.* Heidelberg.

Phillips, J. 1959: *Agriculture and Ecology in Africa; A Study of Actual and Potential Development South of the Sahara.* New York.

Phillips, J. 1961: *The Development of Agriculture and Forestry in the Tropics: Patterns, Problems, and Promise.* New York.

Piekenbrock, P. 1958: *Vegetation und Pflanzenbau in den Tropen.* Deutsche Afrika-Gesellschaft. Series No. 7. Bonn.

Pössinger, H. 1968: *Landwirtschaftliche Entwicklung in Angola und Moçambigue.* No. 31 of *Afrika-Studien,* ed. by Ifo-Institut für Wirtschaftsforschung. Munich.

Prunty, M., Jr. 1951: "Recent Quantitative Changes in the Cotton Regions of the United States," *Economic Geography,* Vol. 27, pp. 189 ff.

Raddatz, E. 1954: "Über den Einfluß der Pflanzenzüchtung auf die Produktivität in der Landwirtschaft," Göttingen Dissertation.

Radulescu, N.A., Velcea, J., and Petrescu, N. 1968: *Geografia Agriculturii Romaniei.* Bucharest.

Rakitnikov, A.K., and Mukomel, I.F. 1962: "Agricultural Regionalization," *Soviet Geography: Review and Translation,* Vol. V, No. 9, pp. 24 ff.

Reeds, L.G. 1964: "Agricultural Geography: Progress and Prospects," *Canadian Geographer,* Vol. 8, pp. 51 ff.

Reeds, L.G. (ed.) 1973: *Agricultural Typology and Land Use.* Proceedings of the Agricultural Typology Commission of the International Geographical Union. Department of Geography, McMaster University, Hamilton, Ontario, Canada. Hamilton.

Rikkinen, K. 1969: „Agricultural Geography in Transition," *Acta Geographica*, Vol. 19, pp. 1 ff.

Rochlin, P., and Hagemann, E. 1971: *Die Kollektivierung der Landwirtschaft in der Sowjetunion und der Volksrepublik China.* Deutsches Institut für Wirtschaftsforschung. Special No. 88. Berlin.

Roubitschek, W. 1964: "Die räumliche Differenzierung der Bodennutzung im Gebiet der DDR," *Mitteilungen für Agrargeographie*, Vol. 4, pp. 967 ff.

Roubitschek, W. 1969: *Standortkräfte in der Landwirtschaft der DDR.* Gotha/Leipzig.

Royen, W. van 1954: *The Agricultural Resources of the World.* Vol. I of *Atlas of the World's Resources*, ed. by W. van Royen. New York.

Rühl, A. 1929: *Das Standortsproblem in der Landwirtschaftsgeographie (Das Neuland Ostaustralien).* No. 6 of *Historisch-volkswirtschaftliche Reihe.* Berlin.

Ruppert, K. 1960: *Die Bedeutung des Weinbaues und seiner Nachfolgekulturen für die sozialgeographische Differenzierung der Agrarlandschaft in Bayern.* No. 28 of *Münchner Geographische Hefte.*

Ruppert, K. 1968: "Die Almwirtschaft der deutschen Alpen in wirtschaftsgeographischer Sicht," *Bayerisches Landwirtschaftliches Jahrbuch*, Special No. 1, pp. 38 ff.

Ruppert, K. 1972: " 'Deagrarisation' in Jugoslawien," *WGI-Berichte zur Regionalforschung 9*, pp. 38 ff. Munich.

Ruppert, K. (ed.) 1973: *Agrargeographie.* Vol. 171 of *Wege der Forschung.* Darmstadt.

Ruppert, K., and Meienberg, P. 1964: "Das Luftbild als Hilfsmittel agrargeographischer Forschung," *Umschau*, Vol. 7, pp. 207 ff.

Ruthenberg, H. 1965: "Probleme des Überganges vom Wanderfeldbau und semipermanenten Feldbau zum permanenten Trockenfeldbau in Afrika südlich der Sahara," *Agrarwirtschaft*, Vol. 14, pp. 25 ff.

Ruthenberg, H. 1967: "Organisationsformen der Bodennutzung und Viehhaltung in den Tropen und Subtropen." In: *Handbuch der Landwirtschaft und Ernährung in Entwicklungsländern*, Vol. 1, pp. 122 ff. Stuttgart.

Ruthenberg, H. 1971: *Farming Systems in the Tropics.* Oxford.

Ruthenberg, H. 1972: *Landwirtschaftliche Entwicklungspolitik.* Materialsammlung, No. 20 of *Zeitschrift für ausländische Landwirtschaft.*

Rutter, W. 1963: *Die Stellung Australiens im Standortssystem der Weltwirtschaft.* Vol. 2 of *Weltwirtschaftliche Studien.* Göttingen.

Sapozhnikova, S.A., and Shashko, S.I. 1960: "Agroclimatic Conditions of the Distribution and Specialization of Agriculture," *Soviet Geography: Review and Translation,* Vol. 1, No. 9, pp. 20 ff.

Sarfalvi, B. 1970: "Regional Differentiation and Geographical Interpretation of the Social Structures of Agrarian Regions," *Geographical Papers,* Vol. 1, pp. 191 ff.

Sauer, C.O. 1952: *Agricultural Origins and Dispersals.* No. 2 of *Bowman Memorial Lectures,* American Geographical Society. New York.

Schendel, U. 1971: *Bericht über eine wasserwirtschaftliche Studien- und Kongreßreise nach Australien, Neuseeland, Hawaii und Kalifornien.* Kiel.

Schickele, R. 1931: "Untersuchungen über die Formen der Weidewirtschaft in den Trockengebieten der Erde." Berlin Dissertation.

Schiffers, H. 1972: *Die Sahara und ihre Randgebiete.* No. 61 of Afrika-Studien, ed. by Ifo-Institut für Wirtschaftsforschung. Munich.

Schiffers, H. 1974: *Dürren in Afrika.* Vol. 47 of *Forschungsberichte der Afrika-Studienstelle.* Munich.

Schmithüsen, J. 1968: *Allgemeine Vegetationsgeographie.* Vol. IV of *Lehrbuch der allgemeinen Geographie,* ed. by E. Obst. 3d ed. rev. Berlin.

Schreiber, D. 1973: *Entwurf einer Klimaeinteilung für landwirtschaftliche Belange.* Vol. 3, Special Series, of *Bochumer Geographische Arbeiten.* Paderborn.

Schröder, K.H. 1953: *Weinbau und Siedlung in Württemberg.* Vol. 73 of *Forschungen zur deutschen Landeskunde.* Remagen.

Schultz, T.W. 1953: *The Economic Organization of Agriculture.* New York.

Schutt, P. 1972: *Weltwirtschaftspflanzen.* Berlin and Hamburg.

Schweinfurth, U. 1966: *Die Teelandschaft im Hochland der Insel Ceylon als Beispiel für den Landschaftswandel.* No. 15 of *Heidelberger Studien zur Kulturgeographie.* Wiesbaden.

Seavoy, R.E. 1973: "The Shading Cycle in Shifting Cultivation," *Annals of the Association of American Geographers,* Vol. 63, pp. 522 ff.

Shand, R.T. (ed.) 1969: *Agricultural Development in Asia.* Berkeley.

Shirlaw, G. 1966: *An Agricultural Geography of Great Britain.* New York.

Siedlungs- und agrargeographische Forschungen in Europa und Afrika. No. 3 of *Braunschweiger Geographische Studien.* Wiesbaden.

Skibbe, B. (ed.) 1958: *Agrarwirtschaftsatlas der Erde.* Gotha.

Smith, R.H.T. (ed.) 1972: *Spatial Structure and Process in Tropical West Africa.* Vol. 48 of *Economic Geography.* Worcester, Mass.

Southworth, H.M., and Johnston, B.F. (eds.) 1967: *Agricultural Development and Economic Growth.* Ithaca, N.Y.

Spencer, J.E. 1966: *Shifting Cultivation in Southeastern Asia.* Berkeley.

Spencer, J.E., and Stewart, N.R. 1973: "The Nature of Agricultural Systems," *Annals of the Association of American Geographers,* Vol. 63, pp. 529 ff.

Spielmann, O. 1969: "Viehwirtschaft in Costa Rica." Hamburg Dissertation.

Stamp, L.D. (ed.) 1961: *A History of Land Use in Arid Regions.* Vol. XVII of *Arid Zone Research,* UNESCO. Paris.

Statistical Office of the United Nations (ed.) 1975: *Statistical Yearbook 1974.* New York.

Statistisches Bundesamt Wiesbaden, Allgemeine Statistik des Auslandes 1972: *Länderkurzberichte Tschad 1972.* Stuttgart and Mainz.

Statistisches Jahrbuch der DDR 1963–1964, 1966, and 1969. Berlin.

Statistisches Jahrbuch über Ernährung, Landwirtschaft und Forsten 1975. Hamburg and Berlin.

Stokes. C.J. 1968: *Transportation and Economic Development in Latin America.* New York.

Stone, K.H. 1967: "Geographic Aspects of Planning for New Rural Settling in the Free World's Northern Lands," In: Cohen, S.B. (ed.): *Problems and Trends in American Geography,* pp. 221 ff. New York.

Stone, K.H. 1979: "Third World Settlement Frontiers: Modest Population Absorbers," *GeoJournal,* Vol. 2, pp. 193 ff.

Stone, K.H. 1979: "World Frontiers of Settlement," *GeoJournal,* Vol. 3, pp. 539 ff.

Stuart, R.C. 1972: *The Collective Farm in Soviet Agriculture.* Lexington, Mass.

Symons, L. 1972: *Russian Agriculture: A Geographical Survey.* New York.

Symons, L. 1979: *Agricultural Geography.* 2d ed. rev. Boulder, Colo.

Tarrant, J.R. 1974: *Agricultural Geography.* New York.

Thiede, G. 1971: *Standorte der EWG-Agrarerzeugung. Schwerpunkte und Entwicklungstendenzen.* Berlin and Hamburg.

Thiede, G. 1975: *Europas grüne Zukunft.* Düsseldorf and Vienna.

Thünen, J.H. von (1826) 1966: *Der isolierte Staat in Beziehung auf Landwirtschaft und Nationalökonomie.* Hamburg (1826) and Darmstadt.

Tosi, J.A., Jr., and Voertman, R.E. 1964: "Some Environmental Factors in the Economic Development of the Tropics," *Economic Geography,* Vol. 40, pp. 189 ff.

Troll, C. 1966: *Die räumliche Differenzierung der Entwicklungsländer in ihrer Bedeutung für die Entwicklungshilfe.* No. 13 of *Erdkundliches Wissen.* Wiesbaden.

Troll, C. 1968: *Die Entwicklungsländer in ihrer kultur- und sozialgeographischen Differenzierung. Bonner Akademische Reden 23.* Bonn

Troll, C. 1969: "Die Landnutzungskartierung in den Rheinlanden," *Erdkunde,* Vol. 2, pp. 81 ff.

Troll, C. 1970: *Argumenta Geographica. Festschrift.* Bonn.

Troll, C. 1975: "Vergleichende Geographie der Hochgebirge der Erde," *Geographische Rundschau,* Vol. 27, pp. 195 ff.

Tschudi, A.B. 1973: "People's Communes in China," *Norsk Geografisk Tidsskrift,* Vol. 27, pp. 5 ff.

Tsuzuki, T. 1963: "Die Fruchtfolgen des japanischen Ackerbaues," *Berichte über Landwirtschaft,* New Series, Vol. 41, pp. 833 ff.

Udo, R.K. 1965: "Sixty Years of Plantation Agriculture in Southern Nigeria, 1902–1962," *Economic Geography,* Vol. 41, pp. 356.

Uexkull, H.R. von 1969: "Reis in Asien – Probleme und Möglichkeiten einer Produktionssteigerung," *Zeitschrift für ausländische Landwirtschaft,* Vol. 8, pp. 248 ff.

Uhlig, H. 1965: "Die geographischen Grundlagen der Weidewirtschaft in den Trockengebieten der Tropen und Subtropen." In: *Weidewirtschaft in Trockengebieten.* Vol. 1, Series I, *Gießener Beiträge zur Entwicklungsforschung.* Stuttgart.

Uhlig, H. (ed.) 1967: *Basic Material for the Terminology of the Agricultural Landscape.* Vol. 1, *Types of Field Patterns.* Gießen.

Uhlig, H. (ed.) 1975: *Südostasien – Austral-pazifischer Raum.* Vol. 3 of *Fischer-Länderkunde.* Frankfurt/M.

Uhlig, H., Manshard, W., and Gerstenhauer, A. 1962: *Beiträge zur Geographie tropischer und subtropischer Entwicklungsländer. Indien – Westafrika – Mexico.* Schriften, No. 2 of *Gießener Geographische Schriften.* Gießen.

U.S. Department of Agriculture 1972: *Indices of Agricultural Production in Africa and the Near East 1962–71.* ERS-Foreign 265. Washington, D.C.

U.S. Department of the Interior 1970: *The National Atlas of the United States of America.* Washington, D.C.

Varjo, U. 1977: *Finnish Farming: Typology and Economics,* Vol. 6 of *Geography of World Agriculture.* Ed. by G. Enyedi. Budapest.

Visher, S.S. 1955: "Comparative Agricultural Potentials of the World's Regions," *Economic Geography,* Vol. 31, pp. 82 ff.

Waibel, L. 1933: *Probleme der Landwirtschaftsgeographie.* Breslau.

Walter, H. 1964 and 1968: Die Vegetation der Erde in ökophysiologischer Betrachtung. 2 vols. Stuttgart.

Walter, H. 1970: *Vegetationszonen und Klima.* Stuttgart.

Walton, K. 1969: *The Arid Zones.* London.

Wang, Y., Nagel, F., and Ruthenberg, H. 1969: *Bodennutzung und technischer Fortschritt auf Taiwan.* Special No. 7 of *Zeitschrift für ausländische Landwirtschaft.* Frankfurt/M.

Weaver, J.C. 1954: "Changing Patterns of Cropland Use in the Middle West," *Economic Geography,* Vol. 30, pp. 1 ff.

Weaver, J.C. 1954: "Crop-Combination Regions in the Middle West," *Geographical Review,* Vol. 44, pp. 175 ff.

Weaver, J.C. 1954: "Crop-Combination Regions for 1919 and 1929 in the Middle West," *Geographical Review,* Vol. 44, pp. 560 ff.

Weaver, J.C. 1958: "A Design for Research in the Geography of Agriculture," *Professional Geographer,* Vol. 10, No. 1, pp. 2 ff.

Webster, C.C., and Wilson, P.N. 1969: *Agriculture in the Tropics.* 3d ed. rev. London.

Weischet, B.G. 1977: *Die ökologische Benachteiligung der Tropen.* Stuttgart.

Whittlesey, D. 1936: "Major Agricultural Regions of the Earth," *Annals of the Association of American Geographers,* Vol. 26, pp. 199 ff.

Whyte, R.O. 1967: *Milk Production in Developing Countries.* London.

Wilhelmy, H. 1970: *Beiträge zur Geographie der Tropen und Subtropen. Festschrift.* Tübingen.

Wilhelmy, H. 1975: *Reisanbau und Nahrungsspielraum in Südostasien.* Kiel.

Wilkens, P.-J. 1974: Wandlungen der Plantagenwirtschaft. Die Entkolonisierung einer Wirtschaftsform." Hamburg Dissertation.

Windhorst, H.-H. 1975: *Spezialisierte Agrarwirtschaft in Südoldenburg. Eine agrargeographische Untersuchung.* Vol. 2 of *Nordwestniedersächsische Regionalforschungen.* Leer.

Winkler, E. 1963: *Agrargeographie der Schweiz. Schweizer Landwirtschaft.* Zürich.

Wirth, E. 1962: *Agrargeographie des Irak.* No. 13 of *Hamburger Geographische Studien.* Hamburg.

Wirth, E. 1969: "Das Problem der Nomaden im heutigen Orient," *Geographische Rundschau,* Vol. 21, pp. 41 ff.

Woerman, E. 1943: *Landwirtschaftliche Anbausysteme in Europa.* Colored map at a scale of 1:5 million. Berlin.

Woerman, E. 1944: *Europäische Nahrungswirtschaft.* Vol. 14, No. 99, New Series, of *Nova Acta Leopoldina.* Ed. by E. Abderhalden. Halle (Saale). Simplified version of colored map noted above.

Wrigley, G. 1971: *Tropical Agriculture — The Development of Production.* London.

Yates, P.L. 1960: *Food, Land and Manpower in Western Europe.* New York 1960.

Zabko-Potopowicz, A. 1957: "The Development of the Geography of Agriculture since World War I," *Przeglad Geograficzny,* Vol. 29, pp. 21 ff.

Zelinsky, W., Kosinski, L.A., and Prothero, R.M. (eds.) 1970: *Geography and a Crowding World: A Symposium on Population Pressures upon Physical and Social Resources in the Developing Lands.* New York.

Zotschew, T. 1957: *Die Entwicklungsprobleme der polnischen Wirtschaft.* In: *Material über die wirtschaftliche Lage und Entwicklung der Volksrepublik Polen.* Institut für Weltwirtschaft, Forschungsabteilung. Kiel.

Addendum

Andreae, B. 1978: *Agrarregionen unter Standortstress. Produktionsverfahren der Bodennutzung in Marginalzonen des Weltagrarraumes.* Kiel.

Andreae, B. 1978: "Planungsstufen von Kleinfarmen im Generationswechsel — Das Hochland Kenias als Modell," *Geographische Rundschau,* Vol. 30, pp. 304 ff.

Andreae, B. 1978: "The Minimum Cost Combination in Agriculture — with Special Reference to Developing Countries in Tropical Areas," *GeoJournal,* Vol. 2, pp. 203 ff.

Andreae, B. 1979: "Agrarprobleme der Dritten Welt. Irrtum und Wahrheit zwischen Tradition und Fortschritt," *Geographische Rundschau,* Vol. 31, pp. 390 ff.

Andreae, B. 1979: "Cultivation of Sugar Beet and Sugar Cane in the Peshawar Basin of Pakistan." In Institute for Scientific Co-operation (ed.), *Plant Research and Development,* Vol. 9, pp. 52 ff., Tübingen.

Andreae, B. 1980: "Stufen der Wildnutzung in Afrika," *Zeitschrift für Jagdwissenschaft,* Vol. 26, pp. 84 ff.

Andreae, B. 1980: *Weltwirtschaftspflanzen im Wettbewerb. Ökonomischer Spielraum in ökologischen Grenzen. Eine produktbezogene Nutzpflanzengeographie.* Berlin and New York.

Bonnamour, J., Brunet, P., and Flatres, P. (eds.) 1980: *L'Aménagement Rurale.* Actes du Ve Colloque Franco-Polonais de Géographie, Cerisy-la-Salle, Septembre 1977. Centre de Recherches sur l'Evolution de la Vie Rurale, Université de Caen. Caen.

Hayami, Y., Ruttan, V.W., and Southworth, H.M. (eds.) 1979: *Agricultural Growth in Japan, Taiwan, Korea, and the Philippines.* Honolulu.

Husain, M. 1979: *Agricultural Geography.* Delhi.

Koutaniemi, L. (ed.) 1978: *Rural Development in Highlands and High-Latitude Zones.* Proceedings of the International Geographical Union's Commission on Rural Development. Department of Geography, University of Oulu, Oulu, Finland. Oulu.

Rostankowski, P. 1979: *Agrarraum und Getreideanbau in der Sowjetunion 1948–1985. Eine agrargeographische Studie.* Berlin.

Späth, H.-J. 1980: *Die agro-ökologische Trockengrenze in den zentralen Great Plains von Nord-Amerika.* Vol. 15 of *Erdwissenschaftliche Forschung.* Wiesbaden.

Figures

Fig. 1: Organization of Farmed Land 18
Fig. 2: Cost Concepts 20
Fig. 3: Derivation of Measurements of the Efficiency of the Farm Enterprise 22
Fig. 4: Climatic Zones of Africa 34
Fig. 5: Average Annual Precipitation on the Earth 35
Fig. 6: Climatic Zones of the Tropics (Diagram) 38
Fig. 7: Northern Cropping Boundaries in Europe 49
Fig. 8: Distribution of the Principal Useful Plants 51
Fig. 9: Effective Altitudinal Zones of Useful Plants in the Equatorial Belt 54
Fig. 10: Altitudinal Zones of Useful Plants in Java Compared with Natural Vegetation 55
Fig. 11: The Northern Limit of Rainfed Farming in Chad 59
Fig. 12: Agronomic Dry Boundaries in the Republic of South Africa 60
Fig. 13: Agronomic Dry Boundaries in North America 61
Fig. 14: Drought Severity and Land Use in Algeria 62
Fig. 15: The Agronomic Dry Boundary in the Sudan-Sahel Natural Realm 63
Fig. 16: Competitive Shifts among the Production Orientations of Livestock Economies with Increasing Market Distance 73
Fig. 17: Changes in Agricultural Enterprises in the Wet-and-Dry Tropics with Increasing Market Distance 73
Fig. 18: Intensity Qualifications for Two-Man Farms 97
Fig. 19: Tractors on Single-Operator Farms 98
Fig. 20: Technical and Structural Arrangements in Livestock Farming 99
Fig. 21: Interfarm Machinery Use in West German Agriculture, 1971 100
Fig. 22: Price-Cost Relationships at the Farm Gate with Increasing Market Distance 103
Fig. 23: Farming Intensity and Farming Diversity with Changing Market Distance 104
Fig. 24: Development of Purchasing Power of Agricultural Products for Wage Labor 108
Fig. 25: Development of Purchasing Power between Wage Labor and Farm Machinery 109
Fig. 26: Farming Regions of the World (colored map) End Papers
Fig. 27: Seasonal Climates of the Tropics and Subtropics 128

Fig. 28: Humid and Arid Months, Vegetation Belts, and Farming Systems in the Tropics and Subtropics 129
Fig. 29: Farming Systems in the Climatic Zones of the Tropics 131
Fig. 30: Cropping Intensity Levels in World Agricultural Space 133
Fig. 31: Decline in Yields with Prolonged Cropping in the Humid Tropics under the System of Shifting Cultivation 134
Fig. 32: Yield Level of Crops under Shifting Cultivation Depending on the Cropping Interval 136
Fig. 33: Ley Farming in African Wet Savannas 138
Fig. 34: Irrigated Areas in World Agricultural Space 142
Fig. 35: Irrigated Rice Monocultures with Varying Degrees of Land Utilization 145
Fig. 36: Wet Rice Rotations with Varying Degrees of Land Use 146
Fig. 37: Cross-Section of a Village Farm Settlement in Zaire 149
Fig. 38: Comparative Productivity of Tea, Coffee, and Cacao Farms 158
Fig. 39: Seasonal Climates in the Dry Areas of the Earth 161
Fig. 40: Rain Curve for Grootfontein/Southwest Africa, 1899/1900 to 1962/1963 162
Fig. 41: Variations in Annual Rainfall in Africa 165
Fig. 42: Carrying Capacity of Natural Pasture in the Republic of South Africa 168
Fig. 43: Stages of Feed Balance in Grassland Farming of Semiarid Climates 172
Fig. 44: The Effect of Fallowing in the Dry Farming System 176
Fig. 45: Zones of Intensity of the Dry Farming System 177
Fig. 46: Agricultural Zones of Australia 192
Fig. 47: The Regionalization of World Nutrition 196
Fig. 48: Climatic Areas of Europe 198
Fig. 49: Duration of a Temperature of at least 5° C in Europe 200
Fig. 50: Soil Map of Europe 203
Fig. 51: Land in Permanent Grassland in the European Community 204
Fig. 52: Production Elasticity of Agricultural Regions in the EEC 206
Fig. 53: Crop Rotation Types in the European Community 211
Fig. 54: Agricultural Land Use Systems in the European Community 215
Fig. 55: Livestock Raising Systems in the European Community 223
Fig. 56: Cattle Farming Types in the European Community 225
Fig. 57: Types of Cattle Raising and their Locational Orientation 226
Fig. 58: Labor Intensity in the Agriculture of the European Community 229
Fig. 59: Annual Precipitation in the United States 233
Fig. 60: The Available Growing Season in the United States 234

Figures

Fig. 61: Distribution of the Principal Soil Types in the United States 235
Fig. 62: Farming Zones of the United States 236
Fig. 63: The Thünen Spatial Pattern in Chicago's Hinterland 239
Fig. 64: Pastureland per Cattle Unit in the Western States of the U.S. 243
Fig. 65: Agricultural Regions in the United States 248
Fig. 66: Agricultural Zones in the Baltic Sea-Adriatic Sea Area 257
Fig. 67: Regional Zones in the Soviet Union 263
Fig. 68: Hydroelectric Power and Irrigation Projects in the Volga-Don Area 265
Fig. 69: Agricultural Zones of China 268
Fig. 70: Long-Term Trends in Prices and Wages in Germany 272
Fig. 71: Factor Costs and Factor Combinations in the Course of National Economic Development 275
Fig. 72: Farmland Shares in World Agricultural Space 276
Fig. 73: Long-Term Development Trends in Farm Size in the Agricultural Domain 281
Fig. 74: Medium- and Long-term Development Alternatives for West European Farms 282
Fig. 75: Adaptation Processes of Various Farm Size Classes in an Expanding Agriculture 283
Fig. 76: Farm Diversification and Specialization in Economic Growth 290
Fig. 77: The Epochal Sequence of Agrarian Land Use Types in the Dry Savannas 294
Fig. 78: Genesis of the Crop Rotations of Central England 296

Index

Abaca, 26, 124
Aborigines, 27
Abruzzi, 226
Absolute costs, 107
Absolute prices, 107
Acorn hog-fattening area, 227
Acquisitive economy, 123
Actual agricultural boundaries, 46 f.
Adaptation processes, 106 ff., 160, 170 ff., 283
Afforestation, 135
Africa 46, 57, 66, 78, 95, 139, 141, 193; East, 167, 181; South-West, 189; Southeast, 138; Southern, 122 f.; Southwest 38, 78, 162, 167, 169. 187 ff.; West, 70, 123, 135, 139, 151, 154
African Highlands, East, 118
Agave landscapes, 155
Agrarian-industrial countries, 273, 275 ff. *See also* Agricultural states; Developing countries
Agrarian structure: defined, 30
Agricultural (agrarian) states, 45 f., 103 f., 143, 145, 271 ff., 280 ff., 290, 294; overpopulated 278, 280. *See also* Agrarian-industrial countries; Developing countries
Agricultural area: defined, 30; 85. *See also* Agricultural space
Agricultural boundaries. *See* Boundaries
Agricultural capacity: expansion, 83 ff.; intensification, 83 ff.
Agricultural evolution theory, 27 ff., 299 ff.
Agricultural geography, 30 ff.; and Aeroboe's evolution theory, 299 ff.; definitions, 30; objectives and significance, 31; of the dry areas, 160 ff.; of the East Bloc countries, 250 ff.; of the humid tropics, 127 ff.; of the middle latitudes, 196 ff.; of North America, 232 ff.; of Western Europe, 197 ff.; work methods, 31 f.
Agricultural Industrial Complex (AIC), 23
Agricultural land (AL), 15, 46, 84, 146 f., 159, 178 f., 182 f., 187, 204, 207 f., 210, 253, 255 ff., 260 f., 263, 267, 287, 301. *See also* Cultivable land; Farmed land
Agricultural Producer's Cooperative (APC), 23, 254, 256
Agricultural products: price-cost relations between, and inputs, 110 f.; price relations between, 107 f. *See also* individual products
Agricultural regions, 214 ff.; 232; complex, 227; defined, 30; in the United States, 248; delimitation, 207 ff. *See also* Agricultural zones; Economic formations

Agricultural space: contraction, 83; defined, 30; delimitation, 45 ff., 85 f.; expansion, 45, 82, 86; expansion as a problem, 45 ff.; farming systems, 113 ff., end papers; structural changes, 271 ff. *See also* Agricultural area
Agricultural surpluses, 45, 193
Agricultural system, 113. *See also* Farming systems
Agricultural zones, 105, 127; defined, 30; French, 212; in the Baltic-Adriatic area, 256 ff.; of Australia, 190 ff.; of China, 266 ff.; of the Soviet Union, 260 ff.; of the United States, 236 ff.; of the World, end papers. *See also* Agricultural regions; Economic formations
Agriculture: origins and evolution of, 27 ff.; farming systems of, 113 ff.
Agro-Chemical Center (ACC), 23
Agronomic dry boundary, 57, 60 ff.; 77 f., 82 ff., 127, 129, 131, 160 ff., 181 ff., 187, 190
Agrospatial structure, 190
Aircraft, 122
Airplane seeding, 147 f.
Alaska, 43, 116
Albania, 253
Alcohol, 75, 257; acetyl, 195
Alfalfa, 18 f., 74, 90, 92, 122, 138, 179, 238, 246, 264
Algeria, 57, 62
Algerian Plateau, 166
Alkalinization, 67
Allgäu, 202, 217, 235
Almonds, 41, 124, 199, 221
Alpine area, 202, 212, 216 f.
Alpine farming, 17
Alpine farms, 69
Alpine ley farming, 19
Alpine pastures, 69, 85
Alpine valleys, 69
Alps, 43, 53, 68, 84, 116, 119, 199, 224
Altes Land, 221, 230
Altitudinal boundaries, 36 f., 46, 52 ff., 141
Altitudinal minimum, 55
Altitudinal range of cultivation, 55
Altitudinal zones, 53 ff.
Amazon Basin, 33, 47, 67, 95, 123
Amazonas, 36
Andes, 36, 39, 43, 53
Angola, 150
Animal breeding, 170, 224, 226
Animal cropping systems, 17, 114, 117 ff., 212 f., end papers

Index

Animal crops, 201, 285
Apennines, 47, 84 f., 224
Appalachian Highlands, 233
Apples, 244
Appropriative economies, 9, 102, 294
Apricots, 124, 221
Apulia, 40 f., 201
Arabia, 57
Arabian Desert, 40
Arabian Peninsula, 163
Arable land. *See* Cultivable land
Arctic coast, 261
Ardennes, 217
Argentina, 57, 74, 176, 275
Arid semideserts, 127
Arid zones, 38 f., 115, 124, 127 f., 160 ff., 190 ff., 235, 263 ff. *See also* Dry climates
Arizona, 115, 235, 243
Arkansas, 147
Ash, 134
Asia, 39, 46, 53, 57, 141; East, 110 f., 171; Southeast, 122; Southeastern, 95, 110, 141
Assam, 69, 135
Aswan Dam, 68
Atlantic coastal area, 201 f., 212, 214, 234
Atlantic Coast: French, 230
Australia, 37, 39 ff., 46, 119, 176, 190 ff., 275
Autarky, 94, 230, 269

Bacteria, 185
Baden, 220
Baden-Württemberg, 213
Bahamas, 46
Bali, 69
Baltic-Adriatic area, 256 f.
Baltic Provinces, 43
Baltic Sea coastal area, 67, 77, 202
Balkan Peninsula, 250
Balkans, 43, 72, 259
Bananas, 25, 33, 36 f., 50, 52 f., 58, 72, 74, 125, 150 ff., 285
Banat, 227, 258
Bangladesh, 66
Bangkok Plain, 66
Bantu tribes, 167
Bare fallow, 58
Barley, 18, 44, 48 f., 52 ff., 56, 58, 62, 67, 71, 77, 81, 86, 90, 100, 111, 119, 122, 138, 141, 164, 176 ff., 199, 261, 264, 293, 296; feeding, 227; hog-fattening area, 227; malting, 230 f.
Bavaria, 201, 205, 213
Bean pickers, 112
Beans, 122, 138 f., 180 f., 245, 262, 285, 296; Bechuana, 138; broad, 90; bush, 36 f., 52, 54 f., 58, 139, 141; Egyptian, 186

Bedouins, 171
Beef, 193 f.; animals, 224; prices and production, 107
Beer, 189 f.
Beet chips, 258
Beet harvester, 77, 288
Beet leaf, 231; silage, 258
Beets, 49, 52 f., 286
Belgium, 84, 110
Benelux countries, 205
Berbers, 166, 171
Biological balance, 27
Biological technological advances, 78 f.
Birth control, 278
Black Forest, 216
Black Sea Coast, 40, 250 f.
Blue lupine, 177
Bog-burning shifting cultivation, 27, 118
Bohemian Mountains, 258
Börde, 202, 218, 220
Borneo, 140
Botswana, 65
Boundaries: actual, 46 f.; agricultural, 47 f.; agronomic dry, 57, 60 ff., 77 f., 82 ff., 127, 129, 131, 160 ff., 181 ff., 187, 190; altitudinal, 36 f., 46, 52 ff., 141; commercialization, 74 ff.; dry, 38 ff., 46, 57 ff., 86, 129, 131, 160, 169, 174 ff., 300; dry, of human habitation, 127, 129, 131; dry, of livestock farming, 129, 131; ecological, 48 ff.; economic, 46 f., 70 ff.; economic dry, 82; effective, 46 f., 86; expansion beyond, 45; geographic, 84; grazing, 36; humid warm-temperate, 40 ff.; in Europe, 49; industrial, 70; livestock grazing, 191 ff.; northern, in Europe, 199 ff.; of tropics and subtropics, 128, 161; polar, 44, 46, 48 ff., 52, 79, 84, 86, 264; precipitation, 115, 169; product-specific transport, 72 f.; profitability, 46 f., 70 ff., 86; settlment, 46, 70; shifting cultivation, 133; shifts, 76 ff.; slope, 68 f.; social, 84; soil, 46, 67 f., 77; subtropical, 185; technological, 46 f., 86; transport, 47, 70 ff.; tropical, 185; wet, 36, 46, 66 f., 86; wet, of grassland farming, 131; wet, of grazing, 129
Boundary shifts, 76 f.
Brabant, 220
Brandenburg, 202; March, 256
Bratislava Lowland, 258
Brazil, 41, 125, 150 f., 182
Brigade, 23
Brittany, 219
Brussel sprouts, 122
Bucket wheels, 147 f.
Bulgaria, 253
Bulls, 80

Bush and tree crop farming, 82 f., 130 ff., 149 ff., 228, 291 f.
Bush and tree crop farms, 87, 114, 122 ff., end papers
Bush and tree crops, 41, 46, 83, 92, 94, 123 ff., 125, 131, 136 f., 163, 199, 201, 205, 214, 216, 220 ff., 244, 256, 269, 278; farm management, 149 ff.; principal locations, 152 ff.; regions with predominantly, 149 ff.; *See also* Perennial crops
Bushmen, 27, 122
Butter, 71, 73, 237
Byelorussia, 262

Cacao, 25, 33, 36, 50, 52 ff., 58, 124 ff., 131, 150 f., 153 f., 158, 291; farms, 114, 158 f.
Calabria, 114, 201
Calcium, 91
California, 40, 53, 67, 89, 122, 163, 244 f.
Caloric yield, 139 f.
Calories, 196
Calves, 65, 71; beet leaf for raising, 231; on feed, and types of cattle raising, 226; production of, 169; raising of, 231
Camels, 66, 80, 166 f., 170
Cameroon, 59
Campania, 40 f., 230
Canada, 43, 79, 110, 116, 176, 190, 232, 237, 242
Canneries, 112, 246
Cape Province, 119
Capital, 21, 94, 173, 186, 271 ff., 283, 291, 295, 302; consumptive, in U.S. agriculture, 249; -cost ratios between labor and machinery, 182; expenditures on U.S. wheat farms; 179; favorable prices for, 244; goods, 123, 249, 271 ff., 290, 302; in the U.S. factor combination, 247 ff.; income, 22; inputs, 82 f., 103 ff., 137, 171 ff., 291, 301; intensity, 110, 151, 173 f., 182 f., 249, 271 ff., 286 ff.; investments, 115, 146, 167, 171 ff., 218, 244, 272, 277 ff., 286, 290, 292; lack of, 123; land, 115; poverty of, 266; productivity, 146, 277 ff.; substitution of, for labor, 154; working, in U.S. agriculture, 249; working, on U.S. wheat farms, 179
Capitalism, 254
Caravan zone, 105
Carbohydrates, 224
Carinthia, 214
Carpathians, 43
Carrot harvesters, 112
Carrots, 122
Cash crop farms, 182
Cash crops, 152, 190, 284
Cash expenditures, 22. *See also* Capital

Cash wages, 102
Cashmere, 75
Cassava, 25, 125, 132, 134, 141. *See also* Manioc
Castor bean, 25
Castor oil, 25
Catch crops (CC), 286; cash, 17; non-cash, 15, 19, 119, 122; winter, 18
Caucasus, 40
Cattle, 24, 38 f., 42, 74, 80 f., 115, 136, 166 f., 170, 173, 188, 226, 244, 246, 260; dairy, 101, 209 f., 225, 285; drives, 194; fattening, 81, 224, 238 f., 293; -fattening farm, 65, 225 f.; grazing, 60; herding, 163; price of slaughter, 226; raising, 94, 104, 115, 166, 169, 192 ff., 199, 224 ff., 240; slaughter, 107 ff., 226, 230, 285, 293; types of, raising, 226; young, 209 f., 222, 225 f. *See also* Farming systems; Livestock raising
Cattle farming. *See* Farming sytems; Livestock raising
Cattle feeding. *See* Cattle fattening
Cattle finishing. *See* Cattle fattening
Cattle Unit (CU), 101. *See also* Large livestock unit
Cauliflower, 122
Celery, 245
Central African Republic, 59, 184, 186
Central America, 274
Central Asia, 171
Central Europe, 77 f., 111, 118, 120, 132, 207, 213 f., 216, 228, 302
Central European loess belt, 101, 202
Central German Highlands, 219
Central Schwerin Marsh, 77
Cereals. *See* Grains
Chaco, 57
Chad, 59, 144, 175
Cheese, 71, 73, 237; making, 85
Chemical fertilization, 219, 272. *See also* Fertilizers, chemical
Cherbourg Peninsula, 116
Cherries, 124, 221
Chickpeas. *See* Garbanzos
Chile, 40, 120, 145, 269, 275
China, 24, 41, 43, 69, 107, 143 f., 153, 181, 250 f., 253 f., 266 ff.; agricultural zones, 268; North, 265
Chinarinde 55
Cinchona, 55
Citrus farming, 89
Citrus fruit, 25, 49, 50, 52 ff., 58, 62, 131, 201, 221, 244, 269; raising, 221
Citrus trees, 150
Clearing, 28 f., 117 f., 132 ff. *See also* Shifting Cultivation
Climates, 33 ff., 231, 290 ff.; and tropical farming regions, 127; Arctic, 250;

(Continuation Climates)
  Central European transitional, 198 f.; constantly humid subtropical, 191, 269; constantly wet tropical rainforest, 33 ff., 38, 127 ff., 291 f., 299, 301; continental cool-summer, 43, 237, 250 f., 297 f.; continental steppe, 258; continental warm-summer, 43, 238, 250; dry, 34, 37 ff., 53, 250 f., 293 ff.; dry savanna, 34, 37, 293 ff.; East European continental, 198 f.; hot, dry summer and warm, wet winter, 198 f., humid cool-temperate, 42 ff., 297 f.; humid savanna, 33 ff., 292 f.; humid warm-temperate, 34, 40 ff., 295 ff.; in Europe, 197 ff.; in the Common Market, 232; marine cool-summer, 34, 42, 295; Mediterranean, 94, 191, 198 f., 222, 228, 235, 244, 259; mid-latitude, 190; mild winter, cool summer, 198 f.; rainforest, 33 f., 38, 127 ff., 291 f., 299, 301; seasonal, 127 f., 161; semidesert, 39 f.; shrub savanna, 38 f.; steppe, 34, 39, 191, 293 ff.; Subarctic, 43, 250 f.; subtropical, 222; subtropical dry-summer, 34, 40 f.; subtropical warm-summer, 41 f., 250; temperate summer and winter, 197 f.; temperate summer, cold winter, 197 f.; tropical highland, 36 f.; tropical rainy, 33 ff., 291 ff.; very mild winter and warm summer, 198 f.; warm-temperate oceanic, 299; West European oceanic, 198 f.; wet-summer subtropical, 41, 190; winter-rain subtropical, 40, 190, 269
Climatic zones, 33 ff., 290 ff., 300; in Europe, 197 ff.; of Africa, 34; of the tropics, 38
Clover, 18 f., 49, 71, 122, 194, 238, 240, 285; red, 37, 55, 90, 238, 258, 296 f.
Cocoa. *See* Cacao
Cocoa Marketing Board, 155
Coconut palms, 25, 33, 50 ff., 58, 124 ff., 126, 131, 150, 153, 155, 269
Coffee, 33, 36 f., 50 ff., 58, 123 ff., 131, 150 ff., 269, 285, 287; farms, 158 f.
Collective farms, 24, 252 ff., 260
Collectivization, 254 f., 266
Cologne-Aachen Bay, 202, 220, 230
Colombia, 53, 74, 275
Colorado, 179, 181
Columbia Basin, 77, 119, 179 f., 242
Columbia Plateau, 57
Combine-harvested crops, 100, 180
Combines, 77, 100, 111, 122, 147 f., 260, 288
Commercial steppe grassland farming, 167 ff. *See also* Farming systems; Grassland farming; Ranching
Commercialization: boundaries, 74 ff.; of farms, 75; stages, 74
Common Market. *See* European Economic Community

Communes, People's, 266
Competition, terms of, 107 ff., 139 ff., 187 ff., 230 ff.
Competitive shifts, 73, 81
Congo, 134
Congo Basin, 33, 36, 47, 67, 95
Continental cool-summer climate, 43, 237, 250 f.; farming changes in, 297 f.
Continental warm-summer climate, 43, 228, 250
Cooperative Organization (COO), 23 f.
Cooperatives, 23 f., 75, 155; Agricultural Producers', 255, 266; wine growers, 124
Coriander, 139
Corn, 18, 25, 33, 36, 49 ff., 58, 60, 69, 72, 74, 78 f., 91 f., 100, 111, 119, 122, 125, 132, 134, 138 f., 141, 164, 174, 178, 184, 186, 188 f., 213 f., 237, 259, 262 ff., 288, 293, 300; Belt, 43, 236 ff.; cropping, 285; U.S., zone 238 ff.
Cornwall, 42, 119, 212
Corsica, 220, 227
Cost-production ratio, 81, 182. *See also* Costs; Input-output ratio; Minimum cost combination; Price-cost relations
Cost relation: between agricultural inputs, 109 f.
Costa Rica, 37, 53 f., 74
Costs, 88 ff., 231, 242, 286 ff.; digressions of, 125; fixed, 115; joint, 288; of release of labor, 88; opportunity, 97; recovered, 150; structure of, 115; types defined, 20. *See also* Cost-production ratio; Input-output ratio; Minimum cost combination; Price-cost relations
Cottage industry, 75 f.
Cotton, 26, 42, 49 f., 52, 55, 58 ff., 64, 91, 122, 134, 138 f., 141, 143 f., 176, 186 f., 220, 240, 262, 264, 268 f., 287, 293, 300; Belt, 236, 240 f.; -seed, 241; -seed cake, 246; Zone, 241
Cowpeas, 175
Cows, 166, 169, 183, 193, 246, 259; dairy, 209, 222, 224, 238; in calf, and types of cattle raising, 226
Cream, 72 f., 115
Crop atlases, 207
Crop compatibility: defined, 18, 20
Crop failure, 190
Crop ratios, 207
Crop rotation regions, 210 ff.
Crop rotations, 42, 90 f., 100 f., 245 f., 258 f., 277, 286, 288, 293, 297; changes, with increasing precipitation, 180 f.; component, 19; cotton, 186; cover, 17; double-crop, 18; extensive, in Pakistan, 164; five-course, 18; field, 18; four-course, 18, 211, 214; in African wet savannas, 138; in altitudinal zones, 56; in China, 269; in dry farming,

(Continuation Crop rotations)
177; in Graubünden, 50; in hoe crop farming, 122; in irrigated cropping, 138, 144 ff., 164; in Lapland, 48; in New York state, 50; in rainfed cropping, 138 f.; in Siberia, 50; in West German farming, 119; in the Australian wheat farming-sheep grazing system, 194; in the Central European loess strip, 101; in the Corn Belt, 240; in the New England states, 238; in the Soviet Union, 262, 264; intensive, in Pakistan, 164; near the agronomic dry boundary, 64; Norfolk, 19, 211, 213 f., 295 f.; normal, on U.S. wheat farms, 179; of central England, 295 f.; on large farms, 138; on small farms, 138; on two-man farms, 100; peanut-millet, 58; peanut-sorghum, 58; peanut-millet-sorghum, 58, 184; period, 19; plan, 19; primary, 19; poorer in leafy crops, 213; pure grain, 78; secondary, 19 f., three-course, 211, 214, 295 ff.; types of, in the European Community, 211; vegetable, 246; wet rice, 146
Crop weight, 208 ff.
Crop years, 18, 135
Crop yields, 139 f.; under shifting cultivation, 136
Cropland: defined, 17 f.
Cuba, 125, 150
Cultivable (arable) land (CL), 260; defined, 15; proportion in, 16, 18, 48, 81, 84, 100, 107, 138, 141, 177 ff., 183, 186, 211 f., 239, 258, 261, 263, 270. *See also* Agricultural land; Farmed land
Cultural landscape, 141
Cumberland, 119, 212
Czechoslovakia (CSSR), 253, 258

Dairy Belt, 43, 92, 236 ff. *See also* Dairy zone
Dairy-cattle fattening farms, 225 f.
Dairy-cattle raising farms, 225 f.
Dairy farming, 20, 58, 104, 213, 222 ff., 226 ff., 256 ff., 265, 273
Dairy-hog farming, 223, 227
Dairy production, 170
Dairy-young cattle farming, 223 f.
Dairy zone: in Australia, 194; in the United States, 237 f. *See also* Dairy Belt
Dams, 292
Danuble Basin, 250
Date palms, 25, 51, 59, 62, 67, 150, 163, southern limit, 63
Dates, 124, 244
Deciduous dry forest, 37
Denmark, 46, 107, 109, 204, 227
Densely settled countries, 271 ff., 277 ff., 281, 290
Desert climate, 34. *See also* Climates

Deserts, 191, 260, 263 f.; dry-hot, 127; semi-, 260, 263 f., 300
Developing countries, 27 ff., 45 f., 80 ff., 104 f., 122, 154, 274 ff., 301. *See also* Agrarian-industrial countries; Agricultural states
Development: forces for, 271 ff.; policy, 84, 195 f.; price-cost, 106 ff., 275 ff.; 290; stages (national economic), 79 ff., 171 ff., 196, 290. *See also* Economic growth; Economic integration
Diesel engines, 147 f.
Digging stick, 117, 147, 175, 299
Dikes, 270
Dill district, 84
Diluvium, 220
Dinaric Mountains, 259
Dingoes, 194
Diseases, 81, 91, 93, 151, 246, 286
Distillery, 257
Diversification, 21, 88 ff., 154; in the pre-industrial era, 286 ff., hindrance of risk to, 187; of farming, 104 ff.; of the production program, 284 ff.; stages in farming, 80 ff., 289 ff., 294
Diversified farms, 87 f., 105 f. *See also* Many-sided farms; Mixed farms
Diversity, farming, 93; stages in, 80 f., 284 ff.
Divided inheritance, 205
Dole Company, 156
Domesticated animals, 114
Don Steppe, 285
Donkeys, 66, 80, 166
Drainage, 29, 77
Draining: marsh soils, 77
Drinking places, 167, 171 ff.
Drought, 62, 160 ff., 187, 226, 242 f., 264, 270
Dry boundaries, 38 ff., 46, 57 ff., 86, 129, 131, 169, 174 ff., 300
Dry climates, 34, 37 ff., 53, 250 f., 293 ff. *See also* Arid zones
Dry farming, 18, 37, 58, 64, 77, 81, 119, 127, 164, 175 ff., 242, 264, 300. *See also* Grain-fallow farming; Dryland cropping
Dry savanna climate, 37 f.; farming changes in, 293 ff.
Dryland cropping, 58 ff., 81, 160, 164, 174 ff., 185 ff., 293 ff. *See also* Dry farming; Grain-fallow farming
Drylot dairy farms, 246
Drylot farms, 225 f.
Dutch, 230

East Bloc, 197, 224; agricultural geography of the, 250 ff.
Ecological boundaries, 48 ff.
Ecological dispersion, 186, 232
Economic boundaries, 46 f., 70 ff.

Economic distance, 47. *See also* Market distance
Economic dry boundaries, 82
Economic formations: defined, 17; growth, 106 ff., 271 ff.; stages, 80 ff., 292; theory, 27 ff., 301 ff. *See also* Agricultural regions; Agricultural zones
Economic growth, 79 ff., 106, 271 ff., 286, 289 ff.; boundary shifts with, 76 ff. *See also* Development; Economic integration
Economic integration, 237, 288 ff. *See also* Development; Economic growth
Economic life, 118, 124, 153, 224; defined, 20; of bush and tree crops, 150; of fodder cropping, 258; of leys, 212; of pineapple, 156
Ecosystem, 68
Ecuador, 275
Educational level, 95 f.
Effective boundaries, 46 f., 86
Efficiency: defined, 21
Egg production, 23, 111
Egypt, 46, 68, 120, 143 ff., 269, 275
Eifel, 221
El Centro area, 246
Elbe Marsh, 114
Electric fence, 112
Embroidery, 75
Emigration, 278
Emilia, 90
Employment opportunities, 178
Ems country, 230
Enclosure farming, 167 ff.
Enclosures, 167, 172 ff.
Energy: content, 139; sources, 237
England, 77, 90, 110, 205, 216, 224, 231, 295 f. *See also* Great Britain
Ermland, 258
Estanciero, 74
Ethiopia, 39, 123, 125, 141, 149 f., 170, 275
Europe, 46, 106, 110, 116, 118, 124, 174, 196 ff.; agricultural geography of Western, 197 ff.; Central, 43; Eastern, 43, 202, 214; Northern, 118, 212, 214; South-eastern, 259; Southern, 213, 216, 228; Western, 197 ff., 237
European Common Market. *See* European Economic Community
European Economic Community (EEC, EC), 15, 42, 94 ff., 197 ff., 230; cattle farming types in the, 225; internal boundaries of the, 230; labor intensity in the, 229; land use systems in the, 215; livestock raising systems in the, 223; market opportunities in the, 231; permanent grassland in the, 204; production elasticity in the, 206; rotation types in the, 211
European farms, 189 f., 282
Exchange value. *See* Purchasing power

Exploitive economies, 27, 294
Export needs, 194
Export policy, 137
Export trade, 137
Exports, 125, 149, 154
Exposure, unfavorable, 69 f.
Extensive farms, 104, 284
Extensive grassland farming, 151, 160, 242 ff., 293 ff., 297 ff.; and dryland crop farming, 187 ff.; sedentary, 115; regions, 165 ff.
External upgrading, 282
Extra-income farming, 282

Factor combinations, 28, 113, 173, 273; changes in, 273 ff.; differences in the United States, 247 ff.; in densely settled countries, 277 ff.; in sparsely settled countries, 274 ff. *See also* Minimum cost combination; Production factors
Factor costs, 28, 247 ff.; in densely settled countries, 277 ff; in sparsely settled countries, 274 ff. *See also* Minimum cost combination; Production factors
Factory processing, 154
Fallow, 19, 186, 297, 301; bare, 122, 130, 164, 176 f., 181, 184; bastard, 261; bush, 27, 130, 151, 186; forest, 27, 130, 134 f., 139, 291; full, 261 f.; grass, 27, 130, 175, 293; old, 264; pasture, 193; years, 135
Fallowing, 184, 272, 294, 296; bare, 164, 176 ff.; forest, 292; grass, 291 f.; in dry farming, 176; legume, 291 f.
Family farms, 21, 146 ff., 224, 288; small-, 114, 207
FAO, 15, 31, 45, 84
Far East, Russian, 265
Farm development forces, 88 ff.
Farm family, 152
Farm gate prices, 102, 106, 126, 137
Farm income, 21 f.
Farm Machinery Association (FMA), 24
Farm organizations, 102
Farm parcels, 205
Farm-produced inputs: and market distance, 103 ff.
Farm production, 22
Farm settlement: in Zaire 149
Farm size, 21, 76, 96 ff., 146, 157 ff., 188 f., 192 ff., 222, 231, 252, 254; and collectivization stages, 266; and precipitation, 179, 183; and production elasticity, 206; and types of cattle raising, 226; classes for tea farms and plantations, 157; classes for West German agriculture, 282 f.; developmental tendencies in, 280 ff.; differences in, and pastureland-cropland ratios, 189; distribution of, in Europe, 205; minimum, 143, 151, 280 ff.; production and, 206; productivity and, 283

Farmed land (FL), 15, 18, 97 f., 100, 102, 114, 206, 219, 226, 239, 274, 276 f.
See also Agricultural land; Cultivable land
Farmers, 27, 252; Central European fruit, 230; collective, 253; European, 189 f.; fodder cropping, 230; hoe cropping, 219; large, 231, 266; medium-scale, 266; native, 189 f.; peasant, 149, 151, 222; potato, 230; small, 254, 266; suitcase, 242; vegetable, 230; West European, 254; West German, 231; wheat, 194
Farming systems, 113 ff.; and market distance, 105; annual cropping, 17, 114, 117 ff., 212 f., end papers; bog-burning shifting cultivation, 27, 118; bush and tree crop, 82 f., 130 ff., 149 ff., 228, 291 f., end papers; cattle, 24, 38 f., 42, 72 ff., 80 f., 115, 136, 166 f., 170, 173, 188, 193, 226, 244, 246, 260; changes in, 290 ff.; commercial steppe grassland, 167 ff.; corn, 285; dairy, 20, 58, 104, 213, 222 ff., 226 ff., 256 ff., 265, 273; dairy-hog, 223, 227; dairy-young cattle, 223 f.; determining, 210, 227, 256 f.; dry, 18, 37, 58, 64, 77, 81, 119, 127, 164, 175 ff., 242, 264, 300; dryland, 58 ff., 81, 160, 164, 174 ff., 185 ff., 293 ff.; enclosure, 167 ff.; evolution of, 199 ff.; extensive grassland, 115, 151, 160, 164 ff., 171 ff., 187 ff., 242 ff., 293 ff., 297 ff.; extensive livestock, 72 f.; extra-income, 282; fenced-range, 167 ff.; fodder crop, 18, 43 f., 68, 71, 81, 91 f., 116 ff., 169, 172, 174, 199, 201 f., 204, 207, 209 f., 214 ff., 232, 237, 240, 256 ff., 293 ff.; fodder crop-grain crop, 215 f.; fodder crop-hoe crop, 215 f.; fodder crop-livestock, 219; fodder crop-specialty crop, 215 f.; forest-burning shifting cultivation, 27, 118, 125, 130, 132, 134 f., 291, 301; fruit, 221, 244; fruit, truck, and mixed, 236, 244 ff.; full-time, 282; grain, 60, 72 f., 92, 104, 111, 119 f., 190, 197, 199, 207, 209 f., 213 f., 218 ff., 256, 258, 285, 298 f., end papers; grain-fallow, 15, 76 f., 110, 114 f., 178, 180, 184 f., 242, 293 ff., 301; grass-clover, 18; grassland, 40, 44, 58, 60 ff., 65, 80 f., 83, 86, 110, 114 ff., 127, 130, 151, 160, 164 ff., 171 ff., 187 ff., 217 f., 242 ff., 293 ff., 297 ff., end papers; grassland (grazing), 114 ff., 131, 188, 301; *Hauberg* shifting cultivation, 19, 118; hoe (root) crop, 86, 91, 104, 107, 120 ff., 189, 197, 199, 207, 209 f., 214 ff., 228, 256 ff., 273, 291 ff., end papers; hoe crop-grain crop, 215 f., 219; improved three-field, 297; in European agriculture, 197 ff., 207; in the dry areas, 160 ff.; in the humid tropics, 127 ff.; in the middle latitudes, 196 ff.; integrated, 81; irrigated fodder crop, 174; irrigation, 29, 38, 41, 60, 68, 82 f., 93, 102, 125, 127, 129 f., 132, 141 ff., 149, 151, 163 f., 300; *Kunstegart,* 19, 68 f.; labor-intensive, 231; latifundia, 206; ley, 18 f., 44, 48, 56, 92, 118 f., 138 f., 141, 211 ff., 261, 291, 295 ff., end papers; livestock, 33, 36 f., 40, 42, 58, 71, 75, 78, 92, 94, 110, 115 f., 119, 123, 129, 131, 179, 190, 217, 246 f., 273, 285, 297; market garden, 297; maritime fodder, 116; millet-sorghum-peanut, 181 ff.; mixed, 21, 139, 152, 236, 240, 244, 273, 285; *molapo,* 65 f.; monocultural, 19, 21, 56, 91, 125, 145, 151, 154, 169, 181, 184, 217, 221, 240, 242, 245, 285, 289 f.; montane fodder, 116; *Naturegart,* 19, 50, 68 f.; nomadic, 66, 75, 80, 114, 166 ff., 171 f., 291, 294 f.; of the Corn Belt, 238; one-field, 297; open-range steppe grassland, 167 ff.; part-time 282; peasant, 56, 155 f.; perennial-cropping, 114, 117 ff., 122 ff., end papers; plantation, 17, 55, 87, 95, 102, 105, 122 ff., 153, 289, end papers; plow, 27 ff., 77, 86, 135, 294; polar fodder, 116; primitive rotation, 117 f.; principal, in world agriculture, 113 ff.; pure dairy, 222 ff.; pure grassland, 217 f.; rainfed, 37, 40, 57, 59 f., 83, 129 ff., 137 ff., 143, 151, 160, 292, 295; ranching, 64, 77 f., 80 f., 115 ff., 167 ff., 129, 167 ff., 182 ff., 236, 293 ff., end papers; rice, 145 ff., 278; savanna grassland, 166 ff.; savanna shifting cultivation, 174 f., 293 ff.; sedentary extensive grassland, 115, 167 ff., 243 f.; sedentary grassland, 167 ff.; sedentary intensive grassland, 115 ff.; sequence of, 291 ff.; sheep, 58, 60, 62, 115, 191 f., 194, 226; sheep, goat, and dairy, 223 f., 226; sheep, goat, and young cattle, 223, 227; shifting cultivation, 27 f., 37, 58, 80 f., 95, 117 ff., 125, 129 ff., 140, 174 ff., 285, 291, 293 ff.; special-crop and general, 236; specialized, 21; specialty crop, 215 f., 221; specialty crop-hoe crop, 215 f., 222; steppe shifting cultivation, 27, 58, 80 f., 117 ff., 174 ff.; subsistence, 187; sugar beet, 77, 100, 110 f., 258; tea monoculture, 56; terrace, 55, 68; three-field, 20, 120, 296 ff.; three-field and fallow, 297; tobacco, 236, 240; traditional, 231; tropical, 131 ff.; 300 f.; two-field, 297; vegetable, 245 f.; wet-rice, 299; wheat-fallow, 182, 190, 193; wheat-sheep, 192 194; wild steppe grassland, 167 ff.; young cattle-dairy, 223
Farms, 17, 87 ff., 289; and force groups, 87 f., 105 f.; bush and tree crop, 87, 114, 122 ff., end papers; cacao, 114, 158 f.; cash crop,

(Continuation Farms)
182; cattle fattening-dairy, 225 f.; cattle raising-fattening, 226; coffee, 158 f.; collective, 252 ff., 260; dairy-cattle fattening, 225 f.; dairy-cattle raising, 225 f.; diversified, 87 f., 105 f.; drylot, 225 f.; drylot dairy, 246; European, 189 f.; 282; expansion of, 178, 280 ff.; extensive, 104, 284; family, 21, 146 ff., 224, 288; feeder stock, 169; fodder cropping, 214 ff., 220 f., 232; forces determining organizational structure, 88 ff., 105 f., 189 f., 191 ff., 252; four-tractor, 98; full-time, 282 f.; fully commercialized, 21, 75; German, 94 f.; grain cropping, 218, 221; hoe-cropping, 220, 222; horticultural, 245; in Congress Poland, 255; income on, 21 ff., 97, 182 f.; industrialized livestock, 247; industrialized poultry, 247; Kikuyu, 285 285 f.; labor-intensive, 104; land-poor family, 21, 220, 283; land-poor intensive, 282; land-rich extensive, 282; land-rich family, 21, 218, 283; large, 21, 75, 89, 112, 124, 151, 218 ff., 252 ff., 256, 283, 302; large family, 21; large scale wage-labor, 21, 220, 224, 283; little commercialized, 75; livestock raising, 116; location of, and supply and demand, 28, 70 ff., 102 ff., 190 f., 238 f., 295; location of, for transport, 102 ff.; many sided, 104 ff., 284; marginal, 77; maritime fodder cropping, 116; mechanized, 281; mixed, 221, 240, 284, 289; monocultural, 89, 104 ff., 117, 151, 217, 240, 244, 284 ff., 289; one-crop, 244; one-sided, 104 ff., 284 f., 289; one-tractor, 98; Ovambo, 190; part-time, 226, 282 f.; peasant, 75, 124, 153 ff., 253, 256, 266; pioneer, 47, 77; pure cattle-fattening, 224 ff.; pure crop, 187; pure grassland, 93, 217 f.; self-supplying cattle fattening, 169; self-supplying dairy, 225 f.; semi-drylot, 225 f.; sheep, 114; single-operator, 98; single-product, 284; size, 96 ff.; small, 97 ff., 285; small family, 114, 207; socialist, large, 251, 254, 257, 260; spatial differentiation of, 94 ff.; specialized, 104, 221, 240, 246, 284, 289; specialty crop, 220; state, 24, 252 ff., 260; strongly commercialized, 21, 75; subsistence, 21, 182, 280; summer feeding, 116; tea, 157 ff.; temporal changes in, 106 ff.; three-crop, 244; three-tractor, 98; two-crop, 244; two-man, 97, 100; two-tractor, 98; wage-labor family, 21, 218 f., 283; weakly commercialized, 21; wheat, 179, 242; wheat-fallow, 178, 180

Fats, 224
Federal Republic of Germany (FRG). *See* Germany, West

Feed, 39, 101, 181, 249, 259, 293, 295; balance, 92, 114 ff., 151, 169 ff., 224, 259, 286, 292, 295; concentrates, 21, 92, 110, 170, 174, 272; corps, 221; purchases, 116; purchasing, 160, 170 ff., 295; reserves, 172 ff.; shortages, 170, 188, 217; supplies, 116; surplus, 224
Feeder stock, 66, 72 f., 101, 107, 167, 169, 183, 224, 226, 243, 246; farms, 169
Feedlots, 246
Feeds, 111
Fehmarn Island, 119, 218
Fenced-range farming, 167 ff. *See also* Farming systems; Grassland farming; Ranching
Fences, 172, 301
Fertilization, 29, 91, 101 f., 147, 162, 189. *See also* Manuring
Fertilizer: balance, 91 f., 221; expenditures, 157; spreaders, 277
Fertilizers, 16, 111, 148, 204, 249, 277, 292; chemical, 21, 91, 109 f., 135, 152, 174, 251, 267, 286, 288; organic, 92, 124, 221 ff., 286
Fields: rational distribution, 219
Figs, 25, 221
Finland, 79, 84, 275
Fish, 122
Fishermen, 299
Fishing, 27, 123, 291, 294, end papers
Fixed costs, 288
Flanders, 202, 220
Flax, 75, 179, 262 f.
Flood control projects, 270
Flooding, 82
Floods, 68, 270
Florida, 235
Flowers, 245
Fodder crop-grain crop farming, 215 f.
Fodder crop-hoe crop farming, 215 f.
Fodder crop-livestock farming, 219
Fodder crop-specialty crop farming, 215 f.
Fodder cropping, 18, 43 f., 68, 71, 81, 91 f., 116 ff., 169, 172, 174, 199, 201 f., 204, 207, 209 f., 214 ff., 232, 237, 240, 256 ff., 293 ff.; farms, 214 ff., 220 f., 232
Fodder crops, 118, 122, 143 f., 145, 179 f., 222, 261, 293, 297; manure-producing, 221
Food, 259, 289, 297; demands, 45; habits, 190; processing, 59; production, 45 ff.; production capacity, 143; surpluses, 45
Forage, 92; crops, 19, 42. *See also* Feed
Force groups: and farm organization, 87 f., 105 f.
Forest-burning shifting cultivation, 27, 118, 125, 130, 132, 134 f., 291, 301
France, 84, 95, 199, 203, 205, 212, 216, 224, 230 f.
Franconia, 219

Free trade, 95
French Central Plateau, 212, 216 f.
Fruit, 25, 95, 122, 199, 230 f., 245, 256, 294; growing, 221, 244; trees, 220
Fula, 171, 167
Full-time farming, 282
Full-time farms, 282 f.

Galicia, 258
Gambia, 135
Game, 122
Ganges, 29
Garbanzos, 122, 180, 185, 293
Garden culture, 291
Gardens, 266
Garonne, 220
Gascony, 224
Gatherers, 40, 299
Gathering, 27, 40, 122 f., 291, 294
Geest, 202, 230
General farming. *See* Mixed farming
Geographic boundaries, 84. *See also* Boundaries
German agriculture, 231. *See also* Germany
German Democratic Republic (GDR). *See* Germany, East
German farms, 94 f.
German marsh belt, 217
Germany, 84, 95, 107, 109 f., 111, 114, 176, 207, 271 ff.; East, 15, 23, 253 ff.; West, 15, 24, 46 f., 69, 84, 107 ff., 111 f., 145, 205, 213, 216, 224, 227, 230 f., 258, 267, 275, 302
Ghana, 114, 125, 149, 155, 275
Goats, 66, 80, 166, 170, 209, 226, 259, 285
Gobabis District, 188
Göta Plain, 50, 199
Grain: feeding, 227; landscapes, 259
Grain-corn zone. *See* Corn Belt
Grain-cropping farms, 218, 221
Grain-fallow farming, 15, 76 f., 110, 114 f., 178, 180, 184 f., 242, 293 ff., 301. *See also* Dry farming; Dryland farming; Grain farming
Grain farming, 60, 72 f., 81, 92, 104, 111, 119 f., 190, 197, 199, 207, 209 f., 213 f., 218 ff., 256, 258, 285, 298 ff., end papers. *See also* Grain-fallow farming
Grain Unit (GU), 15
Grains, 16, 18, 25, 42 f., 50, 59, 69, 107, 117, 120, 122, 139, 144, 147, 176, 178 ff., 186, 212, 218, 220 f., 239, 241, 258, 262 ff., 267, 288, 293, 297
Grape farms, 114
Grapes, 41, 49, 52, 62, 222, 244, 256
Grapevines, 202, 220 f.
Grass, 19, 71, 139, 171, 238, 240 f., 258; -clover farming, 18; cultivated, 138; guinea, 74; natural, 238, 264; renge, 42, 144 f.; seeds, 18, 119; Sudan, 122; years, 19
Grassland, 19, 83, 255; farmers, 174; -farmland ratio, 202 ff., 213; (grazing) systems, 114 ff., 131, 188, 301; marsh, 110; permanent, 18, 44 f., 48, 77, 84, 202 ff., 214, 240, 258, 261, 263, 267, 297; permanent, and types of cattle raising, 226; permanent, in the European Community, 203; pure, farmers, 93; rotation, 50
Grassland farming, 40, 44, 58, 60 ff., 65, 83, 86, 130, 217 f., 293, 295, end papers; extensive, 80, 83, 89, 110, 115, 127, 151, 127, 151, 160, 164 ff., 171 ff., 187 ff., 242 ff., 293 ff., 297 ff.; sedentary extensive, 115; sedentary intensive, 115 ff.; semi-intensive, 81
Graubünden, 50
Grazing boundaries, 36
Grazing systems. *See* Grassland (grazing) farming systems
Great Britain, 116, 119, 199, 202 f., 205, 212, 214, 217, 275. *See also* England
Great Lakes area, 235, 237, 244
Greenhouse crops, 245
Grootfontein District, 188
Gross income: defined, 22 f.
Gross return: defined, 21
Groundnuts. *See* Peanuts
Groundwater, 217, 292, 295
Growing season, 212; and land use intensity, 122; and the polar boundaries, 48; available, in the United States, 234; for tropical field crops, 139; length of, and land use systems, 216; length of, and rice monocultures, 145; length of, and wet rice rotations, 145; short, 261; short, and adaption to aridity, 164
Growth cycles, 151
Guatemala, 134 f.
Guinea Coast, 33
Gulf Coast (U.S.), 235, 244
Gum arabic, 25, 123
Gum plants, 25
Gur, 164
Guyana, 120, 145

Hacendados, 74
Haciendas, 74
Hand processing, 156
Handicraft industries, 84
Hanover, 287; North, 220
Harvest: failure, 160; fluctuations, 143; processing, 124, 155; risk, 77, 152, 289
Harvesting: mechanized, 110; problems, 77, 151, 154; technology, 57, 77, 147 f., 242
Hauberg shifting cultivation, 19, 118
Hawaii, 155

Hay, 81, 92, 170, 174, 218, 220, 293; baler, 100; bean, 138; chopper, 100; mowing wild, 172
Heifers, 183; price of, and types of cattle raising, 226
Hemp, 26, 185 f.
Herbicides, 77 f., 111 f., 122, 288
Herreros, 78, 167
Herzegovina, 259
Hesse, 201, 213, 219
Hessian loess zone, 218
Hides, 72 f.
Highland exodus, 47
Hildesheim-Brunswick *Börde,* 220
Himalayas, 36, 53
Hoe, 28 f.
Hoe crop-grain crop farming, 215 f., 219
Hoe cropping, 86, 91, 104, 107, 120 ff., 189, 197, 199, 207, 209 f., 214 ff., 228, 256 ff., 273, 291 ff.
Hoe cropping farms, 220, 222
Hoe crops, 18, 19, 100, 108, 122, 146, 220, 222, 262, 297, 302. *See also* Root crops
Hoe culture, 27, 28, 291, 299
Hoe farmers, 96
Hoe (root) crop farming, 86, 91, 104, 107, 114, 120 ff., 189, 197, 199, 207, 209 f., 214 ff., 228, 256 ff., 273, 291 ff., end papers
Hoes, 117, 143, 147 f., 175, 299
Hog fattening, 23, 239, 256
Hog finishing. *See* Hog fattening
Hog raising, 89, 227, 240
Hogs, 24, 108, 188, 209 f., 238. *See also* Swine
Hohe Venn, 217
Holland, 95, 109 f., 230
Home processing, 154
Honan, 268
Honduras, British, 134
Honey, 123
Hopeh, 268
Hops, 124
Hopyards, 252
Hormone weed killers, 91
Horses, 166 f., 288
Horticultural farms, 245
"Hub" crop, 185
Humid cool-temperate climates, 42 ff., 297 f.
Humid savanna climate, 33 ff.; farming changes in, 212 f.
Humid temperate climates: and attitudinal crop boundaries, 53
Humid tropics: agricultural geography of, 127 ff.
Humid warm-temperate boundaries, 40 ff.
Humid warm-temperate climates, 34, 40 ff.
Humus, 221, 245, 293; losses, 137; production, 91 f.; root, 205

Hungary, 227, 253
Hunger, 45
Hunsrück, 221
Hunters, 40, 299
Hunting, 27, 40, 80, 123, 291, 294, end papers
Hybrids, 111
Hydroelectric power, 195, 265, 270
HYVs, 79

Iberian Peninsula, 227
Idaho, 242
Ile-de-France, 259
Illinois, 237, 239
Improved three-field farming, 297
Income: expectations, 74, 273; farm, 21 f.; gross, 21 f.; gross, on U.S. wheat farms, 179; levels, 81, 84 f.; mass, 174; maximization, 41, 249
India, 37, 41, 68, 125, 139, 145, 150, 154, 181 f., 267, 278
Indochinese areas, 251
Indonesia, 33, 125, 149
Indus, 28, 163
Industrial-agrarian countries, 273, 275, 277
Industrial boundaries, 70
Industrial countries, 46, 68, 85, 104, 116, 145, 157, 170, 254, 273, 275, 277 f., 293, 302
Industrial processing, 154
Industrial zone, 105, 249
Industrialization, 81, 205, 271, 273, 275 ff., 288 ff., lack of, 266; policy, 137
Industrialized animal production, 246 f.
Industrialized livestock farms, 247
Industrialized poultry farms, 247
Industrially-produced inputs: and market distance, 103 ff.
Infrastructure policy, 137
Inheritance customs, 205
Input-output ratio, 95, 134, 273, 284, 286. *See also* Cost-production ratio; Costs; Minimum cost combination; Price-cost relations
Insecticides, 111, 122, 148, 152
Integrated farming systems, 81
Intensification, 84, 287, 290 ff., 301; and agricultural capacity, 83 ff.; of crop rotations, 296; of cultivation, 137; of yield, 85
Intensity: agricultural, 111, 289; capital, 110, 151, 173 f., 182 f., 249, 271 ff., 286 ff.; cropping, in China, 268; cropping, in world agricultural space, 133; farming, 20, 272, 302; farming, and market distance, 103; high labor, 205, 278, 295; labor, 97, 101, 109 f., 182 f., 205, 222, 229 f., 249, 254 f., 271, 277 ff., 288, 290; labor, in agricultural development,

(Continuation Intensity)
271 ff., 286 ff.; labor, in land use, 228; minimum, 300; of fertilization, 101; of specialization, 20, 101 f.; operational, 20, 45, 104; organizational, 20, 45, 104, 222; orientation of, 301; production, 19, 247 ff.; rings in European agriculture, 205; specialization, 20, 101 f., 272 f.; stages and economic growth, 290 f.; zones of, in the dry farming system, 177
Intermountain states (U.S.), 300
Internal upgrading, 282
Ionian Sea coast, 40
Irak, 163
Iran, 40, 57, 119, 176, 275
Ireland, 224
Iron, 272
Irrigated fodder crop farming, 174. *See also* Irrigation farming
Irrigation, 148, 163, 195, 201, 245, 262, 264, 266, 270, 272; farm management functions and, 143; farming, 29, 38, 41, 60, 68, 82 f., 93, 102, 125, 127, 129 f., 132, 141 ff., 149, 151, 163, 300; ley farming based on, 213; methods compared, 143 f.; projects, 264 f., 270; sprinkler, 96, 112, 292. *See also* Water
Israel, 65
Italy, 41, 84, 95 f., 119 f., 199, 204 f., 213 f., 216, 220 f., 230
Ivory, 84

Jackals, 194
Japan, 41, 43, 69, 120, 143, 146, 153
Java, 53, 69, 95, 125, 135, 148 f., 150, 154, 278
Joint costs, 288. *See also* Costs
Jugoslavia, 214, 253
Jura, French, 217
Jute, 26, 51, 75, 164
Jutland, 117, 202, 212

Kalahari, 38
Kale, 122
Kalenberg country, 218
Kansas, 57, 114, 119, 179 ff., 242
Kapok, 26, 52, 55
Karakul fleece, 72
Karroo, 300
Kazakhstan, 40, 57, 252, 264
Kenaf, 58, 75, 185 f.
Kentucky, 240
Kenya, 53, 57, 64, 141, 150 f., 167, 182 ff., 275, 285; Highlands, 138, 141
Kikuyu country, 287
Kikuyu farms, 285 f.
Kilo starch unit (kStU), 15, 101

Kirghiz, 166
Klagenfurt Basin, 214
"Know-how," 230
Kolkhoz, 24, 252 ff., 260
Kombinat for Industrial Fattening (KIF), 24
Korea, 43; North, 253
Kuban area, 262, 264
Kunstegart farming, 19, 68 f.
Kurds, 171

Labor, 21, 94, 146, 154, 170, 173, 186, 224, 258, 260, 272 ff., 283, 290 f., 300, 302; abundant, 97; capacity of the farm, 222; composition, 21; cost ratios between, and machine capital, 182; costs, 89 f., 286 ff.; demand and types of cattle raising, 226; demands, 167, 221; division of, 104, 126, 136, 169; economics and farmsize, 283; expenditures, 124, 147 ff., 151 f., 178 ff., 301; expenditures per hectare, 256; family, 221 f.; Filipino migrant, 245; gross productivity of, 23; hand, 89, 184, 273; hired, 22; in European agriculture, 205 ff.; in Southeastern Asia, 122; in the Southern states (U.S.), 248 f., in the U.S. factor combination, 247 ff.; income, 22 f., inputs, 137, 139, 244; intensity, 97, 101, 109 f., 182 f., 205, 222, 229 f., 249, 254 f., 271, 277 ff., 288, 290; -intensive farming, 231; mechanized, 89, 96 ff.; Mexican migrant, 245; migratory, 89; net productivity of, 23, 115; peaks, 112, 151 f., 154, 184, 190, 221 f., 240, 244; productivity, 21, 29, 84, 111, 115, 124, 135, 145 ff., 158, 181 ff., 186, 193, 277 ff., 291; saved, 77; seasonal, 89; spacing, 88 ff., 114, 151, 184, 219, 221, 244, 284, 288; wage, costs, 220
Lake Chad area, 167
Lake Constance area, 221
Lambs, 39
Land, 21, 146, 173, 244, 283, 290, 294; and irrigated rice, 141; capital, 115; cheaper, 231; costs of, use, 286; cultivable, 260; cultivated, 178; determination of, use systems, 208; gross, productivity, 23; in the U.S. factor combination, 247 ff.; labor intensity in, use, 228; net, productivity, 23; ownership, 266; permanent and constant, use, 143; -poor intensive farms, 282; -poor family farms, 21, 220, 283; preparation, 147; prices, 174, 248 f., 271 ff.; productivity, 21 ff., 112, 115, 124, 134, 136, 143, 158, 182 f., 278 ff., 292 ff.; reserves, 171; -rich extensive farms, 282; -rich family farms, 21, 218, 283; rising, values, 287; surplus, 247 f.; use, 273, 288; use cycles, 150; use intensity and growing season length, 122; use intensity

Index

(Continuation Land)
and water supplies, 122; use intensity in irrigation farming, 144; use regions, 214 ff.; use systems and cartographic representation, 207; use types in dry savannas, 294
Languedoc, 220
La Plata countries, 41, 269
Lapland, 48, 111, 197
Lapps, 86
Large family farms, 21. *See also* Wage-labor family farms
Large farms, 21, 75, 89, 112, 124, 151, 218 ff., 252 ff., 256, 283, 302
Large livestock unit (LLU), 15, 168. *See also* Cattle unit
Large-scale operations, 151
Latifundia farming, 206
Latifundios, 89
Lead crop, 185, 207, 214
Lead enterprise, 224
Leafy crops, 18, 178, 212 ff., 259, 263, 300
Lease rotation, 246
Legumes, 18 f., 145, 164, 180 f., 262 f., 269, 286, 298
Leningrad Oblast, 262
Lettuce, 245
Ley farming, 18 f., 44, 48, 56, 92, 118 f., 138 f., 141, 211 ff., 261, 291, 295 ff., end papers
Leys, 211, 239 f., 251, 262; useful economic life, 118, 238
Liberia, 135
Libya, 46, 57
Liguria, 201
Linen, 75
Livestock, 21, 23, 114, 160, 213, 218, 248 f., 259, 265, 301 f.; regions, 222 ff.
Livestock raising, 33, 36 f., 40, 42, 58, 71, 75, 78, 92, 94, 110, 115 f., 119, 123, 129, 131, 179, 190, 217, 246 f., 273, 285, 297; and farm size, 283; dry boundary of, 127; extensive, 72 f., 83, 243, 264, 299; in the European Community, 222 ff.; in the South, 241; sedentary, 80 f.; technical and structural arrangements in, 99. *See also* Farming systems
Llama, 39
Local zone, 105
Locational orientation, 94 ff., 191 ff., 226, 252
Loess belt, 101, 202, 219
Lombardy, 201
Los Angeles area, 233, 244
Lucerne. *See* Alfalfa
Lüneburg Heath, 119
Lupine, 18, 64, 262

Machine Tractor Station (MTS), 24
Machinery, 90, 109, 174, 239, 248 f., 277, 279, 288; cooperatives, 99 ff., 283; corn picking, 288; costs, 99 ff.; cotton picking, 288; custom hiring, 99 ff.; interfarm, use, 99 ff.; operator's, 283; pools, 99 ff., 283; use in West German agriculture, 100
Madras, 36
Magdeburg Börde, 122
Maize. *See* Corn
Maize Triangle, 188
Malagasy, 37, 145
Malawi, 28, 96, 138 f.
Malaysia, 125, 150
Mali, 175, 182, 187
Man: -day (MD), 15, 139 f., 184, 186; -hour (MH), 15, 77, 110 f., 124, 147, 153, 158, 164, 179, 182, 219; -year (MY), 15, 22, 101, 182 f., 260, 301
Management, farm, 21, 203, 299; active measures of, 170 ff.; for bush and tree crops, 149 ff.; functions of crop irrigation, 143; in U.S. agriculture, 249; passive measures of, 170 ff.; types, 244 ff.; weight, 259
Manchurian Plain, 267
Mangoes, 25
Manioc, 25, 33, 36, 50, 52 ff., 58, 74, 125, 132, 139, 286. *See also* Cassava
Manufactured inputs, 271 ff.
Manure, 81, 91 f., 151, 221, 245, 258, 293; barn, 222; -demanding specialty crops, 221; green, 181, 184, 269, 297; -producing fodder crops, 221
Manuring, 109; capacity, 221 f.; economy, 240, 273; green, 91, 291 ff., 297. *See also* Fertilization
Many-sided farms, 104 ff., 284. *See also* Diversified farms; Mixed farms
Marginal areas, 86. *See also* Marginal zones
Marginal farms, 77
Marginal locations, 47 ff., 84, 191 ff., end papers
Marginal productivity, 109, 274, 279
Marginal yield, 279
Marginal zones, 47 ff., 75, 80, 115, end papers. *See also* Marginal areas
Marine cool-summer climate, 34, 42; farming changes in, 295 ff.
Maritime climate, 116
Maritime fodder cropping farms, 116
Maritime fodder cropping regions, 217
Maritime fodder farming, 116
Market: dependence, 21; distance, 47, 73, 84 ff., 102 ff., 190 ff., 238 ff., 295; distance and agricultural production zones, 238 ff.; distance and farming diversity, 104; distance and farming intensity, 104; distance and socialist farm production, 256 ff.; distance and types of cattle raising

(Continuation Market)
226; distance and wage levels, 103; expansion, 273; garden farming, 297; position, 156; prices, 137; risks, 93, 169, 289
Marketing boards, 124, 150
Markets, 28, 220; expansion of, 273; great distances to, in Australia, 192; remoteness from, 190
Marsh soils, 77
Marshes: young cultivated, 218
Masai, 167, 171
Mass income, 174
Masuria, 258
Maté, 51
Mauritania, 57
Mayan culture, 28
Meadows, 18 f.; mowing of, 199; permanent, 68 f.
Meat, 139, 170, 193 f., 267; animals, 174
Mechanical technological advances, 76 ff., 111 f.
Mechanization, 77, 97 ff., 106, 110, 120, 125, 180, 220, 237, 266, 277, 282, 300; and farm size, 283; and smaller parcels, 205; difficulties, 240, 249; increasing pressures for, 214; over-, 90
Mechanized farms, 281
Mechanized harvesting, 110
Mecklenburg, 28, 256, 258
Mediterranean area, 40, 62, 68, 78, 163, 199, 201 f., 214, 216, 228
Mediterranean climate, 94, 191, 198 f., 222, 228, 235, 244, 259. *See also* Climates
Mediterranean coast, 181, 202, 227
Mediterranean countries, 201, 218
Mekong Delta, 29
Melons, 123
Meseta, 39, 301
Mesopotamia, 28
Mexico, 37, 53, 275
Michigan, 237
Migrations, 66; out-, 280
Milk, 65, 71, 73, 80, 93, 104, 108, 110, 115, 139, 166, 224, 227, 237, 239, 261, 267, prices and types of cattle raising, 226; production and prices, 107
Millet, 25, 36 f., 52 ff., 58, 74, 77 f., 81, 86, 117, 125, 138 f., 144, 151, 164, 175 ff., 267 f., 286, 288, 293; -sorghum-peanut farming, 181 ff.
Minimum cost combination, 28, 145 ff., 152, 274 ff., 279. *See also* Cost-production ratio; Costs; Input-output ratio; Price-cost ratio; Price-cost relations
Miombo woodland, 37, 129
Mississippi Basin, 234
Mississippi Valley, 241
Missouri Basin, 234

Mixed farming, 21, 139, 152, 236, 240, 244, 273, 285
Mixed farms, 221, 240, 284, 289. *See also* Diversified farms; Many-sided farms
Mixed forest, 260, 262 f.
Molapo farming, 65 f.
Molapos, 65
Molasses, 246
Mongolia, Inner, 267
Mongols, 166
Monocultural farms, 89, 104 ff., 117, 151, 217, 240, 244, 284 ff., 289. *See also* One-crop farms; One-sided farms; Single-product farms
Monocultural landscapes, 72
Monoculture, 19, 21, 56, 91, 125, 131, 145, 151, 154, 169, 181, 184, 217, 221, 240, 242, 245, 285, 289 f.
Montana, 179
Montane fodder cropping regions, 217
Montane fodder farming, 116
Montenegro, 259
Moscow: Oblast, 262; service area, 72
Mosel Valley, 220 f., 230
Mozambique, 66
Münster Bay, 230
Mulberry trees, 269
Mushrooms, 294
Mutton, 193
Mutual aid teams, 266

Nagana disease, 77, 136, 299
Namib, 115
Natural landscape, 141
Naturegart farming, 19, 50, 68 f.
Navigation, 270
Nebraska, 179, 242
Nematodes, 90
Neolithic culture, 28
Nepal, 53
Net return, 22 f.
New York State, 50
New Zealand, 110
Ngamiland, 65
Niger, 175, 182, 187
Nigeria, 46, 59, 125, 135
Nile, 28; Delta, 68, 122, 163
Nilgiri Mountains, 36
Nitrogen, 110, 286; fertilizers, 110
Nomadic grazing (herding), 27, 66, 75, 80, 114, 171 f., 291, 294 f., zones, 166 ff. *See also* Farming systems; Livestock raising
Nomads, 38, 65, 127, 294, 299; desert, 166, 170; mountain, 167, 171; semi-, 166 f., 250; steppe, 166
Norfolk rotation, 19, 211, 213 f., 295 f.
Normandy, 42
North Africa, 39 f., 111, 163
North America, 42, 60, 118, 196 f.; agricultural geography of, 232 ff.

Index

North Carolina, 240
North Dakota, 179, 190, 242
North European Plain, 227
North German Marsh, 101, 116
North German Plain, 219
North Macedonian Mountains, 259
North Sea Coast, 213; German, 230
Northeast (U.S.), 232 f., 237
Northern boundaries: in Europe, 199 ff.
Northern Plains (U.S.), 179; states, 235
Norway, 50, 199, 205, 216, 224, 226
Nug, 141
Nurse crops, 259
Nutrient balance, 91, 109, 286
Nutrient concentration, 286
Nutrient reservoir, 173
Nutrient yield, 173
Nutrients, 117, 134, 141, 170, 177, 251, 301
Nutrition: regionalization of world, 196

Oases, 39, 62, 241
Oats, 18, 48 ff., 54, 56, 71, 100, 119, 138, 180, 238 ff., 262, 296
Oceanic climate: and production risk, 93
Oder-Vistula area, 258
Oil crops, 19, 25, 154, 220, 298. *See also* Oil palms
Oil mills, 87, 155
Oil palms, 25, 33, 36, 50 ff., 55, 58, 124 f., 126, 131, 150 f., 154, 269, 286, 291. *See also* Oil crops
Oil pumpkin, 139
Okavango Swamp, 65
Olives, 25, 41, 49 f., 52, 58, 62, 124, 155, 163, 199, 201, 221
Oman, 46
One: -crop farms, 244; -field farming, 297; -sided farms, 104 ff., 284 f., 289; -tractor farms, 98. *See also* Monocultural farms; Single-product farms
Open-range steppe grassland farming, 167 ff. *See also* Farming systems; Grassland farming; Ranching
Opportunity costs, 97
Optimal productive combination, 247 ff.
Orange Free State, 188
Oranges, 222
Oregon, 242
Organic matter, 141
Organic technological advances, 111
Otavi Valley, 189
Otjiwarongo District, 188
Ovambo farms, 190
Ovamboland, 189
Ovambos, 189 f.
Overgrazing, 27
Overpopulated agrarian state, 278, 280
Overpopulation, 205

Ownership: land, 266
Oxen, 77, 96, 173, 183, 288

Pacific Coastal Lowlands, 234
Pakistan, 40, 68, 154, 163 f.
Palm nuts, 123
Paprika, 122
Paris Basin, 79, 90, 202
Part-time farming, 282
Part-time farms, 226, 282 f.
Passion fruit, 36, 52, 55, 153
Passive adaptation, 106 f., 160, 170 ff.
Pasture, 18, 74, 86, 169, 259, 297; carrying capacity of natural, 168, 193; dryland, 193; fallow, 193; mowing, 217; natural, 164; per cattle unit in western U.S., 243; permanent, 297; requirements on U.S. wheat farms, 179; residual, 231; rotation, 180, 297; zone, 105
Patagonia, 57
Patagonians, 27
Pea viners, 112
Peanut-millet rotation, 58
Peanut-sorghum rotation, 58, 184
Peanuts, 25, 36 f., 42, 50 ff., 55, 58 f., 78, 134, 138 f., 144, 164, 175, 177 f., 181 ff., 220, 241, 286, 293, 300
Peas, 58, 71, 121, 141, 177, 262
Peasant crops, 150, 155 ff.
Peasant family, 151
Peasant farm economies, 56, 155 f.
Peasant farms, 75, 124, 153 ff., 253, 256, 266
Peasants, 124, 252; German, 75
People's Commune (PC), 24, 256
People's Estate (PE), 24
People's Republic of China (PRC). *See* China
Pepper, 36, 150
Perennial cropping systems, 114, 117 ff., 122 f., end papers
Perennial crops, 122, 125 f., 143, 150, 152, 164, 245, 255, 301. *See also* Bush and tree crops
Personality, 29, 95 f.
Peru, 53
Peshawar Basin, 163 f.
Pests, 81, 91, 93, 151, 246, 286
Phenological dates, 199
Philippines, 135
Photoperiodism, 68
Pineapple, 25, 124, 150, 154 f., 269
Pioneer farms, 47, 77
Place-specific characteristics: and spatial differentiation of farms, 94 ff.; of Australia, 191 ff.; of the dry areas, 160 ff.; of the East Bloc countries, 250 f.; of the tropics, 127 ff.; of the United States, 232 ff.; of Western Europe, 197 ff.
Plains states (U.S.), 89

Plant breeding, 78 f., 84, 111, 264
Plant protection, 21, 29, 93, 147, 162, 219, 288, 292 f.
Plantains, 25, 37
Plantation crops, 150, 155 ff.
Plantation farming, 125, 151, 300 f., end papers
Plantation zone, 105
Plantations, 17, 55, 87, 95, 102, 105, 122 ff., 151, 153 f., 157, 287, 289, 300 f., end papers; Indonesian rubber, 123; size, 96 ff.; sugarcane, 145
Planters, 252
Plow, 28 f., 117, 175
Plow cultivation, 27 ff., 77, 86, 135, 291, 294
Plums, 124
Po Plain, 40 f., 79, 96, 119, 201, 212 ff., 220, 224
Poland, 202, 216, 253 ff.
Polar boundaries, 44, 46, 48 ff., 52, 79, 84, 86, 264
Polar fodder cropping regions, 217
Polar fodder farming, 116
Polish United Workers' Party, 255
Pomerania, Farther, 75, 227, 256, 258
Population, 267; consumptive habits, 273; density, 95 f., 146, 173, 175, 205, 248, 267, 294; density and agricultural development, 274 ff.; growth, 28, 45, 82, 105, 175, 178 ff., 244, 286; high, density in the (U.S.) South, 248; over-, 205. *See also* Settlement
Pork, 71, 75, 239, 261
Potash, 91
Potato harvesters, 100, 288
Potato hog-fattening area, 227
Potatoes, 18, 36, 44, 48, 51 ff., 56, 58, 69, 71, 75, 86, 90, 95, 122, 138, 141, 199, 213, 219, 222, 230, 251, 256 ff., 285 f.; English, 138; feed, 256; seed, 104, 219, 256; sweet, 25, 42, 50 ff., 58, 74, 125, 132, 138, 143 ff., 186, 241, 285 ff.
Poultry, 24, 285
Poultry raising, 23, 89, 247
Pre-Alpine areas, 213
Precipitation, 34, 115 f., 118, 160 ff., 168 f., 176 ff., 201 f., 216 ff., 234 f., 242 f.; agricultural effects of a decline in, 191; annual, in the United States, 233; and crop rotations in dry farming, 177; and land use intensity, 122; and land use systems, 216 f.; and ley farming, 118; and ley farming regions, 212; and tropical vegetation formations, 127 ff.; average annual, 35; boundaries, 115, 169; boundaries in Africa, 165; curve for Grootfontein, Southwest Africa, 162, 166; favorable distribution of, 116; fluctuations, 160, 162 f., 165 f.; limits of extensive grazing, 115; mean annual, in Europe, 201 f.; on the agronomic dry boundary, 57; seasonal distribution, 64; uneven distribution of, 160; variations in Africa, 165. *See also* Climates
Preparatory crop, 185
Price-cost development, 106 ff.
Price-cost differences, 106
Price-cost ratio, 74, 110, 144 ff., 170, 173, 247 ff., 297. *See also* Cost-production ratio; Costs; Input-output ratio; Minimum cost combination; Price-cost relations
Price-cost relations, 71, 78, 102 ff., 247 ff., 271 ff., 297; between agricultural inputs, 109 f.; between agricultural products, 107 ff.; between agricultural products and inputs, 110 f. *See also* Cost-production ratio; Costs; Input-output ratio; Minimum cost combination; Price-cost ratio
Price-cost shifts, 86, 88
Prices, 110, 252, 271 ff., 295 ff.; barley, 297; farm gate, 102, 106, 126, 137; in Germany, 272; iron, 272; land, 27 ff., 174, 248 f., 272; potato, 297; rye, 272; wheat, 297; world market, 126
Primary forage area (PFA), 15
Primitive rotation farming, 117 f.
Product-specific transport boundaries, 72 f.
Production, 249; and farm size, 206; Brazilian agricultural, 284; brigades, 266; costs, 88, 231, 242, 286 ff.; deficiency in, 255; diversity, 21, 88 ff., 105 f., 154, 187, 240, 284 ff.; elasticity, 19, 101, 143, 206; factors, 21, 28, 109 f., 146 ff., 183, 247 ff., 255, 274; farm, 22; function, 176; intensity, 19, 247 ff.; methods, 146 ff., 155 f., 169 ff., 247 ff.; physical location of, 94 f.; prices of, factors, 271 ff.; process, 274; program, 19, 21, 88, 101 ff., 106 f., 113 ff., 167 ff., 182 ff., 247, 284 ff.; risk, 169 f., 187; teams, 266
Productivity, 21, 23, 124, 143, 158, 241 ff., 269, 273 f., 282, 292 ff., 302; and farm size, 283; capital, 146, 277 ff.; comparison, 139 f., 158 f.; labor, 21, 29, 84, 111, 115, 124, 135, 145 ff., 158, 181 ff., 186, 193, 277 ff., 291; land, 21 f., 112, 115, 124, 134, 136, 143, 158, 182 f., 278 f., 292 ff.; marginal, 109, 274, 279; market, for sheep and cattle, 167; of livestock farming types, 295; optimal, combination, 247 ff.
Profitability boundaries, 46 f., 70 ff., 86
Protein, 185, 189; animal, 224; content, 139; plant, 224; ratio, 286; yield, 139
Provence, 220
Purchasing power, 271 ff.; between wage labor and farm machinery, 109; low, 136;

Index

(Continuation Purchasing power)
  of agricultural products, 108, 111, 271, 302; of animal products, 170, 174, 295
Pure grassland farmers, 93. *See also* Grassland farmers
Pure cattle-fattening farms, 224 ff.
Pure crop farms, 187
Pure grassland farming, 217 f. *See also* Farming systems; Grassland farming
Pure grassland farms, 217 f.
Pygmies, 27
Pyrethrum, 52, 55, 138

Queensland, 192

Rainfed farming, 37, 40, 57, 59 f., 83, 129 ff., 137 ff., 143, 151, 160, 292, 295; in developing countries, 149; in the humid savannas, 137 ff.; in the tropical highlands, 140 f.; in the tropical rainforest, 132 ff.
Rainforest, 55, 63, 67, 94, 124, 138 ff., 153, 299 ff.; climate, 33 ff., 38, 127 ff., 291 f., 299, 301; rainfed farming in the, 132 ff.
Ramie, 26
Ranchers, 66, 74, 127, 252
Ranches, 17, 38, 74 f., 87, 96 ff., 101, 172, 182 f., 194, 242, end papers
Ranching, 64, 77 f., 80 f., 115 ff., 129, 167 ff., 182 ff., 236, 243, 293, 295, end papers. *See also* Farming systems; Grassland farming
Range: fenced-, farming, 167; open-, farming, 167
Rape, 18, 100, 119 f., 144; -seed, 18, 90; yields, 218
Reclamation: water, 195, 264 f., 270
Reimbursement share, 21
Reindeer, 86, 250
Relief, 52
Rent differential, 159
Residential zone, 105
Rhine: industrial area, 230; -Main area, 79, 119, 201, 220; Plain, 220; Valley, 220, 222, 230
Rhineland, 90; -Pfalz, 84; -Ruhr area, 230
Rhone delta, 20
Rhone Valley, 220
Rice, 25, 33, 37, 42, 49 ff., 55, 59, 66 f., 79, 91, 120, 122, 132, 134, 139, 141, 144 ff., 201, 262, 264, 267, 269, 278; cultivation, 95, 129, 299; farming, 145 ff., 278; yields, 110
Risk, 185, 187 ff., 193, 245, 255; harvest, 77, 152, 289; market, 93, 169, 289; of drought, 264; production, 169 f.; reducing, 221; spreading, 93; to diversified production, 187
Riverside-San Bernardino area, 246
Riviera, 205, 220

Romania, 253
Root crops, 25, 42, 132, 139, 186, 262, 278, 294, 296. *See also* Hoe crops
Root forage crops, 92
Root humus, 118
Root system, 122
Roughages, 18, 204
Row-spraying, 112
Rubber, 25, 33, 36, 50 ff., 55, 58, 95, 124 ff., 131, 150, 153, 286
Rugs, 75, 84
Ruminants, 209, 222, 224, 226
Russia, 232, 251 f., 261 f. *See also* Soviet Union
Ruwer Valley, 221, 230
Rye, 18, 51, 56, 71, 91, 100, 251, 262, 264, 267

Saar, 230
Sacramento Valley, 67, 122, 163, 246
Sago palm, 51
Sahara, 40, 47, 62, 78, 86, 166, 181
Sahel, 27, 37 f., 63, 110
Salinas Valley, 246
Salinization, 68
Salt carbonates, 67
Salt efflorescence, 67
San Francisco Bay Area, 233
San Joaquin Valley, 163
Sao Paulo, state of, 114
Sarawak, 135
Sardinia, 224, 227
Saskatchewan, 57, 242
Sauerland, 217
Savanna climate, 34 ff., 292 f.
Savanna grassland farming, 166 ff. *See also* Grassland farming
Savanna shifting cultivation, 27, 58, 80 f., 117 f., 174 ff., 293 ff.
Savannas: bush, 189; dry, 39, 63, 129, 165, 190 f., 294 f.; high-grass, 130, 137 ff., 194; humid, 63, 129 f., 137, 139, 292; semiarid shrub, 127; shortgrass, 37; shrub, 39, 129, 165, 191, 295
Sawahs, 69
Saxony, Lower, 202, 205, 213
Scania, 117, 212
Scandinavia, 43, 111, 197, 205, 214, 217, 232
Scarcity ratios, 75
Schleswig-Holstein, 79, 202, 205, 213
Scotland, 119, 212, 224, 227
Sedentary extensive grassland farming, 115, 167 ff., 243 f. *See also* Farming systems; Grassland farming
Sedentary grassland farming, 167 ff. *See also* Farming systems; Grassland farming
Sedentary intensive grassland farming, 115 ff. *See also* Farming systems; Grassland farming

Seed dressings, 277
Seed drills, 277
Seed costs, 21
Seedbeds, 147 f., 299
Seeds, 277, 292; stubble, 18; under-, 18
Self-subsistence, 114, 123
Self-sufficiency, 93, 123, 155, 261
Self-supplying cattle fattening farms, 169
Self-supplying dairy farms, 225 f.
Semi-desert, 129, 191; climate, 39 f.
Semi-drylot farms, 225 f.
Senegal, 175, 182 ff.
Serbia, 259
Serengeti National Park, 114
Sesame, 25, 58, 175, 184, 293; cropping, 185
Settlement: boundaries, 46, 70; density, 175, 261, 274 ff., 294. *See also* Population density
Shansi, 268
Shantung, 268
Sheep, 38, 65, 80, 84, 86, 166 f., 170, 192 ff., 209, 244, 259; and goat and dairy farming, 223 f., 226; and goat and young cattle farming, 223, 227; farms, 114; grazing-wheat belt, 193; Karakul, 62; raising, 58, 60, 62, 115, 193 ff., 226; raising-cattle raising belt, 193. *See also* Farming systems; Livestock raising
Shifting cultivation, 27 f., 37, 95, 129, 132 ff., 140, 285, 294 ff., end papers; bog-burning, 27, 118; *Hauberg*, 19, 118; forest-burning, 27, 118, 125, 130, 132, 134 f., 291, 301; savanna, 174 f., 293 ff.; steppe, 27, 58, 80 f., 117 f., 174 ff.; with plow cultivation, 291. *See also* Clearing
Shrub, 127, 129; savanna climate, 38 f., savannas, 39. *See also* Bush
Siberia, 111, 122, 251, 262
Sicily, 201, 220, 222, 230
Sickles, 147 f.
Sierra Leone, 135
Siegen district, 84
Silesia, 251, 256
Silk, 75, 269; cotton trees, 26, 269
Simplification: of farming, 105 f. *See also* Specialization
Single-operator farms, 98
Single-product farms, 284. *See also* Monocultural farms; One-crop farms; One-sided farms
Sinkiang, 268
Sisal, 26, 50, 52 f., 58, 72, 124 f., 150, 154, 163, 269, 287; factories, 87, 155; landscapes, 125
Skins, 169 f.
Slaughtering, 171
Slavs, 28
Slope boundaries, 68 f.
Slovenia, 258 f.

Small family farms, 144, 207. *See also* Farms
Small farms, 97 ff., 285. *See also* Farms
Small-scale processing, 154
Snowy Mountains, 195; Scheme, 195
Social boundaries, 84
Social fallow, 47, 84
Socialism, 253 f.
Socialist farms, large, 251, 254, 257, 260. *See also* Farms
Socialization: of agriculture, 252; of the production factors, 255; policies, 250; stages of, 253 ff.
Soil boundaries, 46, 67 f., 77
Soil-determined grain cropping economies, 218
Soil erosion, 67, 118, 137, 239 f., 242, 264
Soil fertility, 118, 125 f., 130, 132, 137, 141, 150 f., 173, 175, 286, 299 ff.
Soil productivity, 135
Soil types, 202 f., 220, 235
Soils, 202 f., 218 ff.; black, 263; cultivation of marginal, 266; grassland, 242; in central England, 296; in the East Bloc, 251; marsh, 77; of Europe, 202 f.; types in the United States, 235; volcanic, 135
Somalia, 39
Somalis, 171
Sorghum, 25, 36 f., 52 f., 58, 74, 77 f., 81, 117, 125, 138 f., 164, 175 ff., 241 f., 267 f., 288, 293; millet, and peanut farming, 181 ff.
South Africa, Republic of, 42, 57, 60, 70, 168, 176
South America, 42, 46, 53, 274
South Dakota, 57, 190, 242
Southern Plains (U.S.), 179
Soviet Union, 24, 31, 43, 84, 111, 116, 176, 253 ff., 260 ff., 267; crop rotations, 262; large socialist farms, 260; regional zones, 263
Sovkhoz, 24, 252 ff., 260
Soybeans, 25, 42 f., 51 ff., 145, 164, 220, 239 ff., 264, 288, 300
Spain, 199, 205, 214, 221
Sparsely settled countries, 271 ff., 274 ff., 281
Specialization, 21, 92, 94, 105, 117, 125, 152, 247, 277; and farm size, 283; in economic integration, 289 f.; in the industrial era, 287 ff.; in the United States, 236; in vegetable growing, 245; increasing, 238; of the production program, 284 f.; on fodder cropping farms, 217; stages in farming, 289 f.; strong, 246
Specialized farming, 21. *See also* Farming
Specialized farms, 104, 221, 240, 246, 284, 289
Specialty crop farming, 215 f., 221
Specialty crop farms, 220

Index 343

Specialty crop-hoe crop farming, 215 f., 222
Specialty crops, 18, 94, 209 f., 214 f., 218, 220 f., 228, 236, 256 ff.
Spinach, 245; harvesters, 112
Sprinklers, 156
Sri Lanka, 69, 125, 150, 154, 157
Stall feeding, 222, 230
Starch factory, 257
State farms, 24, 252 ff., 260
Steppe, 260, 262 ff., 293 ff.; climate, 34, 39, 191, 293 ff.; hot and dry, 300; salt, 127; shifting cultivation, 27, 58, 80 f., 117 f., 174 ff.; shrub, 129; subtropical grass, 129
Straw, 221
Strawberries, 18
Strip grazing, 112
Strongly commercialized farms, 21, 75. *See also* Farms
Subarctic climate, 43, 250 f.
Subsistence, 93; crops, 132, 152, 182, 186, 190, 289; economies, 28; farming, 187; farms, 21, 182. 280; self-, 114, 123
Subtropical boundaries, 185
Subtropical dry-summer climate, 34, 40 f.
Subtropical warm-summer climate, 41 f., 250
Sudan, 57, 59, 63, 134, 144, 167, 186
Sudanese blacks, 167
Sugar, 150; beet farming, 77, 100, 110 f., 258; beets, 16, 18, 42, 44, 58, 77, 90, 95, 100, 108, 122, 164, 199, 213, 219, 222, 251, 261 ff.; production, 148; refineries, 230; yield in cane, 155
Sugarcane, 25, 33, 37, 50, 52 ff., 58 f., 72, 74, 91, 95, 124 f., 132, 139, 141, 143 ff., 148, 150, 164, 194, 287 f.; landscapes, 125, 155; mills, 87, 102, 155, 164; monoculture, 102; zones, 155
Sumatra, 135
Summer feeding farms, 116
Sunflowers, 177, 263, 293
Supply and demand: and farm location, 28, 70 ff., 102 ff., 190 ff., 238 f., 295
Surpluses: agricultural, 193; land, 247 f.
Sweden, 43, 71, 107, 110, 117, 197, 199, 204, 212
Swine, 209 f., 222, 260, 285. *See also* Hogs

Taiwan, 69, 120, 143, 145, 278
Tanzania, 53, 123, 125, 150, 167
Tariff barriers, 95, 237
Tea, 18, 25, 36, 50, 52 f., 55, 58, 124 ff., 131, 150, 153, 155, 157 f., 268 f., 285, 287; bushes, 269; factories, 87, 153; farms, 157 ff.; landscapes, 56, 125; monoculture, 56
Technical auxiliary enterprises, 75
Technical inputs, 174
Technical progress, 175
Technological advances, 86, 88, 90 ff., 208, 271; biological, 78 f.; mechanical, 76 ff., 111 ff.; organic, 111
Technological boundaries, 46 f., 86
Technology, 180; agricultural, 28, 76 ff., 111ff.; harvesting, 57, 77, 147 f., 242; mechanized, 76 ff., 111 ff., 288; state of, 106
Teff, 52, 54 f., 141
Temperature: in Europe, 201 f.; land use systems and, 216; of at least 5°C, 200. *See also* Climates
Terrace farming, 55, 68
Terraces, 55, 69
Texas, 57, 242
Thailand, 66, 146 ff., 275
Three: -field and fallow farming, 296 ff.; -crop farms, 244; -field system, 20, 120, 297; -stage theory, 27 ff., 290 f.; -tractor farms, 98
Thünen circles, 71
Thünen model, 104 ff., 205; in Chicago's hinterland, 238 ff.; in the Storsjön area, 71; in the vicinity of Moscow, 261, 263
Tibet, 53
Tobacco, 25, 52, 58, 138, 141, 164, 186, 236, 240, 264; farming, 236, 240
Trace elements, 68
Traction power, 16, 288; motorized, 91, 96; unit, 16. *See also* Tractors
Tractor: drivers, 96; four-, farms, 98; one-, farms, 98; plows, 147; three-, farms, 98; two-, farms, 98
Tractors, 28 f., 68, 77 f., 85, 90, 100, 245, 260, 266 f., 288. *See also* Traction power
Trade: foreign, 269, 291, 297
Traditional farming, 231
Transport: access, 76; aerial, 245; boundaries, 47, 70 ff.; consolidation of the, net, 273; costs, 29, 102 ff., 125, 137, 153, 155, 256 f., 273; cotton fiber, 187; density in Europe, 205; density in the East Bloc, 251 f.; facilities, 124; location of farms for, 102 ff.; net and the Australian ranch, 194; network, 29, 93; network in Sweden, 252; nodes, 244; pineapple, 156; problems, 71, 76, 126, 150, 155, 191 ff.; truck, 92; unfavorable, situation, 169; with camel caravans, 166; wool, 192
Transvaal, 138, 188, 285
Tropical boundaries, 185
Tropical farming, 131 ff., 300 f. *See also* Farming systems
Tropical highlands, 38, 124, 130, 163; bush and tree crops in the, 153; climates, 36 f., rainfed farming in the, 140 f.; sisal in the, 163; wheat in the, 181
Tropical rainy climates, 33 f.; and altitudinal crop boundaries, 53; farming changes in, 291 f. *See also* Climates

Tropics, 38, 93, 124, 153; outer, 165 ff.; wet-and-dry, 38
Tsetse fly, 77, 299
Tsumeb District, 188
Tubers, 25, 80, 139, 186. *See also* Hoe crops; Potatoes; Root crops
Tundra, 250, 260, 263; pastures, 86
Tung oil, 269
Tunisia, 57, 180 f.
Turkestan, 40
Turkey, 143, 176, 269
Tuscany, 201
Twelve-stage theory, 297, 302
Two: -crop farms, 244; -field farming, 297; -man farms, 97, 100; -tractor farms, 98
Tyrol, South, 230

Uganda, 150, 186
Ukraine, 202, 235, 251
Umbria, 201
Underemployment, 89, 184
Undernourishment, 301
Undivided inheritance, 205
UNESCO, 31
United States, 31, 39, 41, 84, 89, 110, 122, 146 ff., 176, 178 ff., 220, 232 ff., 247, 249, 275, 277; annual precipitation, 233; farming zones, 236 ff.; growing season, 234; place-specific characteristics, 232 ff.; soil types, 235; Southeastern, 241; Southern, 247 f.; Southwestern, 241; Western, 247 f.
UNO, 31, 45
Upper Volta, 175, 182
Uruguay, 275
Useful economic life, 20, 118, 124, 150, 153, 156, 212, 224, 258
U.S.S.R. *See* Soviet Union
Uttar Pradesh, 144
Uzbekistan, 40, 251

Vaccines, 277
Vegetable canning industry, 112
Vegetable cropping, 112, 244 f.
Vegetables, 18, 36, 42, 53 ff., 60, 62, 93, 95, 120 ff., 138 ff., 141, 144 f., 219 f., 230, 238 f., 244 ff., 261 f., 264
Vegetation: belts, 129; climax, 141; formations, 127 ff.; zones, 171 ff.
Vesuvius, 120
Vetch, 262
Veterinary hygiene, 78, 272
Veterinary services, 21
Vierland area, 220
Vintners, 230
Vineyards: French, 230
Virgin lands, 262; cultivation, 264; reserves, 265
Virginia, 240

Viticulture, 49, 52, 62, 75, 89, 92, 114, 221 f., 256, 258
Vojvodina, 227, 258
Volga-Don area, 265
Vorgebirge, 220, 230
Vosges, 217

Wage-labor family farms, 21, 218 f., 283
Wage levels, 84, 103, 107 ff., 123, 214, 231, 244, 249, 271 f., 297; and market distance, 103; and types of cattle raising, 226; in Germany, 272; on U.S. wheat farms, 179
Wales, 119, 212, 227
Walnuts, 244
Warmia, 258
Washington, 242
Water, 162 f., 175, 178; buffalo, 147 f.; land use intensity and, 122; reclamation projects, 195, 264 f., 270. *See also* Irrigation
Watering places, 39, 192, 295, 301
Weakly commercialized farms, 21
Weaving, hand, 75
Weed control, 91, 111, 137, 219
Weed growth, 118, 134, 137, 301
Weed killers, 277
Weeds, 78, 91, 118, 122, 171, 246, 300 f.
Weighted-index system, 207 ff., 214
Wells, 171 f., 301; tube, 66, 78
Westphalia, 213, 217 f.
Westphalian-Hellweg region, 202
Wet boundaries, 36, 46, 66 f., 86
Wet-rice farming, 299
Wheat, 18, 36, 44, 49, 51 ff., 56, 58, 60, 62, 64, 77 ff., 84, 90, 100, 108, 111, 119, 122, 138, 141, 144, 164, 176 ff., 194, 199, 238, 240, 251, 262 f., 267 ff., 293, 296; Belts, 236, 242; -fallow farming, 176 ff.; fallow farms, 178, 180; farming, 77, 179, 242; regions in the U.S., 242; -sheep farming, 192, 194. *See also* Farming systems
Wickerwork, 189
Wild animals, 80, 114, 194
Wild plants, 123
Wild steppe grassland farming, 167 ff. *See also* Farming systems; Grassland farming; Ranching
Winches, 147 f.
Wine, 231. *See also* Grapes
Wisconsin, 237
Work spacing, 88 ff., 114, 151, 184, 219, 221, 244, 284, 288
Wrocław (Breslau) Platform, 202

Yak, 39
Yakuts, 250
Yams, 25, 33, 36 f., 74, 132, 139, 286
Yangtze, 29

Index

Yaroslavl Oblast, 262
Yellow Plain, 268
Yields, 84 f., 141, 286; fluctuations, 278; for grain crops, 176 f., 179; in different locations, 94; in the peanut-millet-sorghum rotation, 184; increased by capital inputs, 82; increasing gross, 143; of land uses, 90; on U.S. wheat farms, 179; on young cultivated marshes, 218; under shifting cultivation, 134, 136; uniform fodder, 286

Young cattle-dairy farming, 223
Yucatan, 28
Yucca, 74, 139

Zaire, 135, 149 f., 275
Zambesi Basin, 37
Zambia, 135, 275
Zeeland, 219, 230
Zimbabwe, 74